CAMBRIDGE LIBRARY COLLECTION

Books of enduring scholarly value

Life Sciences

Until the nineteenth century, the various subjects now known as the life sciences were regarded either as arcane studies which had little impact on ordinary daily life, or as a genteel hobby for the leisured classes. The increasing academic rigour and systematisation brought to the study of botany, zoology and other disciplines, and their adoption in university curricula, are reflected in the books reissued in this series.

Arctic Zoology

In the 'Advertisement' to this 1784 two-volume work, Thomas Pennant (1726–98), zoologist and traveller, explains that his original intention was to record the zoology of North America 'when the empire of Great Britain was entire'. After the War of Independence, he changed his focus to the zoology (and people, archaeology and geology) of the Arctic regions of America, Europe and Siberia. The content of the volumes, one of the earliest works of systematic zoology published in Britain, is based on the writings of earlier zoologists, information obtained by Pennant from his scientific correspondents all over Europe and America, and his studies in private museums and collections. It is embellished with engravings of animals, birds, landscapes and artefacts. Volume 1 begins with an account of the various Arctic habitats, and describes the quadrupeds of these regions. Other works by Thomas Pennant are also reissued in the Cambridge Library Collection.

Cambridge University Press has long been a pioneer in the reissuing of out-of-print titles from its own backlist, producing digital reprints of books that are still sought after by scholars and students but could not be reprinted economically using traditional technology. The Cambridge Library Collection extends this activity to a wider range of books which are still of importance to researchers and professionals, either for the source material they contain, or as landmarks in the history of their academic discipline.

Drawing from the world-renowned collections in the Cambridge University Library and other partner libraries, and guided by the advice of experts in each subject area, Cambridge University Press is using state-of-the-art scanning machines in its own Printing House to capture the content of each book selected for inclusion. The files are processed to give a consistently clear, crisp image, and the books finished to the high quality standard for which the Press is recognised around the world. The latest print-on-demand technology ensures that the books will remain available indefinitely, and that orders for single or multiple copies can quickly be supplied.

The Cambridge Library Collection brings back to life books of enduring scholarly value (including out-of-copyright works originally issued by other publishers) across a wide range of disciplines in the humanities and social sciences and in science and technology.

Arctic Zoology

VOLUME 1:
CLASS I. QUADRUPEDS

THOMAS PENNANT

CAMBRIDGE
UNIVERSITY PRESS

CAMBRIDGE
UNIVERSITY PRESS

University Printing House, Cambridge, CB2 8BS, United Kingdom

Cambridge University Press is part of the University of Cambridge.
It furthers the University's mission by disseminating knowledge in the pursuit of
education, learning and research at the highest international levels of excellence.

www.cambridge.org
Information on this title: www.cambridge.org/9781108073653

This edition first published 1784
This digitally printed version 2014

ISBN 978-1-108-07365-3 Paperback

P Paillore pinx.   P. Mazell sculp.

ARCTIC ZOOLOGY.

VOL. I.

INTRODUCTION.

CLASS I. QUADRUPEDS.

LONDON:

PRINTED BY HENRY HUGHS.

M.DCC.LXXXIV.

ADVERTISEMENT.

THIS Work was begun a great number of years paſt, when the empire of *Great Britain* was entire, and poſſeſſed the northern part of the New World with envied ſplendor. At that period I formed a deſign of collecting materials for a partial Hiſtory of its Animals ; and with true pains, by various correſpondencies, made far greater progreſs in my plan than my moſt ſanguine expectations had framed. Above a century ago, an illuſtrious predeceſſor in the line of Natural Hiſtory, who as greatly exceeded me in abilities as he did in zeal, meditated a voyage to the New World, in purſuance of a ſimilar deſign. The gentleman alluded to was FRANCIS WILLUGHBY, Eſq; who died in 1672, on the point of putting his deſign in execution. Emulous of ſo illuſtrious an example, I took up the object of his purſuit ; but my many relative duties forbade me from carrying it to the length conceived by that great and good man. What he would have performed, from an actual inſpection in the native country of the ſeveral ſubjects under conſideration, I muſt content myſelf to do, in a leſs perfect manner, from preſerved ſpecimens tranſmitted to me ; and offer to the world their Natural Hiſtory, taken from gentlemen or writers who have paid no ſmall attention to their manners.

Let me repeat, that this Work was deſigned as a ſketch of the Zoology of *North America.* I thought I had a right to

A   the

the attempt, at a time I had the honor of calling myself a fel-
low-subject with that respectable part of our former great empire; but when the fatal and humiliating hour arrived, which
deprived *Britain* of power, strength, and glory, I felt the morti-
fication which must strike every feeling individual at losing his
little share in the boast of ruling over half of the New World. I
could no longer support my clame of entitling myself its humble
Zoologist: yet, unwilling to fling away all my labors, do now
deliver them to the Public under the title of the ARCTIC
ZOOLOGY. I added to them a description of the Quadrupeds
and Birds of the north of *Europe* and of *Asia*, from latitude 60
to the farthest known parts of the *Arctic* World, together
with those of *Kamtschatka*, and the parts of *America* visited
in the last voyage of the illustrious COOK. These additional
parts I have flung into the form of an Appendix to each genus, and distinguished by a *fleur de lis*; and the species by
literal instead of numeral marks, which distinguish those of
North America. These will, in a great measure, shew the
dilatation of Quadrupeds and Birds, and the migrations of the
feathered tribe, within part of the northern hemisphere.

I have, whenever I could get information, given their respective residences, as well as migrations to far more northern
parts, to shew to what very remote places the Author of Nature hath impelled them to retire, to breed in security. This
wise provision preserves the species entire, and enables them
to return by myriads, to contribute to the food or luxuries of
southern climates. Whatever is wanting in the *American* part,
I may foresee, will in time be amply supplied. The powers
of literature will soon arise, with the other strengths of the
new empire, and some native Naturalist give perfection to that

part

part of the undertaking, by obſervations formed on the ſpot, in the uſes, manners, and migrations. Should, at preſent, no one be inclined to take the pen out of my hand, remarks from the other ſide of the *Atlantic*, from any gentlemen of congenial ſtudies, will add peculiar pleaſure to a favorite purſuit, and be gratefully received.

I muſt reckon among my moſt valued correſpondents on the New Continent, Doctor ALEXANDER GARDEN *, who, by his long reſidence in *South Carolina*, was enabled to communicate to me variety of curious remarks and ſubjects, as will appear in the following pages.

To the rich muſeum of *American* Birds, preſerved by Mrs. ANNA BLACKBURN, of *Orford*, near *Warrington*, I am indebted for the opportunity of deſcribing almoſt every one known in the provinces of *Jerſey*, *New York*, and *Connecticut*. They were ſent over to that Lady by her brother, the late Mr. *Aſhton Blackburn*; who added to the ſkill and zeal of a ſportſman, the moſt pertinent remarks on the ſpecimens he collected for his worthy and philoſophical ſiſter.

In the foremoſt rank of the philoſophers of the Old Continent, from whoſe correſpondence I have benefited, I muſt place Doctor PETER SIM. PALLAS, at preſent Profeſſor of Natural Hiſtory in the ſervice of the illuſtrious EMPRESS of *Ruſſia*: he not only favored me with the fulleſt remarks on the Zoological part of that vaſt empire, moſt of which he formed from actual travel and obſervation, but collected for my uſe various other remarks from the manuſcripts of his predeceſſors; eſpecially what related to *Kamtſchatka* from thoſe

---

Now reſident in *London*.

of

ADVERTISEMENT.

of STELLER; which have affifted me in the hiftory of parts hitherto but very flightly underftood.

From the correfpondency and labors of Mr. EBERH. AUG. WILLIAM ZIMMERMAN, Profeffor of Mathematics at *Brunf-wic*, I have collected moft uncommon inftruction. His *Specimen Zoologiæ Geographicæ Quadrupedum* * is a work which gives. a full view of the clafs of Quadrupeds, and the progrefs they have made in fpreading over the face of the earth, according to climates and latitudes. Their limits are defcribed, in general, with uncommon accuracy. Much is faid of the climates themfelves; of the varieties of mankind; of the effects of heat and cold on them and other animals. A moft curious map is joined to the work, in which is given the name of every animal in its proper climate; fo that a view of the whole Quadruped creation is placed before one's eyes, in a manner perfectly new and inftructive †.

To the following foreigners, diftinguifhed for their literary knowlege, I muft pay my beft acknowlegement for variety of moft ufeful communications: Doctor ANDERS SPARMAN, of *Stockholm*; Doctor CHARLES P. THUNBERG, of *Upfal*; MR. AND. J. RETZIUS, Profeffor of Natural Hiftory at *Lund*; Mr. MARTIN THRANE BRUNNICH, Profeffor of Natural Hiftory, and Mr. OTHO MULLER, Author of the *Zoologia Danica*, both of *Copenhagen*: and let me add my great obligations to the labors of the Reverend Mr. OTTO FABRICIUS, for his moft finifhed *Fauna* of *Greenland*.

A quarto in *Latin*, containing 685 pages, printed at *Leyden*, 1777; fold in *London* by Mr. *Faden*, Geographer, *St. Martin's Lane*.

† A new edition of the map has been lately publifhed by the learned Author; the geographical part is corrected according to the late voyages of Captain COOK, and great additions made to the zoological part. An explanation is given, in the third volume of the *Zoologia Geographica*, lately publifhed in *German* by the Author.

To

ADVERTISEMENT.

To many of my countrymen my beſt thanks are due for literary aſſiſtances. Sir JOSEPH BANKS, Baronet, will, I hope, accept my thanks for the free admittance to thoſe parts of his cabinet which more immediately related to the ſubjeČt of the following ſheets.

To Sir ASHTON LEVER, Knight, I am highly indebted, for the more intimate and cloſer examination of his treaſures than was allowed to the common viſitors of his moſt magnificent muſeum.

To Mr. SAMUEL HEARN, the great explorer by land of the *Icy Sea*, I cannot but ſend my moſt particular thanks, for his liberal communication of many zoological remarks, made by him on the bold and fatiguing adventure he undertook from *Hudſon's Bay* to the *ne plus ultra* of the north on that ſide.

Mr. ANDREW GRAHAM, long a reſident in *Hudſon's Bay*, obliged me with numbers of obſervations on the country, and the uſe of multitudes of ſpecimens of animals tranſmitted by him to the late muſeum of the Royal Society, at the inſtance of that liberal patron of ſcience, my reſpeČted friend the Honorable DAINES BARRINGTON.

Let me cloſe the liſt with acknowleging the great aſſiſtance I have found in the Synopſis of Birds by Mr. JOHN LATHAM; a work now brought almoſt to a concluſion, and which contains a far greater number of deſcriptions than any which has gone before. This is owing not only to the aſſiduity of the Author, but alſo to the peculiar ſpirit of the *Engliſh* nation, which has, in its voyages to the moſt remote and moſt oppoſite parts of the globe, payed attention to every branch of ſcience. The advantages are pointed out by the able pen of the Reverend DoČtor DOUGLAS, in his IntroduČtion to the laſt Voyage of

our

ADVERTISEMENT.

our great navigator, publifhed (under the aufpices of the Lords of the Admiralty) in a manner which reflects honor on our country in general, and will prove a moft lafting monument to the memory of the great Officer who fo unfortunately perifhed by favage hands, and his two able conforts, who at length funk beneath the preffure of fatigue, in carrying the glory of difcovery far beyond the attempts of every preceding adventurer.

DOWNING,
February 1, 1785.

THOMAS PENNANT.

PLATES.

P L A T E S.

V O L. I.

For the drawings from which thefe Antiquities were engraven, I am indebted to the Reverend Mr. *Low*, Minifter of *Birfa* in *Orkney*, who, at my requeft, made the voyage of the *Orkney* and *Schetland* ifles in 1778. He hath prepared his journal for the prefs: it is to be hoped, that the liberality of the public will enable him to give this addition to my labors, which will complete the account of the northern part of the *Britifh* dominions.

                        Tab. VII.

P L A T E S.

The Bookbinder is desired to observe, that the Second Volume begins
at p. 187, Class II. Birds.

INTRO-

INTRODUCTION.

OF THE

ARCTIC WORLD.

A KNOWLEGE of the geography, climate, and foil, and a general view of the productions of the countries, whofe Zoologic Hiftory is to be treated of, are points fo neceffary, that no apology need be made for introducing them into a prefatory difcourfe.

It is worthy human curiofity to trace the gradual increafe of the animal world, from the fcanty pittance given to the rocks of *Spitzbergen*, to the fwarms of beings which enliven the vegetating plains of *Senegal:* to point out the caufes of the local niggardnefs of certain places, and the prodigious plenty in others. The Botanift fhould attend the fancied voyage I am about to take, to explain the fcanty herbage of the *Arctic* regions ; or, fhould I at any time hereafter defcend into the lower latitudes, to inveftigate the luxuriancy of plants in the warmer climates.

The Foffilift fhould join company, and point the variations of primæval creation, from the folid rock of *Spitzbergen* through all the degrees of terreftrial matter: the fteps it makes to perfection, from the vileft earth to the precious diamond of *Golconda.* The changes in the face of the globe fhould be attended to ; the deftructions by vulcanoes ; the ravages of the fea on fome coafts, and the recompence it may have made to others, by the retreat of its waters.

The purfuit of thefe enquiries will alfo have a farther and more important object. Hiftory fhould be called in, and a brief account given of the population of the more remote countries—the motives which induced mankind to feek retreats in climates feemingly deftitute of incitements to migration. Particular attention fhould be paid to the means of peopling the new world, and of ftocking it with animals, to contribute to the fupport of mankind, after the firft colonization—the increafe of thofe animals, and their ceffation, and giving place in a certain latitude to genera entirely different.

a

Here

Here the fine ſtudy of Geography ſhould ſtep in to our aſſiſtance. The outline of the terreſtrial globe ſhould be traced; the ſeveral approximations between part and part ſhould be attended to; the nature of the oceans obſerved; the various iſlands pointed out, as the ſteps, the baiting-places where mankind might have reſted in its paſſage from an overcharged continent.

The manners of the people ought not leſs to be attended to; and their changes, both mental and corporeal, by compariſon of the preſent ſtate of remote people with nations with whom they had common anceſtors, and who may have been diſcovered ſtill to retain their primæval ſeats. Some leading cuſtoms may ſtill have been pre-ſerved in both; or ſome monuments of antiquity, proofs of congenial habitudes, poſſibly no longer extant in the ſavage than in the cultivated branches of the common ſtock.

STREIGHTS OF DOVER.

Let me take my departure northward, from the narrow ſtreights of *Dover*, the ſite of the iſthmus of the once peninſulated *Britain*. No certain cauſe can be given for the mighty convulſion which tore us from the continent: whether it was rent by an earthquake, or whether it was worn through by the continual daſhing of the waters, no *Pythagoras* is left to ſolve the *Fortuna locorum* :

> Vidi ego, quod fuerat quondam ſolidiſſima tellus
> Eſſe fretum

But it is moſt probable, that the great philoſopher alluded to the partial deſtruction of the *Atlantica inſula*, mentioned by *Plato* as a diſtant tradition in his days * It was effected by an earthquake and a deluge, which might have rent aſunder the narrow iſthmus in queſtion, and left *Britain*, large as it ſeems at preſent, the mere wreck of its original ſize †. The *Scilly* iſles, the *Hebrides*, *Orknies*, *Schet-lands*, and perhaps the *Feroe* iſlands, may poſſibly be no more than fragments of the once far-extended region. I have no quarrel about the word *iſland*. The little iſthmus, compared to the whole, might have been a junction never attend-ed to in the limited navigations of very early times. The peninſula had never been wholly explored, and it paſſed with the antients for a genuine iſland. The correſpondency of ſtrata on part of the oppoſite ſhores of *Britain* and *France*,

CHALKY STRATA. leaves no room to doubt but that they were once united. The chalky cliffs of *Blanc-nez*, between *Calais* and *Bologne*, and thoſe to the weſtward of *Dover*, ex-actly tally : the laſt are vaſt and continued; the former ſhort, and the termina-tion of the immenſe bed. Between *Bologne* and *Folkſtone* (about ſix miles from

---

* *Plato* died about the year 347 before CHRIST, aged 81. *Pythagoras*, about 497, aged 90.
† See this opinion farther diſcuſſed by Mr. *Somner*, *Ph. Tranſ. Abridg.* iv. 230.

the latter) is another memorial of the junction of the two-countries; a narrow submarine hill, called the *Rip-raps*, about a quarter of a mile broad, and ten miles long, extending eaftwards towards the *Goodwin* Sands. Its materials are boulder-ftones, adventitious to many ftrata. The depth of water on it, in very low fpring-tides, is only fourteen feet. The fifhermen from *Folkftone* have often touched it with a fifteen feet oar; fo that it is juftly the dread of navigators. Many a tall fhip has perifhed on it, and funk inftantly into twenty-one fathoms water. In *July* 1782, the *Belleifle* of fixty-four guns ftruck, and lay on it during three hours; but, by ftarting her beer and water, got clear off.

Thefe celebrated ftreights are only twenty-one miles wide in the narroweft part. From the pier at *Dover* to that at *Calais* is twenty-four. It is conjectured, that their breadth leffens, and that they are two miles narrower than they were in antient times. An accurate obferver of fifty years, remarks to me, that the encreafed height of water, from a decreafe of breadth, has been apparent even in that fpace. The depth of the channel, at a medium, in higheft fpring-tides, is about twenty-five fathoms. The bottom, either coarfe fand or rugged fcars, which have for ages unknown refifted the attrition of the currents. From the ftreights, both eaftward and weftward, is a gradual increafe of depth thorough the channel to a hundred fathoms, till foundings are totally loft or unattended to.

The fpring-tides in the ftreights rife, on an average, twenty-four feet; the neap-tides fifteen. The tide flows from the *German* fea, paffes the ftreights, and meets, with a great rippling, the weftern tide from the ocean, between *Fairleigh*, near *Haftings*, and *Bologne* *; a proof, that if the feparation of the land was effected by the feas, it muft have been by the overpowering weight of thofe of the north.

It is moft certain, that *Britain* was peopled from *Gaul*. Similar cuftoms, as far as can be collected, evince this fact. The period is beyond the reach of hiftory.

* All the intelligence refpecting the tides, &c. in thefe parts, I received from Mr. *James Hammond* of the cuftom-houfe, *Dover*, and Mr. *William Cowly*, a veteran pilot of the fame place.

Beyond the meafure vaft of thought,
The works, the wizard TIME hath wrought!
The *Gaul*, it's held of antique ftory,
Saw *Britain* link'd to his now adverfe ftrand;
No fea between, nor cliff fublime and hoary,
He pafs'd with unwet feet through all our land.
To the blown *Baltic* then, they fay,
The wild waves found another way. *&c.*

COLLINS's Ode to Liberty.

If, after the event by which our ifland was torn from the continent, the migration over fo narrow a ftreight might, in the earlier ages, have been very readily effected in the *vitilia navigia* or coracles, or the *monoxyla* or canoes in ufe in the remote periods; but the numerous fpecies of Quadrupeds never could have fwam into our ifland, even over fuch a contracted water, which at all times muft have been poffeffed by tides fo rapid, as to baffle their utmoft efforts: their paffage, therefore, muft have been over the antient ifthmus; for it is contrary to common fenfe to fuppofe, that our anceftors would have been at the trouble of tranfporting fuch guefts as wolves and bears, and the numerous train of leffer rapacious animals, even had it been practicable for them to have introduced the domeftic and ufeful fpecies.

Would they on board or Bears or Lynxes take,
Feed the She-adder, and the brooding Snake?

PRIOR.

QUADRUPEDS.   Men and beafts found their way into *Great Britain* from the fame quarter. We have no Quadrupeds but what are alfo found in *France*; and among our loft animals may be reckoned the Urus, p. 2; Wolf, Nº 9; Bear, Nº 20; Wild Boar; and the Beaver, Nº 40: all which were once common to both countries. The *Urus* continued among us in a ftate of nature as late at leaft as the year 1466 *: and I have feen fome of their defcendants, fcarcely to be called tame, in confinement in the parks of *Drumlanrig* and *Chillingham* †. The *Caledonian* Bears were exported to *Rome*, and efteemed for their fiercenefs ‡. They continued in *Scotland* till the year 1057. They exifted in *Wales*, perhaps, till the fame period; for our antient laws ranked them among the beafts of chace §. Wolves infefted even the middle counties of *England* as late as the year 1281, and continued their ravages in *North Britain* in the reign of Queen *Elizabeth*; nor were they wholly extirpated till the year 1680. The Wild

---

* Six Wild Bulls were ufed at the inftallation feaft of *George Nevil*, archbifhop of *York*. *Leland's Collect.* vi. 2.   † *Tours in Scotland.*   ‡ *Martial, Plutarch.*   § *Raii Syn. Quad.* 214.

BOARS

Boars were common in the neighborhood of *London* in the reign of *Henry* II. and continued in our kingdom, in a wild ſtate, till 1577 : they were then only to be found in the woods of Lord *Latimer*, who, we are informed by Doĉtor *Moufet*, took great delight in their chace *. Let me add, from the ſame authority, that Roebucks were found at the ſame period in *Wales*, and among the *Cheviot hills*; they are now confined to the Highlands of *Scotland*. Finally, Beavers inhabited *Wales* in 1188, when our hiſtorian, *Giraldus*, made his progreſs through the principality. Every one of theſe animals are at this time to be found in *France*, the *Urus* excepted. *Theodebert*, king of *France*, periſhed in the chace of one about the year 548 †; but it is probable that the ſpecies muſt have exiſted in that vaſt kingdom long after that event.

The Elk, N° 3; Genet, *Hiſt. Quad.* N° 224; Lynx, N° 150; Fat Dormouſe, *Hiſt. Quad.* N° 287; Garden Dormouſe, *Hiſt. Quad.* N° 288; and the Bats *Serotine, Pipiſtrelle*, and *Barbaſtelle, Hiſt. Quad.* N^{ris} 408, 409, 410, either never reached our iſland, or if they did, periſhed ſo early, that even their very names in the *Britiſh* tongue, have periſhed with them. The *Ibex, Hiſt. Quad.* N° 13, and the *Chamois, Hiſt. Quad.* N° 17, inhabitants only of the remote *Gauliſh Alps* and *Pyreneans*, probably never reached us. *France*, therefore, poſſeſſes forty-nine ſpecies of Quadrupeds; we only thirty-nine. I exclude two ſpecies of Seals ‡ in both reckonings; being animals which had at all times powers of making themſelves inhabitants of the coaſts of each kingdom.

Birds, which have the ready means of wafting themſelves from place to place, have notwithſtanding, in numbers of inſtances, their limits. Climate confines ſome within certain bounds, and particular ſorts of food induce others to remain within countries not very remote from us; yet, by wonderful inſtinĉt, birds will follow cultivation, and make themſelves denizens of new regions. The CROSS-BILL has followed the apple into *England*. *Glenco*, in the *Highlands* of *Scotland*, never knew the Partridge, till its farmers of late years introduced corn into their lands: nor did Sparrows ever appear in *Sibiria*, till after the *Ruſſians* had made arable the vaſt waſtes of thoſe parts of their dominions. Finally, the Rice Buntings, p. 360, natives of *Cuba*, after the planting of rice in the *Carolinas*, annually quit the iſland in myriads, and fly over ſea and land, to partake of a harveſt introduced there from the diſtant *India*.

BIRDS.

---

* *Health's Emprovement*.　　　† *Ecole de la Chaſſe*, clxi.

‡ The Common Seal, is common to the ocean and *Mediterranean* ſea. Poſſibly the *Mediterranean* Seal, *Hiſt. Quad.* N° 376, may be ſo likewiſe.—This work is always intended, when the name of the work referred to is not added to the numbers.

FRANCE,

FRANCE, as it exceeds in variation of climate, so it exceeds us in the number of species of birds. We can boast of only one hundred and thirty-one kinds of land-birds, and one hundred and twenty-one of water-fowl. *France*, on the contrary, has one hundred and fifty-six of the first, and one hundred and thirteen of the last. This computation may not be quite accurate; for no one has as yet attempted its *Fauna*, which must be very numerous, in a kingdom which extends from *Calais*, in about lat. 51, to *Collioure* in the south of *Roussillon*, on the *Mediterranean* sea, in about lat. 42. The northern parts possess the birds in common with *England*: and in all probability the provinces in the *Mediterranean* annually are visited by various species from northern *Africa*.

COASTS OF BRITAIN.

Stupendous and precipitous ranges of chalky cliffs attend the coast, from *Dover* eastward, and, from their color, gave the name of *Albion* to our island. Beneath one of them anchored *Cesar*, fifty-five years before CHRIST, and so near as to be capable of being annoyed by the darts of the *Britons*. After weighing anchor, he sailed up a bay, now occupied by meadows, and landed at *Rutupium*, *Richborough*, opposite to the present *Sandwich*. The walls of the former still evince its antient strength; and the vestiges of a quay, now bounded by a ditch, points out the anchorage of the *Roman* commerce. The adjacent *Thanet*, the *Thanatos* of the antients, at present indistinguishable from the main land, was in old times an island, separated by a deep channel, from a mile and a half to four miles in width, the site of *Roman* settlements; and, in 449, celebrated for having been the first landing-place of the invading *Saxons*; to whom it was assigned as a place of security by the imprudent *Vortigern*. But such a change has time effected, that *Thanet* no more exists as an island; and the *Britanniarum Portus*, in which rode the *Roman* navies, is now filled with marshy meads.

After passing the lofty chalky promontory, the *North Foreland*, opens the estuary of the *Thames*, bounded on each side by low shores, and its channels divided by numerous sand-banks; securely passed, by reason of the perfection of navigation, by thousands of ships frequenting annually *London*, our emporium, envied nearly to impending decline.

SUFFOLK AND NORFOLK.

On the projecting coasts of *Suffolk* and *Norfolk*, arise, in certain intervals, eminences of different matter. *Loamy* cliffs appear about *Leostoffe*, *Dunwich*, &c. The *Crag-pits* about *Woodbridge*, are prodigious pits of sea-shells, many of them perfect and quite solid; an inexhaustible fund of manure for arable lands. About *Yarmouth*, and from thence beyond *Wintertoness*, the coast is low, flat, and composed of shingle, backed by sand. From *Hapsburgh* to *Cromer* are a range of lofty clayey precipices, rising from the height of forty to a hundred feet perpendicular; a prey to the ocean, which has effected great changes in these parts. About *Sherringham* and *Cley*, it rises into pretty and gentle hills, sloping down into a

rough

róugh fhore, of little rocks and ftones. At *Holkham, Wells,* and *Wareham,* the fandy fhores terminate in little hillocks of fand, kept together by the *Arundo Arenaria,* or *Bent,* the great prefervative againft the inundations of fand, which would otherwife deftroy whole tracts of country, and in particular foon render ufelefs the range of falt-marfhes which thefe are backed with. *Hunftanton* cliff rifes a diftinguifhed feature in this flat tract. The furface is the ufual vegetable mould, about a foot deep; beneath that are two feet of fmall broken pieces of chalk: the folid ftratum of the fame, after having been loft for numbers of miles, here again makes its appearance, and forms a folid bed thirty feet in thicknefs, refting on a hard red ftone four feet deep, which is often ground and made into a red paint. Seven feet of loofe friable dirty yellow ftone fucceeds, placed on a bafe of iron-colored plumb-pudding-ftone, projecting into the fea, with vaft fragments fcattered over the beach. This cliff is about eighty feet high, lies on the entrance of the wafhes, the *Metaris Eftuarium* of *Ptolemy.* From hence, all the coaft by *Snettifham* to *Lynn* is low, flat, and fhingly.

From *Holm,* the northern promontory of *Norfolk,* the fea advances deeply weftward, and forms the great bay called the *Wafhes,* filled with vaft fand-banks, the fummits of which are dry at low water; but the intervening channels are the means of prodigious commerce to *Lynn* in *Norfolk,* feated on the *Ouze,* which is circulated into the very inland parts of our ifland, through the various rivers which fall into its long courfe. *Lynn* is mentioned in the *Doomfday Book;* but became confiderable for its commerce with *Norway* as early as the year 1284.

The oppofite fhore is that of *Lincolnfhire.* Its great commercial town, *Bofton,* ftands on the *Witham,* a few miles from the head of the bay. Spring-tides rife at the quay fourteen feet, and convey there veffels of above a hundred tons; but greater fhips lie at the *Scap,* the opening of the eftuary. Such is the cafe at *Lynn;* for the fluggifh rivers of thefe tame tracts want force to form a depth of water.

Lincolnfhire, and part of fix other counties, are the *Pais-bas,* the *Low Countries* of *Britain;* the former bounded on the weftern part by a range of elevated land, which, in this humble county, overlooks, as *Alps* would the ocean, the remaining part. This very extenfive tract, from the *Scap* to the northern headland oppofite to *Hull,* prefents to the fea a bow-like and almoft unindented front; and fo low as to be vifible from fea only at a fmall diftance; and churches, inftead of hills, are the only landmarks to feamen. The whole coaft is fronted with falt-marfhes or fand-hills, and fecured by artificial banks againft the fury of the fea. Old *Holinfhead* gives a long lift of ports on this now inhofpitable coaft. *Waynfleet,* once a noted haven, is at prefent a mere creek. *Skegnefs,* once a large walled town, with a good harbour, is now an inconfiderable place a mile from the fea: and the port of
Grimefby,

LINCOLNSHIRE.

Grimeſby, which in the time of *Edward* III. furniſhed him with eleven ſhips, is now totally choaked with ſand.

The *Great Level*, which comprehends *Holland* in this county, with part of *Northamptonſhire*, *Norfolk*, *Suffolk*, *Cambridge*, and *Huntingdon*, a tract of ſixty computed miles in length, and forty in breadth, had been originally a wooded country. Whole foreſts of firs and oaks have been found in digging, far beneath the moor, on the ſolid ground ; oaks fifteen feet in girth, and ſixteen yards long, moſtly burnt at the bottoms, the antient method of falling them : multitudes of others entirely rooted up, as appears, by the force of the ſea burſting in and overwhelming this whole tract, and covering it with *ſilt*, or the mud which it carried with it from time to time. *Ovid*'s beautiful account of the deluge was here verified ; for under *Conington Down*, in *Huntingdonſhire*, was found the ſkeleton of a whale near twenty feet long, which had once ſwam ſecure to this diſtance from its native reſidence.

> Et modo quâ graciles gramen carpſere capellæ,
> Nunc ibi deformes ponunt ſua corpora phocæ.
> —————— ſylvaſque tenent delphines, et altis
> Incurſant ramis, agitataque robora pulſant.

In proceſs of time this tract underwent another revolution. The *ſilt* or mud gained ſo conſiderably as to leave vaſt ſpaces dry, and other parts ſo ſhallow as to encourage the *Romans* to regain theſe fertilized countries from the ſea. Thoſe ſenſible and indefatigable people firſt taught us the art of embanking, and recovered the valuable lands we now poſſeſs. It was the complaint of *Galgacus*, that they exhauſted the ſtrength of the *Britons*, *in ſylvis et paludibus emuniendis* *, ' in clearing woods and draining marſhes.' After the *Romans* deſerted our iſland, another change took place. Neglect of their labors ſucceeded : the drains were neglected, and the whole became fen and ſhallow lake, reſembling the preſent eaſt fen : the haunt of myriads of water-fowl, or the retreat of banditti. *Ely* and many little tracts which had the advantage of elevation, were at that period literally iſlands. Several of theſe in early times became the retreat of religious. *Ely*, *Thorney*, *Ramſey*, *Spiney*, and others, roſe into celebrated abbies, and by the induſtry of their inhabitants firſt began to reſtore the works of the *Romans*. The country above *Thorney* is repreſented by an old hiſtorian † as a paradiſe. Conſtant viſitations, founded on wholeſome laws, preſerved this vaſt recovered country : but on the rapid and rapacious diſſolution, the removal of numbers of the inhabitants, and the neglect of the laws of the *Sewers*, the drains were filled, the cultivated land overflowed, and

---

* *Vita Agricolæ.*   † *Malmſbury*, lib. iv. 294.

the

the country again reduced to a ufelefs morafs *. In the twentieth of *Elizabeth* the ftate of the country was taken into confideration †; no great matters were done till the time of *Francis*, and *William* his fon, earls of *Bedford*, who attempted this *Herculean* work, and reclamed this vaft tract of more than three hundred thoufand acres; and the laft received, under fanction of parlement, the juft reward of ninety thoufand acres. I fpeak not of the reliques of the antient banks which I have feen in *Holland*, *Lincolnfhire*, now remote from the fea, nor yet of the *Roman* tumuli, the coins, and other evidences of the refidence of that nation in thefe parts; they would fwell a mere preface to too great a length: and, it is to be hoped, will be undertaken by the pen of fome native, who will perform it from his actual furvey.

The vaft fenny tracts of thefe counties were in old times the haunts of multitudes of water-fowl; but the happy change, by attention to draining, has fubftituted in their place thoufands of fheep; or, inftead of reeds, made thofe tracts laugh with corn. The Crane, which once abounded in thefe parts, has even deferted our ifland. The Common Wild Duck ftill breeds in multitudes in the unreclamed parts; and thoufands are fent annually to the *London* markets, from the numerous decoys. The Grey Lag Goofe, *Br. Zool.* ii. N° 266, the origin of the Tame, breeds here, and is refident the whole year: a few others of the Duck kind breed here. Ruffs, Redfhanks, Lapwings, Red-breafted Godwits, and Whimbrels, are found here during fummer; but, with their young, in autumn, difperfe about the ifland. The Short-eared Owl migrates here with the Woodcock, and is a welcome gueft to the farmer, by clearing the fields of mice. Knots fwarm on the coafts in winter: are taken in numbers in nets: yet none are feen during fummer ‡. The moft diftant north is probably the retreat of the multitude of water-fowl of each order which ftock our fhores, driven fouthward by the extreme cold: moft of them regularly, others, whofe nature enables them to brave the ufual winters of the frigid zone, are with us only accidental guefts, and in feafons when the froft rages in their native land with unufual feverity.

From *Clea Nefs*, the land retires weftward, and, with the oppofite fhore of *Yorkfhire*, bounds the great eftuary of the *Humber*, which, winding deep into the country, is the receptacle of the *Trent*, and all the confiderable rivers of that vaft province; fome of which arife in its moft remote parts. All thefe coafts of *Lincolnfhire* are flat, and have been gained from the fea. *Barton* and *Barrow* have not at prefent the leaft appearance of ports; yet by *Holinfhed* were ftyled good ones §. Similar

* Compare Sir *W. Dugdale*'s maps of this tract, in its morafly and drained ftate. *Hift. Embank.* p. 375. 416.   † Same, p. 375.
‡ See *Tour in Scotland*, 1769; *Lincolnfhire*, where the fen birds are enumerated.
§ *Defcr. Britain*, 108.

accidents

accidents have befallen the upper part of the low tract of *Holderness*, which faces the congruent shores. *Hedon*, a few miles below *Hull*, several hundred years ago a port of great commerce, is now a mile and a half from the water, and has long given way to the rising fortune of the latter (a creation of *Edward* I. in 1296) on account of the excellency of its port. But in return, the sea has made most ample reprisals on the lands of this hundred : the site, and even the very names of several places, once towns of note upon the *Humber*, are now only recorded in history : and *Ravensper* was at one time a rival to *Hull* * ; and a port so very considerable in 1332, that *Edward Baliol* and the confederated *English* barons sailed from hence with a great fleet to invade *Scotland* ; and *Henry* IV. in 1399, made choice of this port to land at, to effect the deposal of *Richard* II. yet the whole of it has long since been devoured by the merciless ocean: extensive sands, dry at low water, are to be seen in their stead ; except *Sunk Island*, which, till about the year 1666, appeared among them like an elevated shoal, at which period it was regained, by embankments, from the sea ; and now forms a considerable estate, probably restored to its pristine condition.

SPURN HEAD.   *Spurn Head*, the *Ocelum Promontorium* of *Ptolemy*, terminates this side of the *Humber*, at present in form of a sickle, near which the wind-bound ships anchor securely. The place on which the lighthouses stand is a vast beach near two miles long, mixed with sand-hills flung up by the sea within the last seventy years.

The land from hence for some miles is composed of very lofty cliffs of brown clay, perpetually preyed on by the fury of the *German* sea, which devours whole acres at a time, and exposes on the shores considerable quantities of beautiful amber. Fine wheat grows on the clay, even to the edge of the cliffs. A country of the same fertility reaches from *Kilnsey*, near this place, as far as the village of *Sprottly*, extending, in a waved form, for numbers of miles ; and, when I saw it, richly cloathed with wheat and beans.

From near *Kilnsey* the land bends very gently inward, as far as the great promontory of *Flamborough* ; and is a continuance of high clayey cliff, till about the village of *Hornsey*. Near it is a mere, noted for its Eels and Pikes, at present separated from the sea by so small a space as to render its speedy destruction very probable. A street, called *Hornsey Beck*, has long since been swallowed : and of *Hide*, a neighboring town, only the tradition is left.

BRIDLINGTON   The country grows considerably lower ; and, near the base of the promon-
BAY.   tory, retires so far in as to form *Bridlington* bay, antiently called *Gabrantovicorum Sinus*, to which the Geographer adds Ευλιμεν☉, on account of the excellency and

* *Madox. Ant. Exch.* i, 422.

5   safety

safety of its port, where veffels ride in full fecurity under the fhelter of the lofty head-land. *Smithie* fand, the only one between *Flamborough* and *Spurn Head*, ftretches acrofs the entrance into *Bridlington* bay, and, in hard gales from the north and north-eaft, adds to the fecurity of that noble afylum for the coafting veffels. *Sureby*, an adjacent village, feems no more than a tranflation from the old appellation. The *Romans*, in all probability, had a naval ftation here; for here ends the road, vifible in many places between this place and *York*, and named, from its founders, the *Roman ridge*.

The head is formed of lime-ftone, of a fnowy whitenefs *, of a ftupendous height, and vaft magnificence, vifible far at fea. If we may depend on *Richard* of *Cirencefter*, the *Romans* named it *Brigantum Extrema*, and the bay *Portus Felix*. The *Saxons* ftyled the cape *Fleamburg*, perhaps from the lights which directed the great *Ida*, founder of the *Northumberland* kingdom, to land here, in 547, with a great body of their countrymen.

FLAMBOROUGH
HEAD.

The vaft height of the precipices, and the amazing grandeur of the caverns which open on the north fide, giving wide and folemn admiffion, through moft exalted arches, into the body of the mountain; together with the gradual decline of light, the deep filence of the place unlefs interrupted by the ftriking of the oar, the collifion of a fwelling wave againft the fides, or the loud flutter of the pigeons affrighted from their nefts in the diftant roof; afford pleafures of fcenery which fuch formations as this alone can yield. Thefe alfo are wonderfully diverfified. In fome parts the caverns penetrate far, and end in darknefs; in others are pervious, and give a romantic paffage by another opening equally fuperb. Many of the rocks are infulated, of a pyramidal form, and foar to a great height. The bafes of moft are folid; but in fome pierced through and arched. All are covered with the dung of the innumerable flocks of migratory birds which refort here annually to breed, and fill every little projection, every hole, which will give them leave to reft. Multitudes were fwimming about; others fwarmed in the air, and ftunned us with

---

* Soft near the top, and of a crumbling quality when expofed long to the froft. At the foot of the cliff it is hard, folid, and fmooth. Boats are employed every fummer in carrying great quantities to *Sunderland*, where it is burnt into excellent lime. Moft of the lime-ftone ufed at *Scarborough* is made from ftones flung up by the fea. It may be remarked, that whatfoever degree of hardnefs any lime-ftone poffeffes in the quarry, the mortar made from it, by proper management, may be made as hard, but by no means harder. Moft of the houfes in and about *London* are built with lime made of chalk; hence the many miferable cafualties there, by the fall of houfes. The workmen, fenfible of the weaknefs of that kind of mortar, endeavour to keep the walls together by lodging frames of timber in them; which being confumed in cafes of fire, the whole building tumbles fuddenly, and renders all attempts to extinguifh the fire very dangerous.—Mr. *Travis*.

the

the variety of their croaks and fcreams. Kittiwakes and Herring Gulls, Guillemots and Black Guillemots, Auks, Puffins, Shags, and Corvorants, are among the
fpecies which refort hither. The notes of all fea-fowl are moft harfh and inhar-
monious. I have often refted under rocks like thefe, attentive to the various
founds over my head; which, mixed with the deep roar of the waves flowly fwelling,
and retiring from the vaft caverns beneath, have produced a fine effect. The fharp
voice of the Gulls, the frequent chatter of the Guillemots, the loud notes of the
Auks, the fcream of the Herons, together with the deep periodical croak of the
Corvorants, which ferves as a bafs to the reft, have often furnifhed me with a concert, which, joined to the wild fcenery furrounding me, afforded in an high degree
that fpecies of pleafure which refults from the novelty and the gloomy majefty of
the entertainment.

ROCKY COASTS
BEGIN.

At *Flamborough* head commence the hard or rocky coafts of this fide of *Great Britain*, which continue, with the interruption of a few fandy bays and low land, to the
extremity of the kingdom. It often happens, that the bottom of the fea partakes
of the nature of the neighboring element: thus, about the head, and a few miles
to the northward (in places) the fhores are rocky, and the haunts of lobfters
and other cruftaceous animals. From thefe ftrata a tract of fine fand, from one
to five miles in breadth, extends floping eaftward, and from its edge to that of the
Dogger-bank is a deep bottom, rugged, rocky, and cavernous, and in moft parts
overgrown with corallines and fubmarine plants.

This difpofition of fhore gives to the inhabitants of this coaft the advantageous
fifhery which they poffefs; for the fhore on one hand, and the edges of the
Dogger-bank on the other, like the fides of a decoy, give a direction to the immenfe fhoals of the Cod genus, which annually migrate from the northern ocean,
to vifit, refide, and fpawn, in the parts adjacent to our coafts. They find plenty
of food from the plants of the rocks, and the worms of the fand, and fecure
fhelter for their fpawn in the cavernous part of the fcarry bottom. It is in the
channel between the banks and the fhores, in which the Cod are taken, or in
the hollows between the *Doggers* and *Well-bank*; for they do not like the agita-
tion of the water on the fhallows. On the contrary, the Skates, the Holibuts,
Flounders, and other flat fifh, bury themfelves in the fand, and fecure themfelves
from the turbulence of the waves.

An amazing fhoal of Haddocks vifit this coaft periodically, generally about
the tenth of *December*, and extend from the fhore near three miles in breadth,
and in length from *Flamborough* head to *Tinmouth* caftle, perhaps further north.
An army of a fmall fpecies of Shark, the PICKED, *Br. Zool.* iii. N° 40, flanks
the outfide of this fhoal to prey upon it; for when the fifhermen caft their lines

beyond

beyond the diſtance of three miles from land, they never catch any but thoſe voracious fiſh *.

Between *Flamborough* head and *Scarborough* projects *Filey Brig*, a ledge of rocks running far into the ſea, the cauſe of frequent ſhipwrecks. *Scarborough* caſtle, ſeated on a vaſt rock projecting into the water, ſucceeds. The ſpring-tides, at the time of the equinoxes, riſe here twenty-four feet; but at other times only twenty: the neap-tides from twelve to ſixteen. Then *Whitby*, noted for its neighboring allum-works, and more for its fine harbour, the only one on the whole coaſt: the admittance into which is a narrow channel between two high hills: it expands largely within, and is kept clean by the river *Eſk*. From hence to the mouth of the *Tees*, the boundary between this county and that of *Durham*, is a high and rude coaſt, indented with many bays, and varied with little fiſhing villages, built ſtrangely among the cliffs, filling every projecting ledge, in the ſame manner with thoſe of the peaſants in the pichureſque and rocky parts of *China*.

The *Tees*, the northern limit of this great county, opens with a wide mouth and mudded bottom into the ſea. This was the *Dunum Eſtuarium* of *Ptolemy*; and ſerves as a brief entrance for navigators into the country. Almoſt all the northern rivers deſcend with a rapid courſe, from their mountanous riſe and ſupply; and afford but a ſhort navigation. From hence the lead of the mineral parts of *Durham*, and the corn of its more level parts, are imported. In the mud of this eſtuary, more particularly, abounds the *Myxine Glutinoſa* of *Linnæus*, the *Hag* of the neighboring fiſhermen; a worm, which enters the mouths of the fiſh taken on hooks, that remain a tide under water, and devours the whole, leaving only the ſkin and bones. This alſo is the worm which converts water into a ſort of glue.

From *Seaton Snook*, in the biſhoprick of *Durham*, to *Hartlepool*, is a ſeries of ſand-banks, and the ſhore a long-continued ſandy ſhallow. From the *Neſs Point* of *Hartlepool* to *Blackhalls* is a rocky lime-ſtone coaſt, with frequent intervals of ſand-bank, and a ſtony beach; but *Seham* and *Hartlepool* is ſo very rugged, that no enemy could land, or even ſtand off the ſhore, without the moſt imminent danger: in particular, the coaſts about *Hawthorn Hive* are bold, excavated, and formed into groteſque figures, for ſeveral miles, and the ſhores rough with a broken and heavy ſea, by reaſon of the hidden rocks and ſpits of ſands which run out far.

* Conſult vol. iii. of the *Br. Zoology* for an account of the fiſh on this coaſt: alſo the *Tour in Scotland*, 1769. To Mr. *Travis*, Surgeon in *Scarborough*, I am indebted for the moſt curious articles.

from land.   From *Seham* to *Sunderland* are fand-hills and fhallow fandy beaches. From *Weremouth* to near *Cleadon*, low rocks of lime-ftone form the coaft, here and there interfected with fand-hills and ftony beaches. From thence to the mouth of the *Tyne*, and even to *Dunftanbrough* in *Northumberland*, the fhore is fandy, and the land in a few places rocky; but from thence to *Bamborough*, the coafts are high and rocky, in many places run far into the fea, and at low tides fhew their heads above water.

Bamborough caftle ftands on the laft of the range of rocky cliffs.   This fortrefs was founded by the *Saxon* monarch *Ida*.   After various fortunes it, has proved in its difmantled ftate of more ufe to mankind than when it boafted fome potent lord and fierce warders.   A charitable prelate of the fee of *Durham* purchafed the eftate, and left it for the ufe of the diftreffed feamen who might fuffer fhip-wreck on this dangerous coaft, and to unconfined charitable purpofes, at the dif-cretion of certain truftees.   The poor are, in the deareft feafons, fupplied with corn at a cheap rate; the wrecked, found fenfelefs and benumbed with cold, are taken inftantly into thefe hofpitable walls, and reftored to life by the affiftance of food, medicine, and warm beds; and if the fhip is capable of relief, that alfo is faved, by means of machines always ready for the purpofe *.

The *Farn iflands*, or rather rocks, form a group at no great diftance from fhore; the neareft a mile and fixty-eight chains; the fartheft about feven.   Thefe probably, at fome remote period, have been convulfed from the land, but now divided from it by a furious tide, rufhing through a channel from five to twelve fathoms in depth.   The original fea, to the eaft of the *Staples*, the remoteft rocks, fuddenly deepens to forty or fifty †.   *St. Cuthbert* firft made thefe rocks of note: he occafionally made the largeft of them the feat of his devotion and feclufion from the world; expelling, fays fuperftition, the malignant fpirits, the pre-oc-cupants.   Some remains of a chapel are ftill to be feen on it.   For ages paft, the fole tenants are a few cows, wafted over from the main land in the little cobles, or boats of the country; and the Eider Ducks, *Arct. Zool.* ii. N° 480, ftill diftinguifhed here by the name of the Saint.   Numberlefs fea-fowls, and of great variety of kinds, poffefs the remoter rocks, on which they find a more fecure retreat than on the low-cliffed fhores.   To the marine feathered tribe the whole coaft from *Flamborough* head to that of *St. Ebb's* is inhofpitable. They feek the loftieft promontories.   Where you hear of the haunts of the Razor-bills and Guillemots, Corvorants and Shags, you may be well affured, that

---

* *Tour in Scotland*, 1769; and fuller in Mr. *Hutchinfon's Northumberland*, ii. 176.
† *Adair. Hammond. Thompfon.*

the

the cliffs foar to a diftinguifhed height. Where thofe are wanting, they retire to fea-girt rocks, as fpots the left acceffible to mankind. The five fpecies of Auks and Guillemots appear in fpring, and vanifh in autumn : the other birds preferve their native haunts, or fpread along the neighboring fhores.

From *Bamborough* to the mouth of the *Tweed* is a fandy fhore, narrowing as it approaches our fifter kingdom. *Lindesfarn*, or the *Holy ifland*, with its ruined cathedral and caftle, lie remote from fhore, acceffible at every recefs of tide, and poffibly divided from *Northumberland* by the power of the waves in diftant ages. The tides do not fwell over this tract in the ufual manner of apparent flowing and gradual approach; but ooze gently out of every part of the fand, which at firft appears a quaggy extent, then, to the terror of the traveller, furrounds him with a fhining plain of fmooth unruffled water, reflecting the varied land-fcapes of the adjoining fhores *.

The *Tweed*, the antient *Alaunus*, a narrow geographical boundary between us and our fellow-fubjects the *Scottifh* nation, next fucceeds. After a fhort continuance of low land, *St. Ebb's head*, a lofty promontory, projects into the fea (frequented in the feafon by Razor-bills, Guillemots, and all the birds of the *Bafs*, excepting the Gannet) and its lower part is hollowed into moft auguft caverns. This, with *Fifenefs*, about thirty miles diftant, forms the entrance into that magnificent eftuary the firth of *Forth*, which extends inland fixty miles; and, with the canal from *Carron* to the firth of *Clyde*, intirely infulates the antient *Caledonia*. The ifle of *May* appears near the northern fide of the entrance; the vaft towering rock, the *Bafs*, lies near the fouthern. This lofty ifland is the fummer refort of birds innumerable, which, after difcharging the firft duty of nature, feek, with their young, other fhores or other climates. This is one of the few fpots in the northern hemifphere on which the Gannets neftle. Their fize, their fnowy plumage, their eafy flight, and their precipitate plunge after their prey, diftinguifh them at once from all the reft of the feathered tenants of the ifle, the Corvorants and Auks, the flights of whom are rapid, and the Gulls, which move with fluggifh wing.

Near the *Bafs* the entrance narrows, then opens, and bending inwards, forms on each fide a noble bay. The *Firth* contracts to a very narrow ftreight at *Queensferry*; then winds beautifully, till it terminates beyond *Alloa*, in the river to which it owes its name. The fhores are low, in part rocky, in part a pleafant beach; but every where of matchlefs beauty and population. *Edinburgh*, the capital, rifes with true grandeur near the fhore, with its port, the great em-

SCOTLAND.

ST. EBB's HEAD.

FIRTH OF FORTH.

* Mr. *Hutchinfon*, ii. 151.

porium,

porium, *Leith*, beneath, where the spring-tides sometimes rise fifteen and sixteen feet, and to seventeen or eighteen when the water is forced up the firth by a violent wind from the north-east. Almost every league of this great estuary is terminated with towns or villages, the effects of trade and industry. The elegant description of the coast of *Fife*, left us by *Johnston* *, is far from being exaggerated; and may, with equal justice, be applied to each shore.

FIFESHIRE, bounded by the firths of *Forth* and *Tay*, projects far into the sea; a country flourishing by its industry, and happy in numbers of ports, natural, artificial, or improved. Coal and lime, the native productions of the county, are exported in vast quantities. Excepting the unimportant colliery in *Sutherland*, those at *Largo Wood*, midway between the bay and *St. Andrews*, are the last on this side of *North Britain*. The coasts in general of this vast province are rocky and precipitous; but far from being lofty. The bays, particularly the beautiful one of *Largo*, are finely bounded by gravelly or sandy shores; and the land, in most parts, rises high to the middle of the county. Towards the northern end, the river *Edin*, and its little bay, by similarity of sound point out the *Tinna* of the old geographer.

FIRTH OF TAY.   The estuary of the *Tay* limits the north of *Fifeshire*. Before the mouth extends the sand retaining the *British* name of *Aber-tay*, or the place where the *Tay* discharges itself into the sea. The *Romans* preserved the antient name, and Latinized it into *Tava*. The entrance, at *Brough-tay* castle, is about three quarters of a mile wide; after which it expands, and goes about fourteen miles up the country before it assumes the form of a river. At the recess of the tides there appears a vast extent of sands, and a very shallow channel; but the high tides waft, even as high as *Perth*, vessels of a hundred and twenty tons. The shores are low, and the ground rises gently inland on the southern side: on the north it continues low, till it arrives at the foot of the *Grampian* hills, many miles distant. In some remote age the sea extended on the north side far beyond its present bounds. At a considerable distance above the flourishing port of *Dundee*, and remote inland, anchors have been found deep in the soil †. When these parts were deserted by the sea, it is probable that some opposite country was devoured by an inundation, which occasioned this partial desertion.

From thence to *Aberbrothic*, in the shire of *Angus*, noted for the venerable remains of its abbey, is a low and sandy shore. From *Aberbrothic* almost to *Montrose*, arises a bold rocky coast, lofty and precipitous, except where interrupted by the beautiful semicircular bay of *Lunan*. Several of the cliffs are penetrated by

* See *Tour in Scotland*, 1772. part ii. p. 212.   † *Douglas's East Coast of Scotland*, 14.

most

most amazing caverns; some open into the sea with a narrow entrance, and internally instantly rise into high and spacious vaults, and so extensively meandring, that no one as yet has had the courage to explore the end. The entrance of others shame the work of art in the noblest of the *Gothic* cathedrals. A magnificent portal appears divided in the middle by a great column, the basis of which sinks deep in the water. Thus the voyager may pass on one side in his boat, survey the wonders within, and return by the opposite side.

The cavern called the *Geylit-pot*, almost realises in form a fable in the *Persian* Tales. The hardy adventurer may make a long subterraneous voyage, with a picturesque scenery of rock above and on every side. He may be rowed in this solemn scene till he finds himself suddenly restored to the sight of the heavens: he finds himself in a circular chasm, open to the day, with a narrow bottom and extensive top, widening at the margin to the diameter of two hundred feet. On attaining the summit, he finds himself at a distance from the sea, amidst cornfields or verdant pastures, with a fine view of the country, and a gentleman's seat near to the place from which he had emerged. Such may be the amusement of the curious in summer calms! but when the storms are directed from the east, the view from the edge of this hollow is tremendous; for, from the height of above three hundred feet, they may look down on the furious waves, whitened with foam, and swelling from their confined passage.

Peninsulated rocks often jut from the face of the cliffs, precipitous on their sides, and washed by a great depth of water. The isthmus which joins them to the land, is often so extremely narrow as to render it impassable for more than two or three persons a-breast; but the tops spread into verdant areas, containing vestiges of rude fortifications, in antient and barbarous times the retreat of the neighboring inhabitants from the rage of a potent invader *.

Montrose, peninsulated by the sea, and the bason its beautiful harbour, stands on a bed of sand and gravel. The tide rushing furiously through a narrow entrance twice in twenty-four hours, fills the port with a depth of water sufficient to bring in vessels of large burden. Unfortunately, at the ebb they must lie dry; for none exceeding sixty tons can at that period float, and those only in the channel of the *South Esk*, which, near *Montrose*, discharges itself into the sea.

MONTROSE.

A sandy coast is continued for a small distance from *Montrose*. Rude rocky cliffs re-commence in the county of *Merns*, and front the ocean. Among the highest is *Fowls-heugh*, noted for the resort of multitudes of sea-birds. *Bervie* and *Stonehive* are two small ports overhung with rocks; and on the summit of a

---

* These descriptions borrowed from my own *Tours*.

most

moſt exalted one, are the vaſt ruins of *Dunnoter*, once the property of the warlike family of the *Keiths*.   The rocks adjacent to it, like the preceding, aſſume various and grotefque forms.

A little farther the antient *Deva*, or *Dee*, opens into the ſea, after forming a harbour to the fine and flouriſhing town of *Aberdeen*.   A ſandy coaſt continues for numbers of miles, part of which is ſo moveable as almoſt totally to have over-whelmed the pariſh of *Furvie* : two farms only exiſt, out of an eſtate, in 1600, va-lued at five hundred pounds a year.

BULLERS OF
BUCHAN.
A majeſtic rocky coaſt appears again.   The *Bullers* of *Buchan*, and the noble arched rock, ſo finely repreſented by the pencil of the Reverend Mr. *Cordiner* *, are juſtly eſteemed the wonders of this country.   The former is an amazing harbour, with an entrance through a moſt auguſt arch of great height and length. The inſide is a ſecure baſon, environed on every ſide by mural rocks : the whole projects far from the main land, and is bounded on each ſide by deep creeks ; ſo that the traveller who chuſes to walk round the narrow battlements, ought firſt to be well aſſured of the ſtrength of his head.

PETERHEAD.
A little farther is *Peterhead*, the moſt eaſtern port of *Scotland*, the common retreat of wind-bound ſhips ; and a port which fully merits the attention of go-vernment, to reader it more ſecure.   *Kinnaird-head*, the *Taizalum promontorium*, lies a little farther north, and, with the north-eaſtern extremity of *Cathnefs*, forms the firth of *Murray*, the *Tua Æſtuarium*, a bay of vaſt extent.   *Troup head* is an-other vaſt cape, to the weſt of the former.   The caverns and rocks of that pro-montory yield to none in magnificence and ſingularity of ſhape : of the latter, ſome emulate the form of lofty towers, others of inclining pyramids with central arches, pervious to boats.   The figures of theſe are the effect of chance, and owing to the colliſion of the waves, which wearing away the earth and crumbly parts, leave them the juſt ſubjects of our admiration.   Sea-plants, ſhells, and va-rious ſorts of marine exſanguious animals, cloath their baſes, waſhed by a deep and clear ſea ; and their ſummits reſound with the various clang of the feathered tribe.

CAVERNS AND
SINGULAR ROCKS:

HOW FORMED.

From hence the bay is bounded on the ſouth by the extenſive and rich plains of *Murray*.   The ſhore wants not its wild beauties.   The view of the noble cavern, called the rocks of *Cauſſie*, on the ſhore between *Burgh-head* and *Loſſie* mouth, drawn by Mr. *Cordiner*, fully evinces the aſſertion.   The bottom of the bay cloſes with the firth of *Invernefs*, from whence to the *Atlantic* ocean is a chain of rivers, lakes, and bays, with the interruption only of two miles of land between *Loch-oich* and *Loch-lochy*.   Unite thoſe two lakes by a canal, and the reſt of *North Britain* would be completely inſulated.

* *Antiquities and Scenery of Scotland*, letter vi. plates ii. iii.

To

Ch. Cordier pinx.

P. Maivell sculp

CAVES of CAUSSIE.

To the north the firth of *Cromartie*, and the firth of *Tayne*, the *Vara Æstuarium*, penetrate deep into the land. From *Dornoch*, the coast of *Sutherland* is low and sandy, except in a few places : one, at the water of *Brora*, is distinguished by the beauty of the rocky scenery ; in the midst of which the river precipitates itself into the sea, down a lofty precipice. The *Scottish Alps*, which heretofore kept remote from the shore, now approach very near ; and at the great promontory, the *Ripa Alta* of *Ptolemy*, the *Ord*, i. e. *Aird* of *Cathness*, or the *Height of Cathness*, terminate in a most sublime and abrupt manner in the sea. The upper part is covered with gloomy heath ; the lower is a stupendous precipice, excavated into vast caverns, the haunt of Seals and different sea-fowl. On the eastern side of the kingdom, this is the striking termination of the vast mountains of *Scotland*, which form its Highlands, the habitation of the original inhabitants, driven from their antient seats by the ancestors of Lowland *Scots*, descendants of *Saxons*, *French*, and *Normans*, congenerous with the *English*, yet absurdly and invidiously distinguished from them. Language, as well as striking natural boundaries, mark their place. Their mountains face on the west the *Atlantic* ocean ; wind along the west of *Cathness* ; among which *Morvern* and *Scaraben*, *Ben-Hop* and *Ben-Lugal* arise pre-eminent. *Sutherland* is entirely *Alpine*, as are *Ross-shire* and *Inverness-shire*. Their *Summæ Alpes* are, *Meal Fourvounich*, the *Coryarich*, *Benewish*, and *Benevish* near *Fort William* ; the last of which is reported to be fourteen hundred and fifty yards in height. Great part of *Aberdeenshire* lies in this tract. It boasts of another *Morvern*, soaring far beyond the others : this is in the centre of the *Grampian hills*, and perhaps the highest from the sea of any in *Great Britain*. They again comprehend the eastern part of *Perthshire*, and finish on the magnificent shores of *Loch-lomond*, on the western side of which *Ben-lomond* rises, distinguished among its fellows. From hence the rest of *North Britain* forms a chain of humbler hills ; but in *Cumberland*, part of *Westmoreland*, *Yorkshire*, *Lancashire*, and *Derbyshire*, the *Alps* resume their former majesty. A long and tame interval succeeds. The long sublime tract of *Wales* arises, the antient possession of the antient *British* race. From the *Ord*, the great mountains recede inland, and leave a vast flat between their bases and the sea, fronting the waves with a series of lofty rocky precipices, as far as the little creek of *Staxigo* ; the whole a bold, but most inhospitable shore for shipping. *Wick* and *Staxigo* have indeed their creeks, or rather chasms, which open between the cliffs, and may accidentally prove a retreat, unless in an eastern gale.

Sinclair and *Freswick* bays are sandy, and afford safe anchorage : from the last the country rises into lofty cliffs, many composed of small strata of stones, as regular as a mason could lay them ; and before hem rise insulated stacks or co-

ORD OF CATHNESS.

HIGHLAND ALPS.

ENGLISH.

CAMBRIAN.

lumns

lumns of fimilar materials, fome hollowed into arches; others, pillar-like, afpire in heights equal to the land *. Thefe are animated with birds. All their œconomy may be viewed with eafe from the neighboring cliffs; their loves, incubation, exclufion, and nutrition.

Dungfby-head, the antient *Berubium*, terminates the eaftern fide of this kingdom, as *Far out-head*, the old *Tarvedum*, does the weftern. *Strathy-head*, the *Vervedrum* of *Ptolemy*, lies intermediate. The whole tract faces the north, and confifts of various noted headlands, giving fhelter to numerous bays, many of which penetrate deep into the country. Let me make this general remark,—that nature hath, with a niggardly hand, dealt out her harbours to the eaftern coafts of the *Britifh* ifles; but fhewn a profufion on their weftern fides. What numberlefs lochs, with great depth of water, wind into the weftern counties of *Scotland*, overfhadowed and fheltered by lofty mountains! and what multitudes of noble harbours do the weftern provinces of *Ireland* open into the immenfe *Atlantic* ocean!

GERMAN OR NORTH SEA.

The fea which wafhes the fhores of *Britain*, which have paffed under my review, was originally called, by one of the antients †, *Oceanus Britannicus*, forming part of that vaft expanfe which furrounds our iflands. *Pliny* confined that title to the fpace between the mouth of the *Rhine* and that of the *Seine*; and beftowed on this fea the name of *Septentrionalis* ‡; and *Ptolemy* called it *Germanicus*: both which it ftill retains. Its northern extremity lies between *Dungfby-head*, in lat. 58, 35 north, and the fame latitude in the fouth of *Norway*. Before the feparation of *Britain* from *Gaul* it could only be confidered as a vaft

TIDES, THEIR DIRECTION;

bay; but that period is beyond the commencement of record. The tides flow into it from the north-eaft to the fouth-weft, according to the direction of the coaft; but in mid-fea the reflux fets to the north, to difcharge itfelf through the great channel between the *Schetland* ifles and *Norway* §. The depth of

DEPTHS.

water, at higheft fpring-tides, in the ftreights of *Dover*, is twenty-five fathoms: it deepens to thirty-one, between *Lowftoff* and the mouth of the *Maes*: between the *Wells-bank* and *Doggers-bank* gains, in one place only, a few fathoms. Beyond the *Dogger* it deepens from forty-eight to feventy-two: between *Buchan-nefs* and *Schutnefs* in *Norway*, within the *Buchan deeps*, it has from eighty-fix to a hundred fathoms; then decreafes, towards the *Orkney* and *Schetland* ifles, from feventy-five to forty; but between the *Schetlands* and *Bergen*, the northern end of this fea, the depth is from a hundred and twenty to a hundred and fifty fathoms.

---

* See Mr. *Cordiner*'s beautiful view of a ftack of this kind, tab. xv. † *Mela*. ‡ *Plin*. lib. iv. c. 19. § Mr. *William Fergufon*.

The

Ch. Cordiner pinx.

ROCKS near SAND SIDE.

P. Mazell sculp.

The coasts from *Dungsby-head* to *Flamborough-head* are bold and high, and may be seen at sea from seven to fourteen leagues: from the last to *Spurn-head* is also a clear coast; but the rest of the coast of *Norfolk* and *Suffolk* is low, visible at small distance, and rendered dangerous by the number of sand-banks projecting far to sea. After passing the *Spurn-head*, navigators steer between the inner and outer *Dousings*, for the *floating light* kept on board a small vessel (constructed for that purpose) always anchored at the inner edge of a sand called *Dog shon's Shoal*, about eight leagues from the coast of *Lincolnshire*, in about fifteen fathom water. From thence they make for *Cromer* in *Norfolk*; and from that point, till they arrive at the *Nore*, their track is all the way through a number of narrow channels near the most dangerous sands: to which, if we add foggy weather, dark nights, storms, contrary winds, and very near adjacent lee-shores, it may be very fairly reckoned the most dangerous of the much-frequented navigations in the world.

But fortunately, to the north of these, this sea is much more remarkable for sand-banks of utility than of danger, and would never have been observed but for the multitudes of fishes which, at different seasons, according to their species, resort to their sides, from the great northern deeps, either for the sake of variety of food which they yield, or to depose their spawn in security. The first to be taken notice of does not come within the description, yet should not be passed over in silence, as it comes within the natural history of the *North* sea. An anonymous sand runs across the channel between *Buchan-ness* and the north end of *Juts-riff* the left depth of water over it is forty fathoms; so that it would scarcely be thought of, did not the water suddenly deepen again, and form that place which is styled the *Buchan deeps*.

The *Long Bank*, or the *Long Fortys*, bears E. S. E. from *Buchan-ness*, about forty five miles distant, and extends southward as far as opposite to *Newcastle*; is about fifty leagues in length, and seven in breadth; and has on it from thirty-two to forty-five fathoms of water. The ground is a coarse gravel, mixed with marine plants, and is esteemed a good fishing bank.

The *Mar Bank* lies between the former and the shore opposite to *Berwick*; is oval, about fifteen miles long, and has about twenty-six fathom of water, and round it about forty.

The bank called *Montrose Pits* lies a little to the east of the middle of the *Long Fortys*. It is about fifty miles long, and most remarkable for five great pits or hollows, from three to four miles in diameter: on their edges is only forty fathom water; yet they suddenly deepen to seventy, and even a hundred fathom, on a soft muddy bottom: the margins on the contrary are gravelly. I enquired whether the

surface

surface of this wonderful bank appeared in any way agitated, as I had suspicion that the pits might have been productive of whirlpools; but was informed, that the sea there exhibited no uncommon appearance.

DOGGERS BANK.

The noted *Doggers Bank* next succeeds. It commences at the distance of twelve leagues from *Flamborough Head*, and extends across the sea, nearly east, above seventy-two leagues, joining *Horn-riff*, a very narrow strip of sand which ends on the coast of *Jutland*. The greatest breadth is twenty leagues; and in parts it has only on it ten or eleven fathoms of water, in others twenty-four or five. To the south of the *Dogger* is a vast extent of sand-bank, named, in different parts, the

WELL BANK.

Well Bank, the *Swart Bank*, and the *Brown Bank*, all covered with sufficient depth of water; but between them and the *British* coasts are the *Ower* and the *Lemon*, dreaded by mariners, and numbers of others infamous for shipwrecks. The channel between the *Dogger Bank* and the *Well Bank* deepens even to forty fathoms. This hollow is called the *Silver Pits*, and is noted for the cod-fishery which supplies the *London* markets. The cod-fish love the deeps: the flat-fish the shallows. I will not repeat what I have, in another place, so amply treated of *. I must only lament, that the fisheries of this bank are only subservient to the purposes of luxury. Was (according to the plan of my humane friend, Mr. *Travis* of *Scarborough*) a canal formed from any part of the neighboring coast to that at *Leeds*, thousands of manufacturers would receive a cheap and wholesome food; insurrections in times of scarcity of grain be prevented; our manufactures worked at an easy rate; our rivals in trade thereby undersold; and, in defiance of the probably approaching decrease of the *Newfoundland* fishery (since the loss of *America*) contribute to form a nursery of seamen sufficient to preserve the small remnant we have left of respect from foreign nations.

I have, to the best of my abilities, enumerated the *British* fish, in the third volume of the *British Zoology*. The *Faunula* which I have prefixed to Mr. *Lightfoot's Flora Scotica*, contains those which frequent the northern coasts of *Great Britain*; in which will be found wanting many of those of *South Britain*. The Reverend Mr. LIGHTFOOT, in that work, hath given a most elaborate account of the submarine plants of our northern sea.

I will now pursue my voyage from the extreme shores of *North Britain* through a new ocean. Here commences the *Oceanus Caledonius*, or *Deucaledonius*, of *Ptolemy*;

CALEDONIAN OCEAN.

a vast expanse, extending to the west as far as *Greenland*, and northward to the extreme north. This I should call the NORTHERN OCEAN, distinguishing its parts by other names suitable to the coasts. From *Dungsby Head* the *Orkney* islands

* See *Br. Zool.* iii. Articles Haddock, Ling, and Turbot.

‘ appear

'appear spreading along the horizon, and yield a most charming prospect. Some
'of them are so near as distinctly to exhibit the rocky fronts of those bold promon-
'tories which sustain the weight of the vast currents from the *Atlantic*. Others
'shew more faint: their distances finely expressed as they retire from the eye,
'until the mountains of the more remote have scarcely a deeper azure than the
'sky, and are hardly discernible rising over the surface of the ocean *.

Between these and the main land, about two miles from the *Cathness* shore, lies
Stroma, the *Ocetis* of *Ptolemy*, a little island, an appertenance to that county, fertile
by the manual labor of about thirty families; pleasant, and lofty enough for the
resort of the Auk tribe. The noted mummies are now lost, occasioned by the doors
of the caverns in which they were deposited being broken down, and admission
given to cattle, which have trampled them to pieces. This catacomb stands on a
neck of land bounded by the sea on three sides. The salt air and spray expels all
insects, and is the only preservation the bodies have; some of which had been
lodged here a great number of years. In many of the isles, the inhabitants use
no other method for preserving their meat from putrefaction than hanging it in
caves of the sea, and the method is vindicated by the success.

This island lies in the *Pentland Firth*, noted for the violence of the tides; tremendous to the sight, but dangerous only when passed at improper times. They
set in from the north-west: the flood, on the contrary, on the coasts of *Lewis*, pours
in from the south †. The tide of flood upon *Stroma* (and other islands similarly
situated in mid-stream) divides or splits before it reaches it. A current runs with
great violence on both sides, then unites, at some distance from the opposite end,
and forms a single current, running at spring-tides at the rate of nine knots an
hour; at neap, at that of three only. The space between the dividing tides, at
different ends of an island, is quite stagnant, and is called the eddy. Some of
them are a mile or two long, and give room for a ship to tack to and fro, till the
tides are so far spent as to permit it to pursue its voyage.

The most boisterous parts of the streams are at the extremities of the island, and
a little beyond the top of the eddy, where they unite. The collision of these opposite streams excite a circular motion, and, when the tide is very strong, occasion
whirlpools in form of an inverted bell, the largest diameter of which may be about
three feet. In spring-tides they have force enough to turn a vessel round, but not to
do any damage: but there have been instances of small boats being swallowed up.
These whirlpools are largest when first formed; are carried away with the stream,
and disappear, but are quickly followed by others. The spiral motion or suction

---

* Mr. *Cordiner's* elegant description, p. 85.   † *Mackensie's Charts of the Orknies*, p. 4, 5.

does

does not extend far beyond the cavity: a boat may pass within twenty yards of these whirlpools with safety. Fishermen who happen to find themselves within a dangerous distance, fling in an oar, or any bulky body, which breaks the continuity of the surface, and interrupts the vertiginous motion, and forces the water to rush suddenly in on all sides and fill up the cavity. In stormy weather, the waves themselves destroy this phænomenon. A sunk rock near the concourse of these rapid tides occasions a most dreadful appearance. The stream meeting with an interruption, falls over with great violence, reaches the bottom, and brings up with it sand, shells, fishes, or whatsoever else it meets with; which, with boats, or whatsoever it happens to meet, is whirled from the centre of the eruption towards the circumference with amazing velocity, and the troubled surface boils and bubbles like a great cauldron, then darts off with a succession of whirlpools from successive ebullitions. These are called *Rousts*, and are attended with the utmost danger to small boats, which are agitated to such a degree, that (even should they not be overset) the men are flung out of them, to perish without any chance of redemption. It is during the ebb that they are tremendous, and most so in that of a spring-tide with a west wind, and that in the calmest weather; for during flood they are passed with the greatest safety. Vessels in a calm are never in danger of touching on an island or visible rock, when they get into a current, but are always carried safe from all danger.

ROUSTS.

Swona, a little island, the most southern of the *Orknies*, is about four miles beyond *Stroma*, and is noted for its tremendous streams, and in particular the whirlpools called the *Wells* of *Swona*, which in a higher degree exhibit all the appearances of the former. What contributes to encrease the rage of the tides, besides their confinement between so many islands, is the irregular position of the sounds, and their little depth of water. The same shallowness extends to every side of the *Orknies*; an evidence that they had once been part of the mother isle, rent from it by some mighty convulsion. The middle of the channel, between *Stroma* and the main land, has only ten fathom water: the greatest depth around that island is only eighteen. The sounds are from three to forty-six fathom deep: the greater depths are between *South Ronaldsha* and *South Wales*; for in general the other sounds are only from three to thirteen; and the circumambient depth of the whole group very rarely exceeds twenty-five.

SWONA.

DEPTH OF WATER.

About these islands commences a decrease of the tides. They lie in a great ocean, in which the waters have room to expand; therefore never experience that height of flood which is constant in the contracted seas. Here ordinary spring-tides do not exceed eight feet; and very extraordinary spring-tides fourteen, even when acted on by the violence of the winds *.

TIDES.

* *Murdoch Mackensie.*

The

The time of the difcovery and population of the *Orknies* is unknown. Probably it was very early; for we are told that they owe their name to the *Greeks*.

Orcades has memorant dictas a nomine Græco *.

Mela and *Pliny* take notice of them; and the laft defcribes their number and cluftered form with much accuracy †. The fleet of *Agricola* failed round them, and made a conqueft of them; but the *Romans* probably never retained any part of *Caledonia*. I found no marks of them beyond *Orrea* or *Inchtuthel* ‡, excepting at *Fortingal* ‖ in *Breadalbine*, where there is a fmall camp, poffibly no more than a temporary advanced poft. Notwithftanding this, they muft have had, by means of fhipping, a communicated knowlege of the coafts of *North Britain* even to the *Orcades*. *Ptolemy* hath, from information collected by thofe means, given the names of every nation, confiderable river, and head-land, on the eaftern, northern, and weftern coaft. But the *Romans* had forgotten the navigation of thefe feas, otherwise the poet would never have celebrated the courage of his countrymen, in failing in purfuit of the plundering *Saxons* through *unknown ftreights*, and a naval victory obtained off thefe iflands by the forces fent to the relief of the diftreffed *Britons* by *Honorius*.

Quid Sidera profunt?
Ignotumque fretum? Maduerunt <i>Saxone</i> fufo
Orcades §.

The *Orkney* ifles in after times became poffeffed by the *Picts*; and again by the *Scots*. The latter gave way to the *Norwegian* pirates, who were fubdued by *Harold Harfargre* about the year 875 ¶, and the iflands united to the crown of *Norway*. They remained under the *Norwegians* till the year 1263, accepted their laws, and ufed their language. The *Norfe*, or *Norwegian* language was generally ufed in the *Orkney* and *Schetland* iflands even to the laft century: but, except in *Foula*, where a few words are ftill known by the aged people, it is quite loft. The *Englifh* tongue, with a *Norwegian* accent, is that of the iflands; but the appearance of the people, their manners and genius, evidently fhew their northern origin. The iflands vary in their form and height. Great part of *Hoy* is mountanous and lofty. The noted land-mark, the hill of *Hoy*, is faid to be five hundred and forty yards high. The fides of all thefe hills are covered with long heath, in which breed multitudes of Curlews, Green Plovers, Redfhanks, and other Waders. The Short-eared Owl is alfo very frequent here, and neftles in the ground. It is

LANGUAGE.

ROCKS OF THE ORKNIES.

---

* *Claudian.*   † *Mela*, lib. iii. c. 6. *Plin.* lib. iv. c. 16.   ‡ *Tour Scotl.* 1772. p. 70.
‖ *Same* p. 25.   § *Claudian*, de iv. Conf. Honorii.   ¶ *Torfæus Rer. Orcad.* lib. i. c. 3. p. 10.

d   probable

probable that it is from hence, as well as from *Norway*, that it migrates, in the beginning of winter, to the more southern parts of *Britain*. Most of the Waders migrate; but they must receive considerable reinforcements from the most distant parts of the north, to fill the numbers which cover our shores. The cliffs are of a most stupendous height, and quite mural to the very sea. The *Berry Head* is an exalted precipice, with an august cave at the bottom, opening into the sea. The Ern Eagles possess, by distant pairs, the upper part of the rocks: neither these nor any other Falcons will bear society; but, as *Pliny* elegantly expresses it, *Adultos per-sequitur parens et longe fugat, æmulos scilicet rapinæ. Et alioquin unum par aquilarum magno ad populandum tractu, ut satietur, indiget* *. Auks, Corvorants, and all the tribes which love exalted situations, breed by thousands in the other parts. The Tyste, or Black Guillemot, N° 236, secures itself in a crack in the rock, or by scraping a burrow in the little earth it may find; there it lays a single egg, of a dirty olive blotched with a darker. This species never migrates from the *Orknies*. The Foolish Guillemot, N° 436, continues till *November*. The Little Auk, N° 429, a rare bird in other parts of *Britain*, breeds in the holes of the lofty precipices. And the Lyre, or the Sheerwater, N° 462, burrows in the earth among the rocks of *Hoy* and *Eda*, and forms an article of commerce with its feathers, and of food with its flesh, which is salted and kept for the provision of the winter. In that season they are seen skimming the ocean at most surprizing distances from land. The Stormy Petrel, N° 464, breeds frequently among the loose stones; then takes to sea and affrights the superstitious sailor with its appearance. Woodcocks scarcely ever appear here. Fieldfares make this a short baiting-place: and the Snow Bunting, N° 122, often alights and covers whole tracts of country, driven by the frost from the farthest north.

A few Wild Swans breed in some lochs in *Mainland*; but the greatest part of these birds, all the Bernacles, Brent Geese, and several other palmated birds, retire in the spring to more northern latitudes. But to the Swallow-tailed Duck, the Pintail, and a few others, this is a warm climate; for they retire here to pass their winters in the sheltered bays. Any other remarks may be intermixed with those on *Schetland*; for there is great similiarity of subjects in both the groups.

The last lie about sixty miles to the north-east of the most northern *Orkney*. Midway is *Fair Island*, a spot about three miles long, with high and rocky shores, inhabited by about a hundred and seventy people: an industrious race; the men fishers; the women knitters and spinners. The depth of water round varies to twenty-six fathoms. The tide divides at the north end, runs with great velocity, and forms on the east side a considerable eddy.

BIRDS.

FAIR ISLE.

* *Hist. Nat.* lib. x. c. 3.

Schetland

III

G Low del.

P Mazell sculp.

The DOREHOLM.

Schetland confifts of feveral iflands. *Mainland*, the principal, extends from fouth to north twenty-eight leagues, and is moft fingularly formed; confifting of an infinite number of *peninfulæ* connected by very narrow ifthmufes. That called *Mavifgrind*, which unites the parifh of *North Maven*, is only eighty yards broad. But the irregular fhape of this ifland occafions it to abound with the fineft and moft fecure ports, called here *voes*; a moft providential difpenfation in a fea which fwarms with fifhes of the moft general ufe. The adjacent iflands are in general fo near to the mother ifland, and their headlands point fo exactly to its corresponding capes, that it is highly probable that they once made a part of the *Mainland*. The rocks and ftacks affume great variety of forms, fuch as steeples and Gothic cathedrals rifing out of the water, fleets of fhips, and other fancied fhapes. The *Doreholm*, in the parifh of *North Maven*, is very fingular: part is rounded, the reft feems a ruin, compofed of a fingle thin fragment of rock, with a magnificent arch within, feventy feet in height.

To ufe the words of Captain *Thomas Prefton*, to whom we are indebted for an excellent chart of this group, ' the land is wild, barren, and mountanous; nor ' is there fo much as a bufh or a tree to be feen. The fhores are difficult, and in ' many parts inacceffible; rude, fteep, and iron-like; the fight of which ftrikes ' the mind with dread and horror; and fuch monftrous precipices and hideous rocks ' as bring all *Brobdingnag* before your thoughts. Thefe iflands lie between lat. 60 ' to 61. In winter the fun fets foon after it rifes, and in fummer rifes foon after it ' fets; fo that in that feafon the nights are almoft as light as the day; as on the ' contrary, in *December* the day is nearly as dark as the night. About the folftice, ' we fee every night the *aurora borealis*, or, as they are called by the natives, the ' *merry dancers*, which fpread a broad glaring appearance over the whole northern ' hemifphere *.'

They are the conftant attendants of the clear evenings in all thefe northern iflands, and prove great reliefs amidft the gloom of the long winter nights. They commonly appear at twilight near the horizon, of a dun color, approaching to yellow: fometimes continuing in that ftate for feveral hours without any fenfible motion; after which they break out into ftreams of ftronger light, fpreading into columns, and altering flowly into ten thoufand different fhapes, varying their colors from all the tints of yellow to the obfcureft ruffet. They often cover the whole hemifphere, and then make the moft brilliant appearance. Their motions at thefe times are moft amazingly quick; and they aftonifh the fpectator with the rapid change of their form. They break out in places where none were feen before, fkimming

AURORA BORE-
ALIS.

* *Phil. Tranf. abr.* xi. 1328.

d 2

brifkly

briſkly along the heavens: are ſuddenly extinguiſhed, and leave behind an uniform duſky tract. This again is brilliantly illuminated in the ſame manner, and as ſuddenly left a dull blank. In certain nights they aſſume the appearance of vaſt columns, on one ſide of the deepeſt yellow, on the other declining away till it becomes undiſtinguiſhed from the ſky. They have generally a ſtrong tremulous motion from end to end, which continues till the whole vaniſhes. In a word, we, who only ſee the extremities of theſe northern phœnomena, have but a faint idea of their ſplendor, and their motions. According to the ſtate of the atmoſphere they differ in colors. They often put on the color of blood, and make a moſt dreadful appearance. The ruſtic ſages become prophetic, and terrify the gazing ſpectators with the dread of war, peſtilence, and famine. This ſuperſtition was not peculiar to the northern iſlands; nor are theſe appearances of recent date. The antients called them *Chaſmata*, and *Trabes*, and *Bolides*, according to their forms or colors *. In old times they were extremely rare, and on that account were the more taken notice of. From the days of *Plutarch* to thoſe of our ſage hiſtorian Sir *Richard Baker*, they were ſuppoſed to have been portentous of great events: and timid imagination ſhaped them into aerial conflicts.

> Fierce fiery warriors fight upon the clouds
> In ranks and ſquadrons and right form of war.

After, I ſuppoſe, a very long intermiſſion, they appeared with great brilliancy in *England*, on *March* 6rh, 1715-16. The philoſophers paid a proper attention †. The vulgar conſidered them as marking the introduction of a foreign race of princes. The novelty is now ceaſed, and their cauſe perhaps properly attributed to the greater abundance of electrical matter.

STORMS.

HERRINGS.

The tempeſts which reign over theſe iſlands during winter is aſtoniſhing. The cold is moderate; the fogs great and frequent; but the ſtorms agitate the water even to the bottom of theſe comparatively ſhallow ſeas. The fiſh ſeek the bottom of the great deeps: and the Herrings, which appear off the *Schetlands* in amazing columns in *June*, perform the circuit of our iſland, and retire beyond the knowlege of man. When the main body of theſe fiſh approaches from the north, it alters the very appearance of the ocean. It is divided into columns of five or ſix miles in length, and three or four in breadth, and they drive the water before them with a ſort of rippling current. Sometimes they ſink for a ſmall ſpace, then riſe again; and in bright weather reflect a variety of ſplendid colors, like a field of moſt

* *Ariſtol. Meteorolog.* lib. i. c. 5. *Plin. Nat. Hiſt.* lib. ii. c. 26.
† See various accounts of them in the *Phil. Tranſ. abr.* iv. part ii. 138.

precious

precious gems. Birds and fish of prey attend and mark their progress. The Whales of several kinds keep on the outside, and, deliberately opening their vast mouths, take them in by hundreds. Gannets and Gulls dart down upon them; and the diving tribe aid their persecution, with the cetaceous fishes*. Mankind joins in the chace; for this useful species gives food to millions, mediately and immediately. *Dutch, French, Flemings, Danes,* and *English*, rendezvous in *Braffa* found to meet these treasures of the ocean: and return to distribute their booty even to the distant *Antilles*.

Cod, Ling, and *Torsk* †, furnish cargoes to other adventurers. I wish I could speak with the same satisfaction of this as of the free fishery of the Herring; but in these distant islands, the hand of oppression reigns uncontrolled. The poor vassals (in defiance of laws still kept in bondage) are compelled to slave, and hazard their lives in the capture, to deliver their fish to their lords for a trifling sum, who sell them to adventurers from different parts at a high price.

COD-FISH.

Among other scarcer fishes the Opah, *Br. Zool.* iii. N° 101. is found in abundance. It seems a fish of the north as well as the *Torsk*; the last is not found south of the *Orknies*; the former extends even to the banks of *Newfoundland*.

OPAH.

The birds of these islands are the same with those of the *Orknies*, except the Skua, p. 531, which breeds only in *Foula* and *Unst*. Among the few land-birds which migrate to them in summer, is the Golden-crested Wren, N° 153. Its shortest flight must be sixty miles, except it should rest midway on *Fair island*; a surprising flight for so diminutive a bird!

BIRDS.

Multitudes of the inhabitants of each cluster of islands feed, during the season, on the eggs of the birds of the cliffs. The method of taking them is so very hazardous, as to satisfy one of the extremity to which the poor people are driven for want of food. *Copinsha, Hunda, Hoy, Foula,* and *Nofs head*, are the most celebrated rocks; and the neighboring natives the most expert climbers and adventurers after the game of the precipice. The height of some is above fifty fathoms; their faces roughened with shelves or ledges, sufficient only for the birds to rest and lay their eggs. To these the dauntless fowlers will ascend, pass intrepidly from one to the other, collect the eggs and birds, and descend with the same indifference. In most places, the attempt is made from above: they are lowered from the slope contiguous to the brink, by a rope, sometimes made of straw, sometimes of the bristles of the hog: they prefer the last, even to ropes of hemp, as it is not liable to be cut by the sharpness of the rocks; the former is apt to untwist. They trust themselves to a single assistant, who lets his companion down, and holds the rope, depending on his strength alone; which

FOWLING.

---

* See my *Voy. to the Hebrides*, and *Br. Zool.* iii. for the history of the Herring.   † *Br. Zool.* iii. N° 89.

often.

often fails, and the adventurer is sure to be dashed to pieces, or drowned in the subjacent sea. The rope is often shifted from place to place, with the impending weight of the fowler and his booty. The person above receives signals for the purpose, his associate being far out of sight; who, during the operation, by help of a staff, springs from the face of the rocks, to avoid injury from the projecting parts.

In *Foula*, they will trust to a small stake driven into the ground, or to a small dagger, which the natives usually carry about them; and which they will stick into the ground, and, twisting round it a fishing cord, descend by that to climbing places, and, after finishing their business, swarm up by it without fear. Few who make a practice of this come to a natural death. They have a common saying, ' Such a one's *Gutcher* went over the *Sneak*; and my father went over ' the *Sneak* too.' It is a pity that the old *Norwegian* law was not here in force. It considered this kind of death as a species of suicide. The next of kin (in case the body could be seen) was directed to go the same way; if he refused, the corpse was not to be admitted into holy ground *.

But the most singular species of fowling is on the holm of *Noss*, a vast rock severed from the isle of *Noss* by some unknown convulsion, and only about sixteen fathoms distant. It is of the same stupendous height as the opposite precipice †, with a raging sea between; so that the intervening chasm is of matchless horror. Some adventurous climber has reached the rock in a boat, gained the height, and fastened several stakes on the small portion of earth which is to be found on the top: correspondent stakes are placed on the edge of the correspondent cliffs. A rope is fixed to the stakes on both sides, along which a machine, called a cradle, is contrived to slide; and, by the help of a small parallel cord fastened in like manner, the adventurer wafts himself over, and returns with his booty, which is the eggs or young of the Black-backed Gull, N° 451, and the Herring Gull, N° 452.

QUADRUPEDS.

The number of wild Quadrupeds which have reached the *Orkney* and *Schetland* islands are only five; the Otter, Brown Rat, Common Mouse, Fetid Shrew, and Bat. Rabbits are not of *British* origin, but naturalized in every part. In the sandy isles of *Orkney*, they are found in myriads, and their skins are a great article of commerce; but the injury they do in setting the unstable soil in motion, greatly counter-vales the profit.

THESE ISLES ONCE WOODED.

In many parts of these islands are evident marks of their having been a wooded country. In the parish of *St. Andrew* in the *Orknies*, in *North Maven*, and even in *Foula* in the *Schetlands*, often large tracts are discovered filled with the remains of large trees, which are usually found after some violent tempest hath

* *Debes, Hist. Ferroe Isles*, 154. † 480 feet.

3

blown

IV

G. Low del.

BIRD CATCHING at ORKNEY.

P. Mazell sc.

p. XXX.

blown away the incumbent ſtrata of ſand or gravel with which they have been covered. They are lodged in a moraſſy ground, and often ten feet beneath the peat. Some ſtand in the poſition in which they grew ; others lie horizontally, and all the ſame way, as if they had either been blown down, or overturned by a partial deluge. Yet at preſent no kind of wood can be made to grow ; and even the loweſt and moſt common ſhrub is cultivated with the greateſt difficulty. The hazel, the herbaceous, reticulated, creeping, and common willow, are the only ſhrubs of the iſland, and thoſe are ſcattered with a ſparing hand. I ſhall, in another place, conſider the decreaſe of vegetation in this northern progreſs.

The great quantity of turf which Providence hath beſtowed on all theſe iſlands, excepting *Sanda*, is another proof of the abundance of trees and other vegetables, long ſince loſt from the ſurface. The application of this *humus vegetabilis* for the purpoſe of fuel, is ſaid to have been firſt taught the natives by *Einar*, a *Norwegian*, ſurnamed, from that circumſtance, *Torf einar*, *Einar de Ceſpite* *. Had he lived in *Greece*, he could not have eſcaped deification for ſo uſeful a diſcovery.

Before I quit the laſt of *Britiſh* iſles, I ſhall, as ſupplemental to the antiquities mentioned in my *Tours in Scotland*, give a brief account of others found in theſe groups.

ANTIQUITIES.

The *Orknies*, the *Schetlands*, *Cathneſs*, *Sutherland*, and *Roſs-ſhire*, with the *Hebrides*, were, for centuries, poſſeſſed by the *Norwegians*; and, in many inſtances, they adopted their cuſtoms. Of the antient monuments ſtill remaining, ſeveral are common to *Scandinavia* and the old inhabitants of *Britain :* others ſeem peculiar to their northern conquerors. Among thoſe are the circular buildings, known by the names of *Pictiſh* houſes, *Burghs*, and *Duns* . the firſt are of modern date, and to be exploded, as they never were the work of the *Picts* ; the ſecond are aſſuredly right, and point out the founders, who at the ſame time beſtowed on them their natal name of *Borg*, a defence or caſtle †, a *Sueo-Gothic* word ; and the Highlanders univerſally apply to theſe places the *Celtic* name *Dun*, ſignifying a hill defended by a tower ‡. This alſo furniſhes the proof of their uſe, was there no other to be diſcovered. They are confined to the counties once ſubject to the crown of *Norway*. With few exceptions, they are built within ſight of the ſea, and one or more within ſight of the other ; ſo that on a ſignal by fire, by flag, or by trumpet, they could give notice of approaching danger, and yield a mutual ſuccour. In the *Schetland* and *Orkney*

---

* *Torfæus Rer. Orcad.* lib. i. c. 7.　　　† See *Ibre Gloſſarium Sueo-Gothicum*, where the word is defined, *munimentum*, derived from *Berga cuſtodire*, or *Byrgia claudire*.　　　‡ *Baxter. Gloſſ. Antiq. Brit.* 109.

iflands, they are moft frequently called *Wart* or *Wardhills*, which fhews that they were garrifoned. They had their *wardmadher* *, or watchman, a fort of centinel, who ftood on the top, and challenged all who came in fight. The *gackman* † was an officer of the fame kind, who not only was on the watch againft furprize ; but was to give notice if he faw any fhips in diftrefs. He was allowed a large horn of generous liquor, which he had always by him, to keep up his fpirits ‡. Along the *Orkney* and *Schetland* fhores, they almoft form a chain ; and by that means not only kept the natives in fubjection, but were fituated commodioufly for covering the landing of their countrymen, who were perpetually roving on piratical expeditions. Thefe towers were even made ufe of as ftate-prifons ; for we learn from *Torfæus*, that after *Sueno* had furprized *Paul*, count of *Cathnefs*, he carried him into *Sutherland*, and confined him there in a *Norwegian* tower § So much has been faid on this fubject by the Reverend Mr. *Cordiner* and myfelf, that I fhall only refer to the pages, after faying, that out of our kingdom, no buildings fimilar to thefe are to be found, except in *Scandinavia*. On the mountain *Swalberg* ‖ in *Norway* is one ; the *Stir-bifkop* ¶, at *Upfal* in *Sweden*, is another ; and *Umfeborg*, in the fame kingdom, is a third **.

Thefe towers vary in their inner ftructure ; but externally are univerfally the fame ; yet fome have an addition of ftrength on the outfide. The burgh of *Culfwick* in *Schetland*, notwithftanding it is built on the top of a hill, is furrounded with a dry ditch thirteen feet broad; that of *Snaburgh* in *Unft*, has both a wet and a dry ditch ; the firft cut, with great labor, through the live rock. The burgh of *Moura* is furrounded by a wall, now reduced to a heap of ftones, and the infide is cylindrical, not taper, as ufual with others. The burgh of *Hogfeter*, upon an ifle in a loch of the fame name, has alfo its addition of a wall ; a peculiarity in a caufeway, to join it to the main land, and a fingular internal ftructure. Numbers of little burghs, with fingle cells, are fcattered about thefe iflands, in the neighborhood of the greater ; and which probably were built by the poorer fort of people, in order to enjoy their protection. A multitude of places in thefe iflands have the addition of *burgh* to their names, notwithftanding there is not a veftige of a tower near them ; the materials having long fince been carried away, and applied to various ufes. One was, by way of pre-eminence, called *Coningfburgh*, or the *burgh of the king*. I lament its lofs the more, as it might have proved fimilar to its namefake in *Yorkfhire*, and furnifhed additional materials to

* *Ibre Glofs. Sueo-Goth.* 1085. † *Crit. Diff. by John Macpherfon, D. D.* 325. ‡ *Torfæus Rer. Orcad.* 8. § *Baxter, Glofs. Antiq. Brit.* 109. ‖ Information by letter from Mr. *Suhm* of *Copenhagen*. ¶ *Dalhberg*, tab. 64. ** The fame, tab. 300.—For more ample accounts, fee Mr. *Cordiner's Letters*, 73, 105, 118, and my *Tours in Scotland*.

my

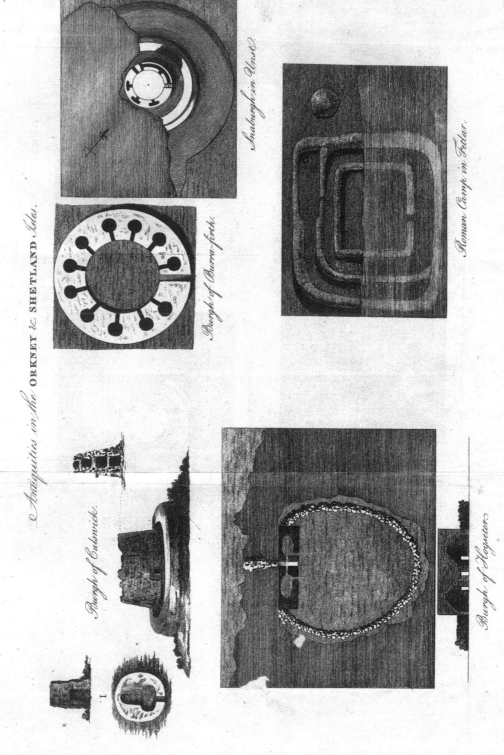

Antiquities in the ORKNEY & SHETLAND Isles.

Snaburgh in Unst.

Burgh of Burra-firth.

Roman Camp in Fetlar.

Burgh of Culswick.

Burgh of Hogsetter.

P. XXIII.

my worthy friend, *Edward King*, Efq; for his moft elaborate hiftory of *Englifh* caftles *. The plates, with explanatory accounts, fhall fupply what farther can elucidate thefe curious antiquities.

After the expulfion of the *Norwegians*, the coafts of *Scotland*, which they poffeffed, were ftill protected by caftles; many of which, fuch as *Oldwick*, ex-hihit very fmall improvements on the model left by the antient *Scandinavian* architects: a few deviated from the original manner, were fquare, had great thick-nefs of wall, furnifhed with cells like thofe in the round towers or *burghs.* *Borve* caftle, in *Cathnefs*, is a little more advanced. This was the refidence of *Thorkel*, a famous freebooter in the tenth century. It is a fmall fquare building, on a rock projecting into the fea, adjoined to the main land by an ifthmus not ten feet wide; and beneath the caftle is a magnificent paffage for boats, which pierces the rock from fide to fide, and is covered by a matchlefs natural arch.

BORVE CASTLE.

I cannot but revert to the former fubject, to mention the *Snaburgh* in *Tet-lor*, one of the moft remote of the *Schetland* ifles. It is in the form of a *Roman* camp; and when entire, had in the middle a rectangular area furrounded by a wall, and that by an earthen rampart of the fame figure, at fome diftance from it. Two fides of the walled area have the additional defence of another rampart of earth; which commences on the infide of one of the narrower fides, and, preferving the fame diftance from the leffer area as the two other fides of the outward fence do, terminates at the latter, near an artificial well. That this was *Roman*, I greatly fufpect. The care for water was a peculiar object with that wife nation; but neglected by barbarians. This is inclofed within the rampart, and at a fmall diftance on the outfide, had the protection of a mount, which once probably had its caftellet, garrifoned for the further fecurity †. The regular *portæ* are wanting; in other refpects it refembles a *Roman* camp. The fea, over which it impends, has deftroyed one half: the entire part is given in the plate, and the reft fupplied with dotted lines.

ROMAN CAMP?

I know but of two periods in which the *Romans* vifited thefe iflands: one at the time when the fleet of *Agricola* fubdued them; the other, when the fleet of *Honorius* defeated the *Saxons* in the feas of *Orkney*. A copper medal of *Vef-pafian*, with *Judæa devicta* ‡ on the reverfe, was found on the fouth fide of *Main-land*, probably loft there by the firft invaders, who might venerate *Vefpafian*, under whom many of them had ferved, and who might naturally carry with them fuch honorable memorials of his reign. The only antiquities found near

---

* See his curious account of *Coningfborough* caftle, which he juftly compares to the *Scottifh Duns*; and judicioufly afcribes to it a very early date. *Archæologia*, vi. 234. tab. xxiii.

† *Vegetius de re Milit.* lib. iv. c. 10.   ‡ Mr. *Low*.

this

this place, were fix pieces of brafs, caft into a form the neareft refembling fetters. They were wrapped in a piece of raw hide; but we cannot pretend to fay that they belonged to the occupiers of the camp.

STONE WEAPONS.   Flint heads of arrows, flint axes, fwords made of the bones of a whale, ftones, beads, and antiquities, muft be referred to the earlieft inhabitants, at a period in which thefe kingdoms were on a level with the natives of new-difcovered iflands

CIRCLES.   in the *South Sea.* Druidical circles of ftones, the temples of primæval religion of our ifland, are not uncommon. The fineft and moft entire are thofe at *Stennis*, in one of the *Orkney* ifles. The diameter of the circle is about a hundred and ten yards. The higheft ftone fourteen feet. The whole is fingularly furrounded with a broad and deep ditch, probably to keep at a diftance the unhallowed vulgar.

SEMICIRCLES.   At the fame place is a noble femicircle, confifting of four vaft ftones entire, and one broken. The higheft are twenty feet high above ground. Behind them is a mound of earth, conformable to their pofition. If there never was a number of ftones to complete a circle, this antiquity was one of the kind which the learned Doctor *Borlafe* calls a theatre, and fuppofes was defigned for the exhibition of dramatical performances *. I fufpect them to have been either for the purpofes of religion, or judicial tranfactions; for the age was probably not fufficiently refined for the former amufements. Upright ftones, either memorials of

PLAIN COLUMNS.   the dead, or victories obtained on the fpot, are very numerous. The moft remarkable is the ftone of *Sator*, in the ifle of *Eda*. It is a flag, fifteen feet high, five and a half broad, and only nine inches thick. Its ftory is quite unknown; but it probably refts over a hero of that name. Notwithftanding the long refidence of the *Norwegians* in thefe iflands, I find only one ftone with a *Runic* infcription, which runs along the fides. The reft of the ftone is plain, and deftitute of the fculptures fo frequent on thofe found in *Scandinavia*.

SCULPTURED CO-LUMNS.   In the wall of the church at *Sandnefs*, is a ftone with three circles, a femicircle, and a fquare figure, engraven on it. This is the only one which bears any refemblance to the elegant carved columns at *Meigle* and *Glames*, and which extend, after a very long interval, as far as the church-yard of *Far*, on the extreme northern coaft of *Cathnefs*. Several of thefe have been before attended to. I can only remark, that they are extremely local, and were, by their fimilarity, only the work of a fhort period. We imagine that the firft, about which we can form any conjecture, was erected in 994, on the defeat of *Camus*, the *Dane:* the laft in 1034, on the murder of *Malcolm* the Second.

* *Antiq. Cornwall*, 195.

In

In the isle of *Unst* are two singular circles, near each other. The largest is fifty feet in diameter, to the outmost ring; for it consists of three, concentrical; the outmost is formed of small stones, the two inner of earth; through all of which is a single narrow entrance to a tumulus which rises in the centre. The other circle is only twenty-two feet in diameter, and has only two rings, formed of earth: in the centre is a barrow, the sides of which are fenced with stones. No marks of their having been places of interment have been found, yet most probably that was their use.

The links or sands of *Skail*, in *Sandwich*, one of the *Orknies*, abound in round *barrows*. Some are formed of earth alone, others of stone covered with earth. In the former was found a coffin, made of six flat stones. They are too short to receive a body at full length: the skeletons found in them lie with the knees pressed to the breast, and the legs doubled along the thighs. A bag, made of rushes, has been found at the feet of some of these skeletons, containing the bones, most probably, of another of the family. In one were to be seen multitudes of small beetles. Whether they were placed there by design, or lodged there by accident, I will not determine; but, as I have discovered similar insects in the bag which inclosed the sacred *Ibis*, we may suppose that the *Egyptians*, and the nation to whom these *tumuli* did belong, might have had the same superstition respecting them. On some of the corpses interred in this island, the mode of burning was observed. The ashes, deposited in an urn which was covered on the top with a flat stone, have been found in the cell of one of the barrows. This coffin or cell was placed on the ground, then covered with a heap of stones, and that again cased with earth and sods. Both barrow and contents evince them to be of a different age from the former. These tumuli were in the nature of family vaults: in them have been found two tiers of coffins *. It is probable, that on the death of any one of the family, the tumulus was opened, and the body interred near its kindred bones.

The violence of the winds have, by blowing away the sands in a certain part of *Westra*, one of the *Schetlands*, discovered an extensive burying-place, once covered with the thickness of twenty feet. This seems to have belonged to different nations. One is marked by the *tumuli* consisting of stones and rubbish; some rounded, others flat at top like truncated cones. Near them are multitudes of graves, which are discoverable only by one, two, three, four, and sometimes even more short upright stones, set in the level sand. The corpse was interred a few feet deep, and covered with a layer of fine clay, to keep the sand from touching it.

* See Mr. *Low's* account, and plate, *Archæologia*, iii. 276. tab. xiii.

Not only human bones, but thofe of oxen, horfes, dogs, and fheep, have been found in thefe graves. Befides, were feveral forts of warlike inftruments, battle-axes, two-handed fwords, broad fwords, brazen daggers and fcull-caps, and fwords made of the bones of the whale : knives and combs : beads, brotches, and chains of ornament : a metal fpoon, and a neat glafs cup greatly corroded : fmall flat circular pieces of marble : ftones fhaped like whetftones, and fpherical ftones perforated, fuch as were in former ufe in *Scotland* for turning of fpindles : but the moft fingular thing was a thigh-bone clofely incircled by a ring of gold. The tumuli feem to have been the places of fepulture of the inhabitants of the ifles : the graves, thofe of fome foreign nation who had landed here, had a conflict, and proved victorious. I found my conjecture on the arms and other matters found in them. The brafs were *Norwegian* *, the iron belonging to the natives ; but the weapons of conquerors and conquered were, with ceremonies refembling thofe at the funeral of *Pallas*, flung into the graves of the victorious party.

> Hine alii fpolia occifis direpta *Latinis*
> Conjiciunt igni, galeas enfefque decoros,
> Frenaque, ferventefque rotas ; pars munera nota;,
> Ipforum clypeos; et non felicia tela :
> Multa boum circa mactantur corpora morti.

In Scandinavia. The antiquities of this clafs found in *Scandinavia* are very numerous, and of a magnitude which evince the extreme population of the country. I difcover only three kinds. The firft may be exemplified in the vaft rounded earthen tumulus in *Smaland*, with a rude monumental upright ftone at top ; and near it a fpherical ftone, beautifully carved, flung up in honor of *Ingo* King of *Sweden*, in the latter end of the ninth century † : others in honor of *Humblus*, and *Laudur* brother to King *Angantyr* ; the laft furrounded at its bafe with a circle of rude ftones ‡. The *Rambora Rolle* is a mount of earth, with three upright pillars, placed fo as to form a triangular fpace ‖. Other *tumuli* confift entirely of vaft heaps of ftones. Several of the fepulchral memorials are formed of ftones difpofed in a circular form : fome of low ftones, like that of the *Danifh* King *Harald Hyldeland*, placed round the edge of the flat area of a low mount. He was flain in battle by *Ringo* King of *Sweden* §, who paid him all funeral honors, burnt his body with great pomp, and placed around his tumulus the numerous bodies of his faithful followers who were flain around their prince ; and their places of reft are marked by multitudes of fmall earthen barrows, with a fingle ftone at the top of each. On

---

* *Wormii Mon. Dan.* 50. *Dalhberg Suecia Antiqua, et Hodierna*, tab. 314.   † *Dahlberg Suecia Antiqua*, tab. 322.   ‡ The fame, 315.   ‖ The fame, 323.   § *Saxo Gramm.* 147.

the

the regal mount is a flat ftone, with five hollows in it, bafons to receive the blood of the victims *. Others confift of fmall ftones with *Maen-hirion*, as the *Welfh* ftyle them, lofty rude pillars, intermixed. In fome the leffer ftones depart from the circular form, are oval or oblong : their edges are often contiguous, and thofe parts are often marked with a lofty pillar †. Two pillars are fometimes found, with an enormous ftone fet from top to top, fo as to form the refemblance of a gateway ‡. Columns of great height are alfo found, furrounded at their bafe with two circles of fmall ftones ¶. Finally, the ftones are difpofed fo as to form wedges, fquares, long rows, as well as circles. The firft denoted that armies of foot and horfe had prevaled : the fecond, troops of warriors : the third, duels of champions : and the laft, the burials of families §. Multitudes of fingle obelifes are fcattered over the country : fome quite plain ; others infcribed with *Runic* characters, memorial of the dead, intermixed with well-fancied ornaments ‖.

In many of the *tumuli* are found the weapons and other matters which had been depofited with the burnt bones of the deceafed. In thofe of the earlieft ages are the ftone weapons, fuch as axes and fpears heads made of flint. In others have been met with a fmall lamp, a key, and fwords of brafs of the fame form with fome of the *Roman* fwords **. A fuperftition attending the fwords was fingular : thofe of higheft temper were fuppofed to have been made by *Duergi*, *dwarfs* or fairies, and were thought to have been irrefiftible. The reader will not be difpleafed with the elegant verfion †† of a *Runic* poem, defcribing the incantations of a fair heroine, to obtain the magical fword out of the tomb of her deceafed father.

The *Runic* INVOCATION of HERVOR, the Daughter of ANGANTYR,

Who demands, at her Father's Tomb, a certain Sword, called *Tirfing*, which was buried with him.

HERVOR.

Awake, *Angantyr* ! To thy tomb,
With fleep-expelling charms, I come.
Break thy drowfy fetters, break !
'Tis *Hervor* calls—Awake ! awake !

Tirfing, made by fairy hands,
Hervor from thy tomb demands.
Hervardur, Hiorvardur, hear !
Lift, oh lift, my father dear !

---

* *Dahlberg,* tab. 315.   † The fame, and tab. 281.   ‡ *Olaus Magnus.*   ¶ *Wormii Mon. Dan.* p. 63.
§ I do not well underftand fome of thefe diftinctions ; but give them from *Olaus Magnus,* lib. i. c. 18. Moft of them are exemplified at *Finfta* in *Sweden.* See *Dahlberg,* tab. 104, and *Perinfkiold Monum. Sueo-Goth.* p. 216.
‖ *Wormii Monum. Dan.* 64, & paffim.   ** *Dahlberg,* tab. 314.   †† By my friend, the Reverend Mr. *Williams* of *Vron.*

Each from his filent tomb I call;
Ghofts of the dead, awaken all!
With helmet, fhield, and coat of mail,
With fword and fpear, I bid ye hail!
Where twifted roots of oak abound,
And undermine the hollow ground,
Each from his narrow cell I call;
Ghofts of the dead, awaken all!
In what darkfome cavern deep,
Do the fons of *Angrym* fleep?
Duft and afhes tho' ye be,
Sons of *Angrym*, anfwer me.
Lift'ning in your clay-cold beds,
Sons of *Eyvor*, lift your heads.
Rife, *Hiorvardur*, rife and fpeak;
Hervardur, thy long filence break.
Duft and afhes tho' ye be,
One and all, oh anfwer me.
Never, oh never may ye reft;
But rot and putrefy unblefs'd,
If ye refufe the magic blade,
And belt, by fairy fingers made!

ANGANTYR.

Ceafe, oh daughter, ceafe to call me;
Didft thou know what will befall thee,
Thou hadft never hither fped,
With Runic fpells to wake the dead:
Thou, that in evil hour art come
To brave the terrors of the tomb.
Nor friend, nor weeping father, gave
Angantyr's reliques to the grave;
And *Tirfing*, that all-conqu'ring fword,
No longer calls *Angantyr* lord.
A living warrior wears it now——

HERVOR.

Tis falfe, *Angantyr*; only thou.
So may great *Odin* ever keep
In peace the turf where thou doft fleep;
As *Tirfing* ftill befide thee lies,
Th' attendant of thy obfequies!
My juft inheritance I claim;
Conjure thee by a daughter's name,
Thy only child!

ANGANTYR.
Too well I knew
Thou wouldft demand what thou fhalt rue.
By *Tirfing*'s fatal point fhall die
The braveft of thy progeny.
A warlike fon fhall *Hervor* bear,
Hervor's pride, and *Tirfing*'s heir;
Already, daughter, I forefee
Heidrek the hero's name will be:
To him, the young, the bold, the ftrong,
Tirfing hereafter will belong.

HERVOR.

Ne'er fhall my inchantments ceafe,
Nor you, ye fpirits, reft in peace,
Until ye grant what I demand,
And *Tirfing* glitters in my hand.

ANGANTYR.

Oh Virgin, more than woman bold!
Of warlike mien and manly mould!
What has induc'd thy feet to tread
The gloomy manfions of the dead,
At this lone hour, devoid of fear,
With fword, and fhield, and magic fpear?

HERVOR.

The caufe thou know'ft, why to thy tomb
I've wander'd thro' the midnight gloom:
Yield then the Fairies work divine;
Thou art no father elfe of mine;
But goblin damn'd.

ANGANTYR.
Then hear me, Maid,
That art not ev'n of death afraid!
Hialmar's bane thou fhalt command;
The fatal fword is in my hand:
But fee the flames that round it rife!
Doft thou the furious fire defpife?

HERVOR.

Yes; I dare feize, amidft the fire,
The object of my foul's defire;
Nor do thefe eyes behold with dread,
The flame that plays around the dead.

ANGANTYR.

Rafh Maid! will nothing then controul
The purpofe of thy daring foul?

But

But hold—ere thou fhouldft fall a prey
To thefe fierce flames that round it play,
The fword from out the tomb I'll bring;
Go, and the fong of triumph fing.

H E R V O R.

Offspring of kings! I know thee now,
And thus before thy prefence bow;
Father, Hero, Prince, and Friend!
To thee my grateful knees I bend.
Not half fo happy had I been,
Tho' *Scandinavia* hail'd me queen.

A N G A N T Y R.

How art thou to thy int'reft blind,
Weak woman, tho' of dauntlefs mind!
Tirfing, the objeft of thy joy,
Thy future offspring fhall deftroy.

H E R V O R.

My feamen call; I muft away:
Adieu, O King! I cannot ftay.
Fate, do thy worft! in times to come
Be what it may, my children's doom!

A N G A N T Y R.

Take then, and keep *Hialmar*'s bane,
Dy'd in the blood of heroes flain.

Long fhall the fatal pledge be thine,
Hervor, if truly I divine;
The fell, devouring, poifon'd blade,
For death and for deftruction made.

H E R V O R.

With joy the two-edg'd fword I take,
Nor reck the havock it will make;
Poffeffing which, I little rue
Whate'er my frantic fons may do.

A N G A N T Y R.

Daughter, farewell! as thou doft live,
To thee the death of twelve I give:
To thee, O maid of warlike mind,
What *Angrym*'s fons have left behind.

H E R V O R.

Angantyr, reft in peace! and all
Ye ghofts, who have obey'd my call;
Reft in your mould'ring vaults below!
While from this houfe of death I go,
Where, burfting from the vap'rous ground,
Meteors fhoot, and blaze around.

I fhall juft mention, that the antient *Scandinavians* had alfo their *Cromlehs* *. I
can trace but one inftance, and that on the top of a tumulus in *Zealand*; which,
with two other barrows, is included in a fquare of ftones.

Circles, for the purpofe of religious rites, were not wanting here. The *Etteftupa*,
or circle of lofty rude columns in *Weft Gothland*, was celebrated for the facrifices
of the heathens †; and the great ftones at *Finftad*, difpofed in form of a cell,
and called *St. Birgitta's Oratory* ‡, was no other than a temple of worfhip, analogous, probably, to that of the Druids.

The next ftep is to the F E R O E iflands, a group about two hundred and ten miles to
the north-weft of the northern *Schetland*, between lat. 61, 15. and 62, 30. There are
feventeen which are habitable, each of which is a lofty mountain arifing out of the
waves, divided from the others by deep and rapid currents. Some of them are
deeply indented with fecure harbours; providence feeming to have favored mankind
with the fafeft retreats in the moft boifterous feas. All are very fteep, and moft of

FEROE ISLES.

* *Wormii Mon. Dan.* p. 8.   † *Dalhberg*, tab. 280.   ‡ The fame, 105.

them

them faced with moſt tremendous precipices. The ſurface of the mountains conſiſts of a ſhallow ſoil of remarkable fertility; for barley, the only corn ſown here,
yields above twenty for one ; and the graſs affords abundant paſturage for ſheep.
The exports are ſalted mutton and tallow, gooſe quils, feathers, and Eider down ;
and, by the induſtry of the inhabitants, knit woollen waiſtcoats, caps, and ſtock-
ings. No trees beyond the ſize of juniper, or ſtunted willows, will grow here :
nor are any wild quadrupeds to be met with, except rats and mice, originally
eſcaped from the ſhipping.

LAND BIRDS.

The liſt of land birds is very ſmall :—The Cinereous Eagle, p. 214. B ; the
Lanner, p. 225. K ; the Sparrow Hawk, p. 226. N * ; a ſpecies of Owl; the
Raven, N° 134; and Hooded Crow, p. 251. B. are the pernicious ſpecies.
Ravens were ſo deſtructive to the Lambs and Sheep, that in old times every boat-
man was obliged to bring into the ſeſſions-houſe, on *St. Olaus*'s day, the beak of
one of thoſe birds, or pay one ſkin, which was called the *Raven-fine*, in caſe of
neglect. The remaining land fowl are Wild Pigeons and Stares, White Wagtails,
Wrens, and ſometimes the Swallow. The Snow Bunting only reſts here in
ſpring, on its paſſage northward. The Heron is ſometimes met with. The Spoon-
Bill is Common †. The Sea Pie, Water Rail, and Lapwing, are ſeen here. The
birds of the rocks, ſuch as Puffins, Razor Bills, and Little Auks, Fooliſh and
Black Guillemots, ſwarm here ; and the *Geyir-fugl*, or Great Auk, at certain
periods viſits theſe iſlands. The laſt, by reaſon of its ſhort wing incapable of flight,
neſtles at the foot of the cliffs. The Skua, Arctic, Black-backed, and Herring
Gulls, Fulmars, Manks, Stormy Petrels, Imber and Northern Divers, Wild
Swans and Geeſe, (the Swans only vernal paſſengers towards the north) Eider
Ducks, Havelda or Long-tailed Ducks, Corvorants, and the Sula Gannet, form
the ſum of the palmated fowl of theſe inhoſpitable ſpots.

FOWLING.

FROM ABOVE.

The manner of fowling is ſo very ſtrange and hazardous that the deſcription
ſhould by no means be omitted. Neceſſity compels mankind to wonderful attempts.
The cliffs which contain the objects of their ſearch are often two hundred fathoms
in height, and are attempted from above and below. In the firſt caſe, the fowlers
provide themſelves with a rope eighty or a hundred fathoms in length. The
fowler faſtens one end about his waiſt and between his legs, recommends himſelf
to the protection of the Almighty, and is lowered down by ſix others, who place a
piece of timber on the margin of the rock, to preſerve the rope from wearing againſt
the ſharp edge. They have beſides a ſmall line faſtened to the body of the adventurer, by which he gives ſignals that they may lower or raiſe him, or ſhift him

* Theſe on the authority of Mr. *Debes*, who wrote the hiſtory of theſe iſles in 1670.
† *Brunnich*, p. 46.

from

from place to place. The laſt operation is attended with great danger, by the looſening of the ſtones, which often fall on his head, and would infallibly deſtroy him, was it not protected by a ſtrong thick cap; but even that is found unequal to ſave him againſt the weight of the larger fragments of rock. The dexterity of the fowlers is amazing; they will place their feet againſt the front of the preci-pice, and dart themſelves ſome fathoms from it, with a cool eye ſurvey the places where the birds neſtle, and again ſhoot into their haunts. In ſome places the birds lodge in deep receſſes. The fowler will alight there, diſengage himſelf from the rope, fix it to a ſtone, and at his leiſure collect the booty, faſten it to his girdle, and reſume his pendulous ſeat. At times he will again ſpring from the rock, and in that attitude, with a fowling net placed at the end of a ſtaff, catch the old birds which are flying to and from their retreats. When he hath finiſhed his dreadful employ, he gives a ſignal to his friends above, who pull him up, and ſhare the hard-earned profit. The feathers are preſerved for exportation: the fleſh is partly eaten freſh: but the greater portion dried for winter's proviſion.

The fowling from below has its ſhare of danger. The party goes on the expe-dition in a boat; and when it has attained the baſe of the precipice, one of the moſt daring, having faſtened a rope about his waiſt, and furniſhed himſelf with a long pole with an iron hook at one end, either climbs, or is thruſt up by his companions, who place a pole under his breech, to the next footing ſpot he can reach *. He, by means of the rope, brings up one of the boats crew; the reſt are drawn up in the ſame manner, and each is furniſhed with his rope and fowling-ſtaff. They then continue their progreſs upwards in the ſame manner, till they arrive at the region of birds; and wander about the face of the cliff in ſearch of them. They then act in pairs; one faſtens himſelf to the end of his aſſociate's rope, and, in places where birds have neſtled beneath his footing, he permits himſelf to be lowered down, depending for his ſecurity to the ſtrength of his companion, who is to haul him up again; but it ſometimes happens that the perſon above is overpowered by the weight, and both inevitably periſh. They fling the fowl down to the boat, which attends their motions, and receives the booty. They often paſs ſeven or eight days in this tremendous employ, and lodge in the crannies which they find in the face of the precipice.

The ſea which ſurrounds theſe iſlands is extremely turbulent. The tides vary greatly on the weſtern and eaſtern ſides. On the firſt, where is received the un-interrupted flood of the ocean from the remote *Greenland*, the tide riſes ſeven fa-thoms: on the eaſtern ſide it riſes only three. Dreadful whirlwinds, called by the *Danes*, *oes*, agitate the ſea to a ſtrange degree; catch up a vaſt quantity of water,

FROM BELOW.

* In *Pontoppidan's Hiſt. Norway*, ii. 61. is a plate expreſſive of this manner of fowling.

ſo

fo as to leave a great temporary chafm in the fpot on which it falls, and carries away with it, to an amazing diftance, any fifhes which may happen to be within reach of its fury. Thus great fhoals of Herrings have been found on the higheft mountains of *Feroe*. It is equally refiftlefs on land, tearing up trees, ftones, and animals, and carrying them to very diftant places. We muft no longer laugh at the good archbifhop *, who gravely tells us, that at times, the Rats called *Lemming* are poured down from the clouds in great fhowers on the *Alps* of *Norway*. We affent to the fact; but muft folve the phœnomenon by afcribing it to a whirlwind, as he does in one place; yet immediately fuppofes they may be bred in the upper regions *out of feculent* matter.

Among the numerous whirlpools of thefe feas, that of *Suderoe*, near the ifland of the fame name, is the moft noted. It is occafioned by a crater, fixty-one fathoms in depth in the centre, and from fifty to fifty-five on the fides. The water forms four fierce circumgirations. The point they begin at is on the fide of a large bafon, where commences a range of rocks running fpirally, and terminating at the verge of the crater. This range is extremely rugged, and covered with water from the depth of twelve to eight fathoms only. It forms four equidiftant wreaths, with a channel from thirty-five to twenty fathoms in depth between each. On the outfide, beyond that depth, the fea fuddenly finks to eighty and ninety. On the fouth border of the bafon is a lofty rock, called *Sumboe Munk*, noted for the multitude of birds which frequent it. On one fide, the water is only three or four fathoms deep; on the other fifteen. The danger at moft times, efpecially in ftorms, is very great. Ships are irrefiftibly drawn in: the rudder lofes its power, and the waves beat as high as the mafts; fo that an efcape is almoft miraculous: yet at the reflux, and in very ftill weather, the inhabitants will venture in boats, for the fake of fifhing. Mr. *Debes* omits the times of greateft danger. It is to be hoped that attention will be paid to the various periodical appearances of a phœnomenon, the caufe of which is very fatisfactorily explained by the worthy paftor †.

Mankind found their way to thefe iflands fome time before the difcovery of *Iceland*. *Naddodd*, a *Norwegian* pirate, had retired here, as the only place of fecurity he could find ‡. About this time, *Harold Harfagre* poffeffed himfelf of *Norway*, and flung off the *Danifh* yoke. A party was formed againft him; but it was foon fubdued, and the malecontents quitting the kingdom, retired to the *Hebrides*, *Orknies*, *Schetland*, and *Feroe*, and gave rife to the *Norwegian* reign in all thofe iflands.

ICELAND.

From the *Feroe* iflands, the hardy *Scandinavians* made the next ftep, in their northern migrations, to ICELAND. I muft premife, that there is the higheft probability that this ifland was difcovered in an age moft remote to theirs: and that it was the *Thule* of *Pytheas*, an illuftrious *Marfeillian*, at left cotemporary with

Ariſtotle *, and who puſhed his diſcoveries towards the north, as his countryman *Euthymenes* did beyond the line. *Pytheas* arrived at *Thule,* an iſland, ſays he, ſix days ſailing northward from *Britain,* where, he informs us, was continual day and night for ſix months alternately †. He does not exactly hit on the length of day and night; but he could have been at no other, at that diſtance from *Britain,* but *Iceland,* in which there was a moſt remarkable abſence of light. As to *Naddodd,* in 861, he was accidentally driven by a tempeſt to the eaſtern ſide of *Iceland,* to a place now called *Reidarfiall.* He found the country covered with ſnow, and therefore named it *Snæland;* yet he returned home full of its praiſes. Soon after, *Gardar,* a *Swede,* experienced the ſame fortune. On a voyage to the *Hebrides,* he was tempeſt-driven to the ſame iſland; on which, by the advice of his mother, who was a ſort of diviner, he landed at *Horn.* At this period *Iceland* was cloathed with wood from the ſhore to the very tops of the mountains. He wintered there, and likewiſe returned full of its praiſes ‡.

FLOKE, a celebrated pirate, was the next adventurer. He took with him three Ravens, and, like another *Noah,* made them the augury of the land. Before he ſailed, he performed a great ſacrifice for the ſucceſs, upon a vaſt pile of ſtones, which he raiſed for the purpoſe. This points out another origin of the vaſt tumuli we ſo frequently ſee. He made the *Schetland* and the *Feroe* iſles his firſt ſteps; and looſed from the laſt for *Iceland,* the neareſt point of which is about five hundred and forty miles diſtant. His firſt Raven returned to *Feroe:* the ſecond flew back to the ſhip: the third directed him to the wiſhed-for land §. He wintered there. The cattle he brought with him periſhed through want. The ſpring proved unuſually cold, and the ſea appeared full of ice; for which reaſon he beſtowed on the iſland the name it at preſent bears. *Floke* was ſick of his voyage: returned full of diſpraiſes of the country. This did not diſcourage other adventurers, all of them *Scandinavians,* thruſt out of the exuberant northern hive. The reſt of the world, which their countrymen ravaged, was aſſuredly too ſmall for them, otherwiſe they never would have colonized almoſt the moſt wretched ſpot in the northern hemiſphere. Ambition poſſibly actuated the leaders, who might think it

> Better to reign in hell than ſerve in heaven.

Colony after colony arrived. They confederated, and formed a republic, which exiſted near four hundred years; but with as many feuds and ſlaughters as could

---

* The works of *Pytheas* had been read by *Dicæarchus,* a diſciple of *Ariſtotle*'s. See *Strabo,* lib. ii. p. 163.

† *Plin.* lib. ii. c. 75. ‡ Same, p. 7. § Same, p. 8. *Torfæus. Hiſt. Norveg.* ii. 97.

happen

happen in a climate where luxury might pamper and corrupt the inhabitants. In 1261, wearied with their diffenfions, they voluntarily re-united themfelves to their mother-country, *Norway*, under the reign of its monarch *Haquin*. It is remarkable, that the poetic genius of their aboriginal country flourifhed with equal fublimity in every climate. The *Scalds*, or bards, retained their fire in the inhofpitable climate of *Iceland*, as vigoroufly as when they attended on their chieftains to the mild air of *Spain*, or *Sicily*, and fung their valiant deeds.

Every thing which furnifhed topics to the poets of other countries, was, in the moft remote period, wanting here. Groves, verdant meadows, purling ftreams, and gentle zephyrs, were totally unknown; and in their ftead, ftunted fhrubs, a thin herbage, rude torrents, and fierce gales, reign in every part. We admit the apology of the learned *Torfæus* for the prefent ftate of his country *. Violent tempefts might cover whole tracts with the unftable fand, eruptions of water from the mountains defolate fome parts, earthquakes bury vaft extents of fertile land with fragments of rocks, and inundations of the fea change the face of others. But foft fcenery was not requifite to infpire poets who were to fing only the preparations for warlike exploits, the flaughter of a battle, the deeds of their heroes, and the magic folemnities of fuperftitions.

The ifland, at prefent, exhibits to the traveller amazing flopes of lava, which once ftreamed from the vulcanoes, and terminated in the fea. Such is the appearance, about three miles from *Hafnaifoird*, in lat. 64. 4. of vaft maffes of lava piled to a montanous height upon each other, broken, vitrified, fharp, rude, and black. In parts, fandy tracts intervene: in others, a foil peculiar to the place, a tufa, originated from the violent eruptions of impure water which rufh from the mountains, attendant on the fiery eruptions. Vallies compofed of a very thin foil, afford grafs for a numerous breed of cattle and fheep. Here is found variety of fpecies of the beft graffes; of the *aira, poa, feftuca*, and *carex*. Part is harvefted againft winter; but not in fuch plenty, but that the farmer is obliged often to feed his ftock with the wolf-fifh, or the heads of cod-fifh beaten fmall, and mixed with a quarter part of hay. To what food will not neceffity compel both man and beaft to recur!

WOODS LONG LOST.

DRIFT-WOOD.

The woods of *Iceland* have long fince vanifhed, unlefs we except a few ftunted birch, fcarcely ten feet high, and four inches in diameter; and a few fpecies of willow, fo fmall and fo rare as fcarcely to be of ufe to the inhabitants. But they are abundantly fupplied with drift-wood from *Europe* and *America*, as appears by the fpecies found on the fhores, efpecially on all the northern coaft, as *Langanefs* on the north-eaft, and *Hornftrandt* on the north-weft. That woods were found here

* *Hift. Norweg.* i. 12.

in

in very remote periods, is very evident, from the quantity of *futurbrand* met with in several parts; which still retains traces of its vegetable origin; the marks of branches, and circles of the annual growth of the wood: some pieces are even capable of being planed. It is found in the fissures of the rocks, much compressed by their weight, and in pieces sometimes big enough to make a middle-sized table. This is sometimes used as fuel; but the want of it is supplied, in some measure, by the drift-wood, by peat, and by several strange substitutes, the effect of necessity. Smiths prefer the *futurbrand* to sea-coal in their business. The beds of this fossil strongly refute the notion of *Iceland* having been entirely formed by vulcanic violence, since the original creation; and raised out of the sea in later times, as others have been known to have done. *Delos* and *Rhodos*, in very remote ages; *Thera*, the modern *Santorini*, and *Therasia*, in the 135th *Olympiad*; *Thia*, in the time of *Pliny* *; and in the beginning of this century another sprung from the sea, by the force of subterraneous fires, near to *Santorini* †: and, while I am now writing, an island is forming by the same cause, not remote from the *Reickenes*, part of the very island in question. But these *futur* or *forte brands* are certainly the remains of antient forests, overturned and buried by earthquakes, after the golden age of the island. Let me add to this another proof, from the number of its vegetables: there being found on it not PLANTS, NUMBER OF. fewer than three hundred and nine perfect, and two hundred and thirty-three cryptogamous plants. On the isle of *Ascension*, which is totally and aboriginally vulcanic, a *Flora* of not more than seven plants is to be seen ‡.

THIS vast island extends from 63. 15. to about 67. 18. north latitude: is reckoned to be five hundred and sixty *English* miles long, and about two hundred and fifty broad ||. It has a rugged coast, indented deeply with secure bays; but faced with very few isles. It lies in the *Hyperborean* ocean, divided from *Greenland* by a sea about thirty-five leagues wide §. The whole is traversed with great ridges of mountains; the highest naked, and usually free from snow, by reason of the saline and sulphurous particles with which they abound. The lower, called *Jokkeler*, are cased with eternal ice and snow; and are the *glacieres* of *Iceland*. Of these, *Snæfiæll Jokkel*, which hangs over the sea in the west part of the island, is far the highest ¶. Out of these, at different periods, have been tremendous eruptions of fire and water, the burst of which is attended with a most terrific noise: flames and balls of fire issue out with the smoke: and showers of stones are vomited up; of which there has been an instance of one weighing near three hundred pounds being flung to the

---

* *Hist. Nat.* lib. ii. c. 87.   † Most admirably described in the *Ph. Transf. Abridg.* v. 196, &c.
‡ *Osbeck's Voy.* ii. 98. *Forster's Voy.* ii. 575. 576. || *Mallet,* i. 15.   § *Kerguelin,* 175.
¶ See *Olaffen,* i. tab. xvii.

distance

diſtance of four miles. The heights of the mountains have not been taken; but that of the *Hecla-fiall* is not far ſhort of ſeventeen hundred yards. Of this ſpecies of mountain, *Hecla* has been moſt celebrated : the records of *Iceland* enumerate ten of its eruptions ſince the arrival of the *Norwegians*. It was the hell of the northern nations; but they ſeem divided in their opinions, whether the pains of the damned aroſe from fire, or, what was more tremendous to the natives of theſe countries, from the cold *.

> To bathe in fiery floods, or to reſide
> In thrilling regions of thick-ribbed ice.

Hecla has been known to have had only ten eruptions between the years 1104 and 1693; from the laſt to 1766, when it burſt out in flames and *lava*. It emitted flames in 1771 and 1772 ; but did not overflow with *Stenna*, or a ſtone flood. But other vulcanoes have, in the preſent century, proved the ſpiracles to the internal fires of *Iceland*. The vallies between the mountains are in general ſandy and ſterile. Fiery eruptions are not confined to the mountains. Laſt year they burſt out of the ſulphureous ſoil of the low parts of *Skaftafield Syſſel* or province ; and the lava has overflown the country for the ſpace of thirty miles, and has at laſt reached the ſea, deſtroying every thing in its progreſs. It dries up the rivers, and fills their beds with lava. Moors in ſome places ſtop its courſe; but it totally changes their nature. It has taken to the deſerts of the ſame province, and begins to ſpread to the eaſt, or *Mulé Syſſel*, the moſt populous and fruitful part of the iſland; nor were there any ſigns of its ceaſing at the time when this account was ſent to me †.

HUERS, OR BOIL-
ING JETS D'EAUX.

THE FOUNTAINS of many of the vallies are of a moſt extraordinary nature; are called *Huers*, and form at times *jets d'eaux* of ſcalding water, ninety-four feet high, and thirty in diameter, creating the moſt magnificent *gerbes* in nature! eſpecially when backed by the ſetting ſun. They ariſe out of cylindrical tubes of unknown depths : near the ſurface they expand into apertures of a funnel ſhape, and the mouths ſpread into large extent of ſtalactitical matter, formed of ſucceſſive ſcaly concentric undùlations. The playing of theſe ſtupendous ſpouts is foretold by noiſes roaring like the cataract of *Niagara*. The cylinder begins to fill : it riſes gradually to the ſurface, and gradually encreaſes its height, ſmoking amazingly, and flinging up great ſtones. After attaining its greateſt height, it gradually ſinks, till it totally diſappears. Boiling *jets d'eaux*, and boiling ſprings, are frequent in moſt parts of the iſland. In many parts they are applied to the culinary uſes of the natives. The moſt capital is that which is

* *Bartholinus de Contemptu Mortis*, 359.   † Letter from Mr. *Brunnich*, dated *October* 31, 1783.

called

called *Geyer*, in a plain rifing into fmall hills, and in the midft of an amphi-
theatre, bounded by the moft magnificent and various-fhaped icy mountains ;
among which the three-headed *Hecla* foars pre-eminent.

Thefe *Huers* are not confined to the land. They rife in the very fea, and form
fcalding fountains amidft the waves. Their diftance from the land is unknown ;
but the new vulcanic ifle, twelve miles off the point of *Reickenes*, emitting fire and
fmoke, proves that the fubterraneous fires and waters extend to that fpace ; for thofe
aweful effects arife from the united fury of thefe two elements *. The depth of
water between this new creation and the *Geir-fugl Skier*, is forty-four fathoms ;
ten leagues to the weft, two hundred and five : and the bottom compofed of
black fand † ; doubtlefs no other than the *Pumex arenaceus*, the frequent evomition
of *vulcanoes*. How much paft human comprehenfion muft the powers have been,
that could force up materials for an ifland, even from the medium depth I have
given ! and how deep beneath the bottom of the ocean muft have been the caufes
which could fupply ftone, or pumice, or lava, to fill the fpace which this ifland occupies, many miles in circumference, and poffibly above a hundred fathoms in
depth !

If fome iflands fpring out of thefe feas, others are fwallowed by the force of
earthquakes. Their foundations are undermined by the fury of the fubterraneous
elements, which carries off the materials of their bafis, and difcharges it in
lava, or different forms, through the vulcanic *fpiracula*. The earthquakes fhatter
the cruft on which they ftand, and they tumble into the great abyfs. Such
was the fate of the nine ifles of *Gouberman*, which lay about four leagues from
Sandanefs, between *Patrixfiord* and *Cape Nort*, all which fuddenly difappeared.
Their names ftill exift in feveral maps ; but their place is only diftinguifhable
by the fuperior depth of water in the fpot on which they ftood ‡.

The number of inhabitants in *Iceland* is computed not to exceed fixty thoufand.
Confidering the ungenial furface of this vaft ifland, probably the number is
equal to the means of fupport. Writers apologize for the fewnefs of inhabitants, by attributing it to the almoft depopulation of the place by the *forte diod*,
or black death, a peftilence which commenced in *Cathay*, or *China*, in 1346, fpread
over all *Afia*, and *Africa*, reached the fouth of *Europe* in 1347, and in 1348
fpread itfelf over *Britain*, *Germany*, and northern *Europe*, even to the extremity
of the inhabited north. The fmall-pox, and other epidemics, are mentioned as
contributing to thin the ifland. During the time of the plague, tradition relates,

---

* See Mr. *Whiteburft's* Theory.   † *Sable noir comme la poudre a canon. Voyage au Nord,*
par M. *de Kerguelin,* 69.   ‡ The fame, 65, 66.

in

in terms moſt graphically horrid, that the perſons who eſcaped to the mountains, ſaw the whole low country covered with a thick peſtiferous fog. A gueſs
may be made at the number of inhabitants in the eleventh century; for a biſhop
of *Schalholt* cauſed, in 1090, all who were liable to pay tribute to be numbered:
four thouſand of that rank were found; ſo that, giving five to a family, the
ſum is twenty thouſand *. Much of the labor in the northern world falls to the
female part of the family; and in thoſe patriarchal times, the ſons alſo ſhared
the toil. I cannot therefore under-rate the number of commonalty, or untaxable
people, heads of families, at ten thouſand; which, by the ſame rule, will give
fifty thouſand of the lower rank. Beſides the dearth of food in this rude iſland,
other cauſes contribute to prevent the increaſe of inhabitants. Neceſſity forces
the men to ſeek from the ſea ſubſiſtence, denied by their niggardly land. Conſtant wet, cold, and hard labor, abridge the days of thouſands; and that labor is in-
creaſed tenfold, to ſupply the rapacity of their maſters. Incredible as it may ſeem,
a late king of *Denmark* ſold the whole iſland, and its inhabitants, to a company of
merchants, for the annual rent of one thouſand pounds. This company en-
ſlave the poor natives; who are bound to ſell their fiſh, the ſtaple of the iſland,
at a low price to theſe monopolizers; who, dreading reſiſtance, even have taken
from them the uſe of fire-arms! Here is given a ſtronger cauſe of depopulation,
perhaps, than the others; for *Hymen* can have but faint votaries in the land from
whence liberty is baniſhed. But for theſe cauſes, here ought to be found the
genuine ſpecies of the *Norman* race, unmixed with foreign blood; as muſt be
the caſe with every place remote from the reſt of the world. Here are to be
ſought the antient cuſtoms and diet of their original ſtock, which are now probably worn out in the land of their diſtant anceſtors. The luxury of food has
ſo little crept in among them, that their meat and drink in general is peculiar
to themſelves; and much of the former compoſed of herbs neglected in other
places.

DRESS.

The dreſs of the natives ſeems unchanged for a very conſiderable time: that of
the men is ſimple, not unlike that of the *Norwegian* peaſants †; that of the females is graceful, elegant, and peculiar to them, and perhaps ſome very old-
faſhioned *Norwegian* lady. They ornament themſelves with ſilver chains and
rich plates of ſilver, beautifully wrought. On their head is a lofty ſlender dreſs,
not unlike a *Phrygian* bonnet. I cannot compare this to any antient *European*
faſhion. *Iſabel* of *France*, queen to *Edward* II. wore a head-dreſs of an enor-

---

* *Arngrim Jonas's Comment. Iceland.* in *Hackluyt*, i. 556.   † See *Olaffen*, i. tab. iii. *Pon-*
toppidan, ii. tab. p. 272.

mous

mous height, of a slender conic form * ; but which, for want of the flexure at top, gave place in elegance to the taste of the *Icelandic* fair.

Mr. *Troil* awakens our curiosity about the *Icelandic* antiquities; speaks of castles, and heathen temples, and burying-places, and upright stones, and mounts. Of the first I am solicitous to gain some further knowlege, for possibly they might direct to the origin of the round buildings in the *Hebrides*, *Orknies*, *Schetland*, and the north of *Scotland* † : others seem to me the various *Scandinavian* antiquities, admirably exemplified in Baron *Dahlberg's Suecia Antiqua et Moderna*.

The species of quadrupeds of this island are very few. Small horses of a hardy kind ; cows in great abundance, and mostly hornless, the flesh and hides of which are considerable articles of exportation. Sheep are met with in great flocks in every farm ; the wool is manufactured at home, the meat salted, and, with the skins, much of it is sold to the Company, at the twenty-two ports allotted for the purposes of traffic. It is remarkable, that the climate disposes their horns to grow very large, and even to exceed the number of those of the sheep of other countries ; examples of three, four, and five, being extremely frequent. Goats and swine are very scarce ; the first, for want of shrubs to brouze, the last through deficiency of their usual food, and the supply which the farm-yards of other countries afford.

margin: DOMESTIC QUADRUPEDS.

The dogs are sharp-nosed, have short and sharp upright ears, bushy tails, and are full of hair. Here are domestic cats; but numbers are grown wild, and multiply among the rocks, so as to become noxious. The reader need not be reminded, that these, and every species of domestic animals, were originally introduced into *Iceland* by the *Norwegians*.

An attempt has been made to introduce the Rein Deer, *Arct. Zool.* N° 4. Those which survived the voyage have bred frequently. There can be little doubt of their succeeding, as *Iceland* has, in common with *Lapland*, most of the plants for their summer food ‡, and abundance of the Rein Deer *lichen* for their winter provision.

Rats and Mice seem to have been involuntarily transported. Both the domestic species are found here ; and the white variety of the Mouse, called in the *Icelandic*, *Skogar Mys*, is common in the bushes. I suspect that there is a native species, allied, as Doctor PALLAS imagines, to the Œconomic, *Arct. Zool.* p. 134, A. ; for, like that, it lays in a great magazine of berries by way of winter-stores. This species is particularly plentiful in the wood of *Husafels*. In a country where

margin: RATS.

---

* *Monfaucon Monum. de la Monarchie Fr.* ii. tab. xlii.   † *Voy. Hebrides.*   ‡ *Confer.* *Olaffen.* ii. 234. and *Amœn. Acad.* iv. 151.

berries

berries are but thinly difperfed, thefe little animals are obliged to crofs rivers to make
their diftant forages.   In their return with the booty to their magazines, they are
obliged to repafs the ftream; of which Mr. *Olaffen* gives the following account :—
" The party, which confifts of from fix to ten, felect a flat piece of dried cow-dung,
on which they place the berries in a heap in the middle; then, by their united force,
bring it to the water's edge, and after launching it, embark, and place themfelves
round the heap, with their heads joined over it, and their backs to the water, their
tails pendent in the ftream, ferving the purpofe of rudders *."   When I confider the
wonderful fagacity of Beavers, and think of the management of the Squirrel,
which, in cafes of fimilar neceffity, make a piece of bark their boat, and tail their
fail †, I no longer hefitate to credit the relation.

FOXES.   The Common Fox, *Arct. Zool.* N° 11, and the Arctic, N° 10, are frequent;
are profcribed, and killed for the fake of a reward, in order to prevent the
havock they would make among the fheep.

BEARS.   The Polar Bear, N° 18, is often transported from *Greenland*, on the iflands of
ice; but no fooner is its landing difcovered, than a general alarm is fpread, and
purfuit made till it is deftroyed.   The *Icelanders* are very intrepid in their attack
on this animal; and a fingle man, armed only with a fpear, frequently enters the
lifts with this tremendous beaft, and never fails of victory.   A perfon who lived
near *Langenefs*, the extreme northern point, where the Bears moft frequently land,
is ftill celebrated for having flain not fewer than twenty in fingle combat.   There
is a reward for every fkin, which muft be delivered to the next magiftrate.

The Common Bat, p. 185, A. is fometimes found in this ifland, and finifhes the
lift of the land-animals of the country.

The amphibious quadrupeds, or Seals, are very numerous.   *Iceland*, being
bleffed with domeftic animals, has lefs ufe of this race than other *Arctic* countries; yet they are of confiderable advantage.   The fkins are ufed for cloathing;
a good one is equal in value to the fkin of a fheep, or the hide of a cow; and
the fat fupplies the lamps in the long nights with oil.   The Common, during
winter, is exceffively fat, and will yield fixty pounds.

SEALS.   The *Icelanders* have two fpecies of native Seals: the Common, N° 72, called
by them *Land-Sælur*, becaufe it keeps near the coaft; the other, the Great,
N° 73, or *Ut-Sælur*.   They are taken in nets placed in the creeks and narrow
bays, which they pafs through to get on fhore.   When it begins to grow dark
the hunters make a fire, and fling into it the fhavings of horns, or any thing that
fmells ftrong; this allures the Seals, who ftrike into the nets, and are taken.

---

* *Olaffen*, as related to him.   † *Linnæus, Klein, Rzaczinfki, Scheffer.*

At other times, a *koder* or lure is tied to a rope, and placed before the nets ; to which the Seals, suppofing it to be fome ftrange animal, will eagerly fwim, and ftrike into the nets, paying with their lives for their curiofity. This carries them fometimes fo far, that they will ftray to a confiderable diftance inland, attracted by a candle, or the fire in a fmith's forge. If they are taken young, they are capable of being tamed : they will follow their mafter, and come to him like a dog, when called by the name which is given them. The *Icelanders* have a ftrange fuperftition about thefe animals : they believe they refemble the human fpecies more than any other, and that they are the offspring of *Pharaoh* and his hoft, who were converted into Seals when they were overwhelmed in the *Red Sea.*

Other fpecies of Seals are migratory. Among them is the Harp, N° 77, or *Vade-Selur.* Thefe quit the feas of *Iceland* in *March,* and fwim through the ftreights of *Davies,* by fome unknown opening, to the fartheft north ; bring forth their young, and return, by the north of *Greenland,* in *May,* extremely lean, to the north of *Iceland* ; continue tneir route, and return to that ifland about *Chrift- mas,* chiefly upon the drift-ice, on which they are either fhot, or harpooned. The Hooded Seal, N° 76, or *Bladru* Seal, is rarely taken here. The Walrus, N° 71, or *Roft-unger,* is fometimes wafted here from *Greenland* on the ice.

It cannot be expected, that many of the feathered tribe fhould inhabit an ifland fo very fevere in its climate, and fo remote from the more fouthern continent and iflands. It is, like all other *Arctic* countries, the afylum to water-fowl, to breed and educate their young ; but, being an inhabited place, fewer refort here than to the untrodden waftes of the more diftant north. The Guland Duck, p. 572. E. may poffibly be a local bird. The reft, whether land or water, are common to *Norway,* and many other parts of *Europe.* The Great Auks, N° 428, are found here in greater numbers than elfewhere : they inhabit and breed on the rocks, called from them *Geir-fugl Skier,* off the point of *Rækenes,* the moft fouthern part of the ifland. Notwithftanding they are furrounded with a fwelling fea, and tremendous breakers, the *Icelanders* venture there annually, in order to collect the eggs, to contribute to the provifion of the year. I can only reckon fixteen land-birds * : twenty cloven-footed water-fowl ; four with pinnated

BIRDS.

* Sea Eagle, N° 87.

Cinereous Eagle, p. 214, B.	Raven, - - N° 134.	Leffer Field-Lark, - p. 395.
Iceland Falcon, - 216, D.	White Grous, - N° 183.	Snow Bunting, - - N° 222.
Gyrfalcon, - - 221, F.	Hazel Grous ? - - p. 317, F.	White Wagtail, - - p. 396.
Lanner, - - 225, K.	Stare, - - p. 331. A.	Wheat-ear, - - p. 420.
Short-ear'd Owl? N° 116 †.	Red-wing Thrufh, p. 342. D.	Wren, - - - N° 382.

† *Olaffen,* ii. tab. xlvi. gives the figure of an Owl refembling his fpecies.

feet,

feet, and forty-three with webbed feet, natives or frequenters of the island. I have omitted, in the Zoologic part, the Lesser Guillemot, *Br. Zool.* ii. N° 235, which is a native of *Iceland*, and called there *Ringuia*. It ought to have had a place in an appendage to the Guillemots, p. 517.

The Raven holds the first rank among the land-birds in the *Scandinavian* mythology. We see the use made of them by the chieftain *Floke*. The Bards, in their songs, give them the classical attribute of the power of presage. Thus they make *Thromundr* and *Thorbiorn*, before a feudal battle, explain the foreboding voice of this bird, and its interest in the field of battle *.

THR. Hark! the Raven's croak I hear,
Lo! the bird of Fate is near.
In the dawn, with dusky wings,
Hoarse the song of death she sings.

Thus in days of yore she sang,
When the din of battle rang;
When the hour of death drew nigh,
And mighty chiefs were doom'd to die.

THOR. The Raven croaks: the warriors slain,
With blood her dusky wings distain;
Tir'd her morning prey she seeks,
And with blood and carnage reeks.

Thus, perch'd upon an aged oak,
The boding bird was heard to croak;
When all the plain with blood was spread,
Thirsting for the mighty dead.

R. W.

The Raven had still higher honors in the northern nations. It was sacred to *Odin*, the hero and god of the north. On the sacred flag of the *Danes* was embroidered this bird. *Odin* was said to have been always attended by two, which sate on his shoulders; whence he was called the *God of Ravens*: one was styled *Huginn*, or *Thought*; the other *Muninn*, or *Memory*. They whispered in his ear all they saw or heard. In the earliest dawn, he sent them to fly round the world, and they returned before dinner, fraught with intelligence. *Odin* thus sang their importance:

Huginn and *Muninn*, my delight!
Speed thro' the world their daily flight:
From their fond lord they both are flown,
Perhaps eternally are gone.
Tho' *Huginn's* loss I should deplore
Yet *Muninn's* would afflict me more †.

R. W.

I have already spoken of the excellent Falcons of this island: let me add, that Falcons were among the animals sacrificed to *Odin* ‡, being birds of the first courage, and which delighted in blood.

---

* *Island's Landnamabok*, 172. † *Bartholinus de Causis contemptæ Mortis*, &c. 429. ‡ *Mallot's Northern Antiq.* ii. 132.

The

The fea which furrounds *Iceland* is faid to be more falt than ufual in other countries. It leaves great faline incruftations on the rocks, which the natives fcrape off and ufe. I can, with no certainty, give the depth of the water, except where Mr. *Kerguelin* founded, ten leagues to the weft of *Geir-fugl Skier*, where he found it to be two hundred and five fathoms *. The equinoctial tides rife as high as fixteen feet : the ordinary tides twelve †. The coafts almoft univerfally bold, thofe of the inlets excepted, where there appears a fmall ftrand.

The bays, efpecially thofe of the fouth, which lie under the influence of the cold of *Greenland*, are annually frozen over; that of *Patrixfiord* was fhut up even as late as the 14th of *May* ‡: but the fea near the coafts never feels the influence of the froft. It is in thofe places deep, and agitated by a moft turbulent motion.

The dreaded ice is what floats from *Greenland* and *Spitzbergen*, and often fills, during the whole fummer, the ftreight between the former and this ifland ‖, and even extends along the northern coaft, covering the fea to a vaft diftance from land. It confifts of the two fpecies, the mountanous ice, called *Fiæl-jakar*; and the fmooth ice of inconfiderable thicknefs, ftyled *Hellu-is*. Thefe arrive generally in *January*, and go away in *March*. Sometimes it does not touch the land till *April*, when it fixes for a confiderable time, and brings to the *Icelanders* the moft tremendous evils ; a multitude of polar bears, which fpread their ravages far and wide among the cattle ; and a cold of incredible violence, which chills the air for many miles, and even caufes the horfes and fheep to drop down dead §. To this is attributed the ftunted ftate of the miferable woods of the country ; which caufe muft have exifted from the commencement of its iron age ; for there feems to have been a period in which there had been confiderable wooded tracts ¶.

The bottom of the fea is probably rocky ; for it abounds with greater variety of *fuci* than *Great Britain*, which give fhelter to fifhes innumerable ; a fource of wealth to the natives (were they permitted the free ufe) as they are of food to diftant nations, the veffels of which annually refort here to fifh, but without any commerce with the *Icelanders*, which is ftrictly prohibited. In 1767, two hundred *Dutch*, and eighty *French* doggers, of about a hundred tons each, were employed, thofe of each nation under the orders and protection of a frigate. They keep from four to fix leagues from fhore, and fifh with hooks baited commonly with large muffels, in forty or fifty fathoms water. Others go to the diftance of fifteen leagues, and fifh in the depth of a hundred fathoms. The great cap-

---

* *Voyage au Mer du Nord*, 69. † *Horrebow*, 101. ‡ *Kerguelin*, 31.
‖ *Troill*, 48, 49. § *Kerguelin*, 20, 175. ¶ See p. xlv.

ture

ture is Cod. As foon as the fifhermen take one, they cut off the head, wafh, gut, and falt it in cafks, with either rock-falt or that of *Lifbon*. The fifhery commences in *March*, and ends in *September*. It begins at the point of *Breder-wick*, and extends round the *North Cape*, by the ifle of *Grim*, to the point of *Langenefs*.

The *Englifh* have entirely deferted this fifhery fince they have been in poffeffion of *Newfoundland*. It had been, in very early times, the refort of our veffels, as is evident by the proclamation of *Henry* V. in order to give fatisfaction for the ill conduct of fome of his fubjects, in 1415, on the coafts of this ifland *, in which he forbids them to refort to the ifles of *Denmark* and *Norway*, efpecially to *Iceland*, otherwife than had been antiently cuftomary. In 1429, the *Englifh* parlement enforced this order, by making it penal for any of our fubjects to trade in the *Danifh* ports, except in *North Earn* or *Bergen*. At length, the *Danifh* monarch wifely refolved to referve the benefits of the fifheries to his own fubjects; and in 1465 made it capital for any *Englifhman* to trade in the ports of *Iceland* †. Even thofe of *Helgeland* and *Finmark* were fhut againft them, unlefs they were driven in by a ftorm. I imagine that this feverity muft have arifen from fome glaring infolence of our countrymen. But the antient treaties were revived, which were renewable by a frefh grant every feven years ‡. In later times, even Queen *Elizabeth* deigned to afk leave of *Chriftian* IV. to fifh in thofe feas; but afterwards inftructed her ambaffador to infift on the right of a free and univerfal fifhery. The anfwer does not appear: but in the reign of her fucceffor, we had not fewer than a hundred and fifty veffels employed in this fifhery. Poffibly we might comply with the regulations infifted on by the king of *Denmark*; or perhaps a greater indulgence was given, by reafon of the marriage of *James* with his fifter *Anne*. I obferve, that the *Danifh* prince excepts the port of *Weftmony*, it being referved for the peculiar fupply of the royal court ‖.

The oppreffed natives fifh in the bays in boats, containing one, and never more than four men. If they venture to fea, which they feldom do to above eight miles diftance, they have larger boats, manned with twelve or fixteen hands; in thefe they flave for the benefit of the monopolifts, to whom they are com-pelled to fell their fifh at a trifling price. How weak muft be the feelings of that government which can add mifery to mifery; and not attempt rather to be-ftow comforts on fubjects condemned to fuch a dreadful abode!

The fpecies of fifh in thefe feas are few; but the multitudes, under feveral of the moft ufeful kinds, are amazing; thofe of Cod in particular. Herrings pafs by

---

* *Rymer's Fœd.* ix. 322.   † Ibid. xvi. 443.   ‡ Ibid. xv. 443.   ‖ *Cambden's Life of Queen Elizabeth*, in the *Complete Hift. of England*, ii. 550.

this

this island in their annual migrations from the north, and for a short space fill every bay. Poverty and want of salt make these riches of other nations a tantalizing appearance to the unfortunate natives. This is the most northern place in which the Herring is seen: they are not found in the shallow water of *Spitzbergen*; neither is it probable that they double *Greenland*, and retire to the frozen ocean, equally wanting in depth of water;—are they not rather lost in the vast profundity of these very seas, in the depth of six hundred and eighty-three fathoms, in lat. 65, between this island and the north of *Norway*; or in the unfathomable depths a little farther north, where the water was found bottomless with seven hundred and eighty fathoms *? The other fishes of *Iceland* are in general common to *Greenland*: my remarks respecting them shall be deferred till I treat of that icy region.

VAST DEPTHS OF WATER.

In order to view the correspondent shores of the tract I have passed over, I shall return to the streights of *Dover*. *Calais* is seated in a low wet tract; and the whole coast, from thence to the extremity of *Holland*, is sandy, and fronted with sand-hills; providentially highest in that lowest of countries, in which the strongest protection against the fury of the sea is necessary. The coast of *Flanders*, the rich bait of ambition, stained with blood, is dangerous by reason of frequent narrow sand-banks, disposed in parallel rows, according to the direction of the land. The coasts of *Holland* are also greatly infested with sands; but between them and the land is a clear channel. From between *Dunkirk* and *Calais*, even to the *Scar*, at the extremity of *Jutland*, is low land, not to be seen but at a small distance, unless at *Camperden* in *Holland*; *Heilegeland*, off the mouths of the *Elbe* and *Weser*; and *Robsnout*, and *Hartshal*, in *Jutland*. While the opposite coasts of *England* are comparatively high, and the channel deep, these are universally obstructed with sand: the great *German* rivers bring down by their floods amazing quantities of sand and mud, the course of which is impeded at sea by the violence of the winds, blowing at south and west two-thirds of the year †. These, with the help of the tides, arrest the progress of the sand into the open sea, and form the numerous banks which, fatal as they may be to mariners, are the security of *Holland*, in particular, from naval invasions. The spring-tides at *Calais* rise twenty feet; at the pier head at *Dover*, to twenty-five; the cause of the variation is supposed, by Mr. *Cowley*, to be the different distances of the two piers from low-water mark, the first being half a mile, the last only a hundred yards; at *Ostend* it rises to eighteen; at *Flushing*, sixteen and a half; at *Helvoetsluys* and the *Texel*, twelve; and on the coasts of *Holstein* and *Jutland*, where the sea expands to a more considerable breadth, the tides grow more irregular, and weaken both in height and strength; at the *Elbe* they do not ex-

STREIGHTS OF DOVER.

SAND-BANKS OFF FLANDERS AND HOLLAND.

TIDES.

---

* *Lord Mulgrave's Voy. towards the North Pole.* † *Yarranton's England's Improvement,* 4, 5.

ceed

ceed feven or eight feet; on the coaft of *Jutland* only two or three; a fingular *phænomenon*, as they are fo greatly higher on the correfpondent coafts of *England*. The flood on the weft coaft of *Holland* fets to the northward, contrary to the courfe of the tides on the eaft coafts of *England* and *Scotland*.

ANTIENT FLAN-
DERS AND HOL-
LAND.

Flanders and *Brabant* formed part of the *Gallia Belgica* of *Cefar*; and *Holland* the *Batavorum Infula*. The rivers are the *Scaldis*, *Mofa*, and *Rhenus*, the modern *Scheld*, *Maefe*, and *Rhine*. The two firft probably do not vary greatly in their difcharge into the fea: the laft has experienced a moft confiderable change. The right branch of this river runs, for fome fpace, as it did in antient times, when it formed the lake *Flevo*, then refumed the form of a ftream, and difcharged itfelf into the fea at a place ftill called the *Flie-ftroom*, between the ifles of *Flie-landt* and *Schelling*, at the mouth of the *Zuyder-zee*. Long after that period the country was dry, firm, and well inhabited; a mighty inundation totally changed the face of it, and enlarged the *Flevo lacus* into the prefent *Zuyder-zee*, and broke the coaft into the chain of iflands which now front the fhore, even as far as the mouth of the *Wefer*. The *Dutch* hiftorians date this accident in 1421: it feems to have been the operation of a length of time; for the paffage through the *Texel* was forced open in 1400, and gave rife to the profperity of *Amfterdam* *. This country was firft peopled by the *Catti*, a *German* nation; thefe were thinned almoft to extirpation by the fwarms from the great northern hive, in their expeditions by land to other parts of *Europe*. For a very long fpace *Flanders* and *Holland* were a feat of banditti: the vaft foreft of *Ardennes* gave protection to them in one country; the moraffes fecured them in the other. Government at length took place, in *Holland* under its counts, in *Flanders* under its forefters. Thefe provinces fell at laft under the dominion of the dukes of *Burgundy*; from them to the houfe of *Auftria* and crown of *Spain*. The revolutions from that are well known. *Holland* received its fecond population from *Germany*, happily (for a country whofe exiftence depends on induftry) a moft induftrious race. The *Rhine* annually brings down multitudes of people, to repair the lofs of men occafioned by diftant voyages, and by the moft unwholefome colonies in the *Eaft* and *Weft Indies*. *Holland* is, from its climate, unfavorable to the encreafe of mankind: it cannot depend on itfelf for the reparation of the lofs of people, but muft look elfewhere for fupplies.

ANIMALS.

FLANDERS has many of the fame fpecies of animals with *Great Britain*; but, from the nature of its coaft, wants moft of the water-fowl, a few cloven-footed birds excepted, which breed on fandy fhores. *Holland* has ftill fewer quadrupeds and birds. Of the quadrupeds which we want, are a few Beavers in the *Rhine* and *Maefe*. The Wolf is common in *Flanders*, and is found

* *Anderfon's Dict.* i. 225.

in the parts of *Holland* bordering on *Germany*. Both countries have a few birds which never appear in *Britain*, except forced by the violence of weather or purfuit of fome bird of prey.

The antient *Germany* next fucceeds. *Holland* was a fort of neutral country, a retreat of the *German Catti*, and not *Germany* itfelf. As at prefent, the bordering parts were divided into petty ftates. The rivers which derive their origin far up the country, are the *Ems*, the *Wefer*, and the *Elb*, the antient *Amifius*, *Vifurgis*, and *Albis*.

Oppofite to the mouth of the eftuary of the *Wefer* and the *Elb*, is the remnant of the *Infula*, *Caftum Nemus*, celebrated by *Tacitus*, with his ufual elegance, for the worfhip of HERTHUM, or MOTHER *Earth*, by the neighboring nations. *Eft in infula oceani,* CASTUM NEMUS, *dicatum in eo vehiculum vefte contectum, attingere uni facerdoti conceffum. Is adeffe penetrali* DEAM *intelligit, vectamque bubus feminis multa cum veneratione profequitur. Læti tunc dies, fefta loca, quæcumque adventu hofpitioque dignatur. Non bella ineunt, non arma fumunt, claufum omne ferrum. Pax et quies tunc tantùm nota, tunc tantùm amata. Donec idem facerdos fatiatam converfatione mortalium Deam templo reddat. Mox vehiculum et veftes, et, fi credere velis, numen ipfum, fecreto lacu abluitur. Servi miniftrant, quos ftatim idem lacus haurit. Arcanus hinc terror, fanctaque ignorantia, quid fit illud quod tantum perituri vidit* .* The worfhip was continued very long after that period, and the ifland was diftinguifhed by the name of *Foftaland, Farria, Infula Sacra*, or *Heilgeland*, or the *Holy ifle*, from

the facrifices made there to the goddefs *Fofta*, or *Fofeta*, the fame with *Vefta*, *Herthum*, or the EARTH. She was called by the *Scandinavians, Goya*. The victims to her were precipitated into a pit : if they funk at once, the facrifice was thought to be accepted : the reverfe if they fwam any time on the furface †. This ifland was vifited, out of refpect to the goddefs, by people of high rank. *Radbothus* I. king of the *Frifians*, was here in 690, when *Winbertus*, and other Chriftian miffionaries, landed, overthrew the temples, and put an end to the pagan rites ‡. It had been an ifland of great extent; but by different inundations, between the years 800 and 1649, was reduced to its prefent contemptible fize §. The great ifland of *Nordftrandt* (one of the *Infulæ Saxonum*) not remote from this, in 1634 was reduced, by the fame caufe, from twenty parifhes to one : fifty thoufand head of cattle, and between fix and feven thoufand fouls, were fwept away. Such are the calamities to which thefe low countries are liable.

---

* De Mor. German. c. 40.　　† Mallei's North. Antiq. Tranfl. i. 136.　　‡ Emmii Hift. Rer. Friz. 129. ed. Franck.　　§ Bufching Geogr. i. 157. 167.

JUTLAND.

Jutland and *Holstein*, the antient *Cimbrica Cherfonefus* *, and *Cartris* †, terminating in the low point called the *Skagen*, or *Scaw*, stretches out in form of a peninsula, bounded by the North sea and the *Kattegatte*, the oblique approach into the *Baltic*. It is a very narrow tract, and only the resting-place of birds in their way from *Scandinavia*, and the farther north, the residence of numerous species. The rich marshes, in a climate mild from its situation between two seas, afford numbers of wholesome plants, the food of a remarkably fine breed of cattle. Besides the home consumption, these provinces send out annually thirty-two thousand head. The nobility do not think it beneath them to preside over the dairy: and their number of cows is princely. M. *De Rantzau* had not fewer than six hundred milch cows.

What the extent of this country might have been in very early times is unknown: it must have been prodigiously great, otherwise it never could have poured out that amazing number of people it did, in their eruption into *France*, when they were defeated by *Marius*, in 101 before CHRIST. Their army was computed to consist of three hundred thousand fighting men (including the *Teutoni*) besides women and children. About seven years before, they had suffered a great calamity from an inundation of the sea, which had destroyed great part of their country; and compelled the survivors, then crouded in the narrow *Cherfonefus*, to apply to the *Romans* for other lands. *Tacitus* speaks of the vestiges of this once mighty people, in the lines, visible in his time, on each shore. I presume that the inundations to which this coast is subject from the sea, hath utterly destroyed every trace of them. The charts plainly point out their overwhelmed territories in *Juts-riff*, and the neighboring sand-banks. The first might have been the continuation of land from the end of *Jutland*, beginning at the *Skaw*, and running out into the North sea in form of a scythe, not very remote from land, and terminating a little south of *Bergen* in *Norway*, leaving between its banks and that kingdom a deeper channel into the *Baltic*.

The *Kattegatte* lies between part of *Jutland* and the coast of *Sweden:* the last covered with isles innumerable. It is almost closed at the extremity, by the low *Danish* islands of *Seland* and *Funen*, which had in old times been (with *Sweden*) the seat of the *Suiones*. Between the first and the coast of *Sweden*, is the famous Sound, the passage tributary to the *Danes* by thousands of ships. These isles were of old called *Codonania* §, and gave to the *Kattegatte* the name of *Sinus Codanus*. The proper *Baltic* seems to have been the *Mare Suevicum* of the antients; and the farthest part, the *Mare Sarmaticum*, and part of the *Mare Scythicum*. As a na-

* *Ptolem.* lib. ii. c. 11.   † *Plin. Nat. Hist.* lib. iv. c. 13.   § *Mela*, lib. iii. c. 3. 8.

turalist,

turalift, I muft mention, that when LINNÆUS fpeaks of the *Mare Occidentale*, he intends the *Kattegatte*. Its greateft depth is thirty-five fathoms. It decreafes as it approaches the Sound; which begins with fixteen fathoms, and near *Copenhagen* fhallows to even four.

The *Roman* fleet, under the command of *Germanicus*, failed, according to *Pliny*, round *Germany*, and even doubled the *Cimbricum Promontorium*, and arrived at the iflands which fill the bottom of the *Kattegatte* * : either by obfervation or information, the *Romans* were acquainted with twenty-three. One they called *Gleffaria*, from its amber, a foffil abundant to this day on part of the fouth fide of the *Baltic*. A *Roman* knight was employed by *Nero*'s mafter of the gladiators, to collect, in thefe parts, that precious production, by which he came perfectly acquainted with this country †. I cannot fuppofe that the *Romans* ever fettled in any part of the neighborhood, yet there was fome commerce between them, either direct, or by the intervention of merchants. Many filver coins have been found at *Kivikke*, in *Schonen* in *Sweden*, of *Hadrian*, *Antoninus Pius*, *Commodus*, and *Albinus* ‡. Among the iflands, *Pliny* makes *Norway* one, under the name of *Scandinavia incompertæ magnitudinis*, and *Baltia* another, *immenfæ magnitudinis*, probably part of the fame, and which might give name to the Sounds called the *Belts*, and to the *Baltic* itfelf. The geographer *Mela* had the jufteft information of this great water, which he defcribes with great elegance, ' *Hac re mare* (CODANUS ' SINUS) *quod gremio littorum accipitur, nunquam latè patet, nec* USQUAM MARI SI- ' MILE *verum aquis paffim interfluentibus ac fæpe tranfgreffis vagum atque diffufum* ' *facie amnium fpargitur, qua littora attingit, ripis contentum infularum non longè dif-* ' *tantibus, et ubique pœne tantundem, it anguftum et par* FRETO *curvanfque fe fubinde,* ' *longo fupercilio inflexum eft.*' The different nations which inhabited its coafts fhall hereafter be mentioned.

VOYAGE OF THE ROMAN FLEET.

I would, like *Mela*, prefer giving to the *Baltic* the name of a gulph rather than a fea; for it wants many requifites to merit that title. It wants depth, having in no one place more than a hundred and ten fathoms. From the eaftern mouth of the Sound to the ifle of *Bornholm* it has from nine to thirty: from thence to *Stockholm*, from fifteen to fifty: and a little fouth of *Lindo*, fixty. It has in this courfe many fand-banks, but all in great depths of water. Between *Alands Haff*, amidft the great archipelago, the *Aland* ifles, and the ifle of *Ofel* in the gulph of *Riga*, the depths are various, from fixty to a hundred and ten ‖. Many frefh-water lakes exceed it in that refpect.

THE BALTIC A GULPH.

DEPTH.

---

* *Plin.* lib. ii. c. 67. lib. iv. c. 13.   † Lib. xxxvii. c. 3.   ‡ *Forffenius de Monum. Kivikenfe*, p. 27.   ‖ *Ruffian* and other charts.

It

No tides. It wants tides, therefore experiences no difference of height, except when the winds are violent. At such times there is a current in and out of the *Baltic*, according to the points they blow from ; which forces the water through the Sound with the velocity of two or three *Danish* miles in the hour. When the wind blows violently from the *German* sea, the water rises in the several *Baltic* harbours, and **Not salt.** gives those in the western part a temporary saltness : otherwise the *Baltic* loses that other property of a sea, by reason of the want of tide, and the quantity of vast rivers it receives, which sweeten it so much as to render it, in many places, fit for domestic uses. In all the *Baltic*, *Linnæus* enumerates but three *fuci**, plants of the sea : in the gulph of *Bothnia*, which is beyond the reach of salt water, not one †.

Few species of fish. The fewness of species of fish in the *Baltic* is another difference between it and a genuine sea. I can enumerate only nineteen ‡ which are found in this vast extent of water : and may add one cetaceous fish, the Porpesse. No others venture beyond the narrow streights which divide the *Baltic* from the *Kattegatte*; yet the great *Swedish Faunist* reckons eighty-seven belonging to his country, which is washed only by those two waters. Let me mention the Herring as a species which has from very early times enriched the neighboring cities. There was, between the years 1169 and 1203, a vast resort of *Christian* ships to fish off the isle of *Rugen*, the seat of the antient *Rugii*, insomuch that the *Danes* cloathed themselves with scarlet and purple, and fine linen.

The *Hornsimpa*, or COTTUS QUADRICORNIS, *Faun. Suec.* N° 321, and the SYNGNATHUS TYPHLE, or Blind Pipe-fish, N° 377, are unknown in the *British* seas : the first seems peculiar to the gulph of *Bothnia*, and is a fish of singular figure, with four flat hornlike processes on the head ‖.

Length and breadth of the Baltic. The extent of the *Baltic* in length is very great. From *Helsingor*, where it properly begins, to *Cronstadt*, at the end of the gulph of *Finland*, is eight hundred and ten *English* sea miles. Its breadth, between *Saltwic*, in *Smaland*, and the oppo-**Of the Gulph of Bothnia.** site shore, two hundred and thirty-seven. The gulph of *Bothnia*, which runs due north, forms an extent almost equal to the first, being, from *Tornea* in *Lapland*, to

---

* *Flora Suec.*   † *Flora Lapp.*

‡ Porpesse,	Striated Cod-fish,	Turbot,	Herring,
Sea Lamprey,	Viviparous Blenny,	Flounder,	Sprat,
Sturgeon,	Beardless Ophidion,	Salmon,	Little Pipe-fish,
Lannee,	Lump,	Gar-fish,	Shorter P.
Sword-fish,	Hornsimpa,	Smelt,	Blind P.

I find that the *Asinus Callarias* is common to the *Baltic* and our seas, therefore must be added to the list of *British* fish.

‖ *Mus. Fr. Adolph.* i. 70. tab. xxxii. fig. 4.

the

the shore near *Dantzic*, not less than seven hundred and seventy-eight : an amazing space, to be so ill stocked with fishy inhabitants.

From the isle of *Rugen*, the course of the *Baltic* is strait and open, except where interrupted by the famous isle of *Gottland*, the place of rendezvous from whence the *Goths* made their naval excursions. In 811, on this island, was founded the famous town of *Wisbuy*, the great emporium of the north : it was, for ages, the resort of every Christian nation. The *English* long traded here, before they ventured on the distant voyage of the *Mediterranean*. It became an independent city, and made its maritime laws the standard of all *Europe* to the north of *Spain*. In 1361, *Waldemar* III. of *Denmark*, attacked, ravaged, and plundered it of immense riches ; all which perished at sea after they were shipped *. Its present inhabitants are husbandmen and fishermen, secure from the calamities of war by the happy want of exuberant wealth.

ISLE OF GOTT-LAND.

Beyond *Stockholm* the *Baltic* divides into the gulphs of *Bothnia* and *Finland* : the first runs deeply to the north, and the country is composed chiefly of granite rock, or strewed over with detached masses of the same. Its greatest breadth is between *Gefle* and *Abo*, in *Finland*, where it measures a hundred and sixty-two miles : its greatest depth a hundred and ninety-five yards †. It terminates in *Lapland*, a country divided by the river *Tornea*, which runs navigable far up between a continued mountanous forest. It is supposed to have been peopled in the eleventh century by the *Finni* : a fact not easy to be admitted ; for the *Finni*, or *Fennones*, are a brawny race, with long yellow hair, and brown irides. The *Laplanders* are, on the contrary, small in body, have short black hair, and black irides. It is certain that a party of *Fins* deserted their native country, *Finland*, in the age before mentioned, rather than relinquish the brutality of heathenism. Their offspring remain converted, and in some measure reclamed, between *Norway* and *Sweden* ‖ ; but are a most distinct race from the *Laplanders*, who possessed their country long before. In the ninth century, the hero *Regner* slew its king or leader in battle § : at that period it was in a savage state ; nor was its conquest attempted by *Sweden* till 1277, when *Waldemar* added it to his kingdom, and in vain attempted its conversion ¶. Scarcely two centuries have elapsed since it has sincerely embraced the doctrines of Christianity. In consequence of which, cultivation and civilization have so well succeeded in the southern parts, that many deserts are peopled, morasses drained, and the reason of the natives so greatly improved, that they have united with the *Swedes*, and even sent their representatives to the

LAPLAND.

---

* *Hist. Abregè de Nord.* i. 206.   † *Prof. Ritzius of Lund.*   ‖ *Ph. Tr. Abr.* vii. part iv.
P. 44.   § *Hist. Abregè du Nord.* ii. 59.   ¶ The same, p. 3.

House of Peafants in the national diet * But thefe were at all times the moft cultivated of this diftinct race. They trained the Rein-deer to the fledge, domefticated it from its wild ftate, and made it the fubftitute for the Cow.

Their country, which penetrates even to the Northern ocean, confifts of favage mountains, woods, vaft marfhes, rivers, and lakes, the haunts of myriads of waterfowl, which refort here in fummer to breed, free from the difturbance of mankind. LINNÆUS, the great explorer of thefe deferts, my venerated example! mentions them as exceeding in numbers the armies of *Xerxes*; re-migrating, with him, in autumn, eight entire days and nights, to feek fuftenance on the fhores and waters of more favorable climates †.

Their lakes and rivers abound in fifh; yet the number of fpecies are few. Thefe are the Ten-fpined Stickle-back, *Br. Zool.* iii. N° 130; Salmon, N° 143, in great abundance, which force their way to the very heads of the furious rivers of *Tornea* and *Kiemi*, to depofit their fpawn; Char, N° 149, are found in the lakes in great abundance; and Graylings, N° 150, in the rivers; Gwiniads, N° 152, are taken of eight or ten pounds weight; Pikes, N° 153, fometimes eight feet long; and Perch, N° 124, of an incredible fize §; and the Salmo Albula, *Faun. Suec.* N° 353, clofes the lift of thofe of the *Lapland* lakes and rivers.

The mouth of the gulph of *Bothnia* is filled with a prodigious clufter of little iflands and rocks, dangerous to mariners. *Aland* is the chief, an ifland of furprifing rockinefs, and with all the other afpects as if torn from the continent by fome mighty convulfion. The gulph of *Finland* extends from thence due eaft, and has, on its northern coaft, a chain of fimilar iflands, and a few fprinkled over the channel. All the coaft and all its ifles are compofed of red or grey granite; and all the coafts of *Sweden* are the fame, mixed in places with fand-ftones. *Finland* and *Carelia* are the bounds of the gulph on this fide: *Livonia*, the granary of the north, and *Ingria*, on the other. Thefe countries, with *Ruffia*, made part of the *European Scythia*, or *Sarmatia*; and this part of the *Baltic* has been fometimes ftyled *Mare Scythicum*, and *Mare Sarmaticum* ‖. The gulph decreafes in depth from fixty to five fathoms, as you advance towards *Cronftadt*, the great naval arfenal of *Ruffia*. From thence is twelve miles of fhallow water to *Peterfburg*, that glorious creation of PETER the GREAT; the inlet of wealth and fcience into his vaft dominions, before his time inacceffible to the reft of *Europe*, unlefs by the tedious voyage of the White fea; and a country unknown, but by the report of the fplendid barbarifm of its tyrants. Peter was formed with a fingular mixture of

* *Anderfon*, ii. 419.   † *Amœn. Acad.* iv. 570. *Fl. Lap.* 273.   § *Scheffer's Lapland.*
‖ *Ptolemy.*

endowments

endowments for the purpose of civilizing a rude and barbarous people : his mind was pregnant with great defigns, obftinate perfeverance, and unrelenting feverity in the exertion of punifhment on all who dared to oppofe the execution of his fyftem for the good of the whole.   A mind filled with the milkinefs of human nature, would never have been able to deal with the favage uninformed *Ruffians*.   *Peter* hewed his work into fhape : for the laft polifh, Heaven formed another CATHE-RINE, the admiration of *Europe*, the bleffing of an empire which forms at leaft one eleventh of the globe, extending from the northern point of *Nova Zemlja*, in the frozen latitude of near 78, to the influx of the *Terek* into the *Caspian* fea, in the warm latitude of about 43 and a half; or, to give it the fhorteft breadth, from the coaft of the Frozen ocean, at the extremity of the country of the *Tschutki*, lat. 73, to the mouth of the *Aimakan*, in the gulph of *Ochotz*, in lat. 54.   Its length is ftill more prodigious, from *Peterfburgh* as far as the *Afiatic* fide of the ftreights of *Bering*.

In the following work, I have, by the affiftance of that celebrated naturalift Doctor PALLAS, given a defcription of the Quadrupeds and Birds of this vaft empire, as far as was compatible with my plan, which was confined between the higheft known latitudes of the northern hemifphere, as low as that of 60.   The remainder will be comprehended in the great defign formed by the Imperial Academy, and executed by profeffors whofe glory it is to prove themfelves worthy of their illuftrious and munificent patronefs, under whofe aufpices they have pervaded every part of her extenfive dominions in fearch of ufeful knowledge.

To *Peterfburg*, this corner of the empire, is brought, as to a vaft emporium, the commerce of the moft diftant parts; and from hence are circulated the *European* articles to fupply even the remote *China*.   The place of traffic is on the *Chinefe* borders, at *Kjackta*, a town without women ; for none are allowed to attend their hufbands.   By this route the furs of *Hudfon's-Bay* find their way to warm the luxurious inhabitants of *Pekin*, the animals of the neighboring *Tartary* and *Sibiria* being inadequate to the increafed demand.   The want of a maritime intercourfe is no obftacle to this enterprifing nation to the carrying on a trade with *India*.   It has encouraged above a hundred *Banians*, all males, from *Multan*, to fettle at *Aftracan*; and their number is kept up by a fupply of young unmarried relations from home. Thefe fupport the moft important trade of *Aftracan*, by carrying through *Aftrabad* to the inland parts of the *Mogol* empire.   I ftray a little from my plan ; but it may be excufed on account of the novelty of the relation, and becaufe it points out a more fouthern inland road than was known in the middle ages, when the merchants went by the way of *Bochara* and *Samarcand* to the northern cities of *India*, *Candabar* and *Cabul*.

In

SARMATÆ.

In my return to the *German* fea, let me review the antient inhabitants of the *Baltic*. The wandering *Sarmatæ*, of *Scythian* defcent, poffeffed all the country from lake *Onega* to the *Viftula*; and part of the vaft *Hercynian* foreft, famous of old for its wild beafts, occupied moft of this country. Bifons with their great manes: Uri with their enormous horns, which the natives bound with filver and quaffed at their great feafts: the Alces, or Elk, then fabled to have jointlefs legs: and Wild Horfes, were among the quadrupeds of this tract *. I fmile at the defcription of certain birds of the *Hercynian* wood, whofe feathers fhone in the night, and often proved the guide to the bewildered traveller †. The refplendent plumage of the Strix Nyctea, the Snowy Owl, N° 121, might probably have ftruck the eye of the benighted wanderer, and given rife to the ftrange relation.

ENINGIA.

Eningia was the oppofite fhore, and the fame with the modern *Finland*, inhabited by people of amazing favagenefs and fqualid poverty; who lived by the chace, headed their arrows with bones, cloathed themfelves with fkins, lay on the ground, and had no other fhelter for their infants than a few interwoven boughs ‡. They were then, what the people of *Terra del Fuego* are now. There is no certainty

OONÆ.

refpecting the *Oonæ*; iflanders, who fed, as many do at prefent, on the eggs of wild fowl and on oats ‖; but moft probably they were the natives of the ifles of *Aland*, and the adjacent archipelago; for *Mela* exprefsly places them oppofite to the *Sarmatæ*. We may add, that the *Hippopodæ* and *Panoti* might be the inhabitants

HIPPOPODÆ.

of the northern part of the *Bothnian* gulph; the firft fabled to have hoofs like horfes, the laft ears fo large as to ferve inftead of cloaks. The *Hippopodæ* were certainly the fame fort of people as the *Finni Lignipedes* of *Olaus*, and the *Skride Finnus* of *Ohthere*. They wore fnow-fhoes, which might fairly give the idea of their being, like horfes, hoofed and fhod. As to the *Panoti*, they baffle my imagination.

The *Bothnian* and *Finland* gulphs feem to me to have been, in the time of *Tacitus*, part of his *Mare pigrum ac immotum*, which, with part of the *Hyperborean* ocean, really infulated *Scandinavia*, and which he places beyond the *Suiones*, or modern *Sweden*. *Pliny* gives, I fuppofe from the relation of *Britifh* or other voyagers, to part of this fea, probably the moft northern, the title of *Morimarufa*, or Dead Sea, and *Cronium*. The learned *Forfter*, with great ingenuity, derives the word from the *Gaelic* and *Celtic* language. The firft, from the *Welfh*, *môr*, fea, and *marw*, dead; the other from the *Irifh*, *muir-croinn*, the coagulated, *i. e. congealed fea*. *Tacitus* adds to his account, that it was believed to encir-

---

* Cefar Bell. Gall. lib. iv. Plin. lib. viii. c. 15.   † Solinus, c. 32. Plin. x. c. 47.
‡ Tacitus de Mor. Germ. .   ‖ Forfter's Obf. 96.

cle the whole globe, and that the laſt light of the ſetting ſun continued ſo very vivid as to obſcure the ſtars themſelves. There is not a ſingle circumſtance of exaggeration in all this: every winter the gulph is frozen, and becomes motionleſs. Many inſtances may be adduced even of the *Baltic* itſelf being frozen *. The ſtars are frequently loſt in the amazing ſplendor and various colors of the *aurora borealis.* The *Hilleviones,* an antient people of *Sweden,* ſtyled *Scandinavia, alterum orbem terrarum,* and their deſcendants, long carolled the junction of the *Bothnian* gulph with the northern ocean, traditionally rehearſed in old *Swediſh* ſongs. *Tacitus* uſes the two laſt words to expreſs the world ſurrounded by this ſea. In the days of the geographer *Mela,* there certainly was a ſtrong tide in this upper part of the *Baltic;* for, ſpeaking of the iſlands off *Finland,* he ſays, " Quæ *Sarmatis* adverſa ſunt, ob alternos acceſſus recurſuſque pelagi, et quod " ſpatia queis diſtant, modò operiuntur undis, modò nuda ſunt; alias inſulæ " videntur, aliàs una et continens terra." With propriety, therefore, in another place, does he compare it to a ſtreight, *par freto,* notwithſtanding he was ignorant of its other entrance. Doctor *Pallas* moſt juſtly aſcribes the formation of not only the *Baltic,* but its former communication with the *White Sea,* to the effects of a deluge. The whole intermediate country is a proof; the foundation being what is called the old rock, and that covered with variety of matter; ſuch as beds of pebble and gravel, and fragments of granite, torn from the great maſs. Parts of the channel which formed the inſulation of *Scandinavia,* are the chain of lakes, from that of *Ladoga* to the *White Sea,* ſuch as *Onega,* and others, often connected by rivers, and lying in a low country, filled with the proofs above-mentioned. This was the ſtreight through which the tide poured itſelf from the *Hyperborean* ocean, and covered, at its flux, the iſlands deſcribed by *Mela.* This, like the other northern ſeas, was annually frozen over, and could be no obſtacle to the ſtocking of *Scandinavia* with quadrupeds. There is no fixing the period in which this paſſage was obſtructed. An influx of ſand, or an earthquake, might cloſe it up. As ſoon as this event took place, the *Baltic* felt the want of its uſual feed: it loſt the property of a ſea; and, by a conſtant exhalation, from that time decreaſed in the quantity of water. Modern philoſophers have proved the great loſs it has ſuſtained, and that it decreaſes from forty to fifty inches in a century: that, near *Pithea,* the gulph of *Bothnia* has retired from the land half a mile in forty-five years; and near *Lulea,* a mile in twenty-eight. Notwithſtanding its preſent ſtate, when we conſider the accounts given by the antients, the old *Swediſh* traditions, and the preſent veſtiges of the former channel, we can, without any

---

* *Forſter's Obſ.* 80.

i

force

force of fancy, give full credit to the infulated form of *Scandinavia*, given in one of *Cluverius*'s maps *; which, he fays, is drawn from the erroneous accounts of the antients.

SUIONES.

The *Suiones* poffeffed the modern *Sweden*, and extended even to the ocean, and were a potent naval power. Their fhips were fo conftructed, with prows at each end, that they were always ready to advance. Thefe people, in after times, proved, under the common name of *Nortmans*, the peft and conquerors of great part of fouthern *Europe*; their fkill in maritime affairs fitting them for diftant expeditions. In the fixth century they were called *Suethans*, and were famous for their cavalry. In their time, the Sable, N° 30, was common in their country: *Jornandes*, therefore, obferves, that notwithftanding they lived poorly, they were moft richly cloathed: he alfo informs us, that they fupplied the *Romans* with thefe precious furs, through the means of numbers of intervening nations †. *Scandinavia*, in that period, had got the name of *Scanzia*; and as it was then called an ifland, and by *Jornandes* ‡, a native of the country, there is all the reafon to imagine, that the paffage into the *Hyperborean* ocean was not in his time clofed.

After repaffing the Sound, appear *Schonen*, *Halland*, and *Bohufland*, *Swedifh* provinces, bounded by the *Kattegatte*. *Halland*, from fome fimilitude of found, is fuppofed to have been the feat of the *Hilleviones*, a moft populous nation; perhaps the fame with the *Suiones* of *Tacitus*; for beyond them he places the *Sitones*, or the country of *Norway*, who were a great naval people; as the hiftorian fays that they differed not from the *Suiones*, except in being under a female government. The promontory of the *Naze*, vifible at eight or ten leagues diftance, with the low land of *Bevenbergen* in *Jutland*, forms the entrance into the *German* fea. The *Bommel*, and the *Drommel*, high mountains to the eaft of it; and the high land of *Left*, a vaft mountain, gradually rifing from the fhore, to the weft, are noted guides to mariners. It is reafonably fuppofed, that *Pliny* intended this vaft region by his ifland of *Nerigon*, from whence, fays he, was a paffage to *Thule*. He fpeaks alfo of *Bergos*, which, from agreement of found, is thought to be the prefent province of *Bergen*. The *promontorium Rubeas* is gueffed to be the *North Cape*, between which and the *Cimbri*, *Philæmon* § places the *Mare Morimarufa*, or the Dead Sea, fo called from the clouded fky that ufually reigned there.

NORWAY.

THE NAZE.

Our firft certain knowlege of the inhabitants of this country, was from the defolation they brought on the fouthern nations by their piratical invafions.

---

* At the end of his fecond vol. of *Germania Antiqua*.			† *Jornandes de Reb. Geticis*, c. iii.
‡ The fame, c. iv.			§ As quoted by *Pliny*, lib. iv. c. 13,

Their

Their country had, before that period, the name of *Nortmannaland*, and the inhabitants *Nortmans*; a title which included other adjacent people. *Great Britain* and *Ireland* were ravaged by them in 845; and they continued their invasion till they effected the conquest of *England*, under their leader, *Canute* the Great. They went up the *Seine* as far as *Paris*, burnt the town, and forced its weak monarch to purchase their absence at the price of fourteen thousand marks. They plundered *Spain*, and at length carried their excursions through the *Mediterranean* to *Italy*, and even into *Sicily*. They used narrow vessels, like their ancestors the *Sitones*; and, besides oars, added the improvement of two sails: and victualled them with salted provisions, biscuit, cheese, and beer. Their ships were at first small; but in after times they were large enough to hold a hundred or a hundred and twenty men. But the multitude of vessels was amazing. The fleet of *Harold Blaatand* consisted of seven hundred *. A hundred thousand of these savages have at once sallied from *Scandinavia*, so justly styled *Officina Gentium*, *aut certè velut vagina nationum* †. Probably necessity, more than ambition, caused them to discharge their country of its exuberant numbers. Multitudes were destroyed; but multitudes remained, and peopled more favorable climes.

Their king, *Olaus*, was a convert to Christianity in 994; *Bernard*, an Englishman, had the honor of baptizing him, when *Olaus* happened to touch at one of the *Scilly* islands. He plundered with great spirit during several years; and in 1006 received the crown of martyrdom from his pagan subjects. But religious zeal first gave the rest of *Europe* a knowlege of their country, and the sweets of its commerce. The *Hanse* towns poured in their missionaries, and reaped a temporal harvest. By the year 1204, the merchants obtained from the wise prince *Suer* every encouragement to commerce; and by that means introduced wealth and civilization into his barren kingdom. *England*, by every method, cherished the advantages resulting from an intercourse with *Norway*; and *Bergen* was the *emporium*. *Henry* III. in 1217, entered into a league with its monarch *Haquin*, by which both princes stipulated for free access for their subjects into their respective kingdoms, free trade and security to their persons. In 1269, *Henry* entered into another treaty with *Magnus*, in which it was agreed, that no goods should be exported from either kingdom except they had been paid for; and there is besides a humane provision on both sides, for the security of the persons and effects of the subjects who should suffer shipwreck on their several coasts.

This country extends above fifteen hundred miles in length, and exhibits a most wonderful appearance of coast. It runs due north to Cape *Staff*, the

---

* *Mallet's Introd.* i. 257.   † *Jornandes*, c. 4.

western

SEA.

weſtern point of *Sondmor*, then winds north-eaſt to its extremity at the *North Cape*. High and precipitous rocks compoſe the front, with a ſea generally from one to three hundred fathoms deep waſhing their baſe *. Multitudes of narrow creeks penetrate deep into the land, overſhadowed, by ſtupendous mountains. The ſides of theſe chaſms have depth equal to that of the adjacent ſea ; but in the

DYBRENDES.

middle is a channel called *Dybrendes*, i. e. deep courſes, from fifty to a hundred fathoms broad, and of the diſproportionable depth of four hundred †, ſeemingly time-worn by the ſtrength of the current from the torrent-rivers which pour into them. Fiſh innumerable reſort to their edges. Theſe creeks are, in many places, the roads of the country ; for the vallies which traverſe it are often ſo precipitous as to be impervious, unleſs by water. Some, which want theſe conveniences, are left uninhabited by reaſon of the impoſſibility of conveying to and from them the articles of commerce.

CHAIN OF IS-LANDS.

Millions of iſlands, large and ſmall, ſkerries, or rocks, follow the greateſt part of this wondrous coaſt. The iſlands are rude and mountanous, and ſoar correſpondent to the *Alps* of the oppoſite continent. Thoſe of *Loeffort*, on the north ſide of the dreadful whirlpool *Maelſtrom*, engraven by *Le Bruyn*, give a full idea of the nature of the coaſts ‡. The ſea near the iſlands is ſo deep and rocky, that the *Norwegian* kings cauſed vaſt iron rings to be faſtened with lead § to the ſides, to enable ſhips to moor in ſecurity, or to aſſiſt them in warping out. A few of the former give ſhelter to the fiſhermen and their ſmall ſtock of cattle ; the reſt riſe in columns of groteſque forms. On the outſide of theſe natural counterſcarps, are multitudes of *haubroe*, or ſea-breakers, longitudinal banks of ſand, running north and ſouth, from the diſtance of four to ſixteen leagues from the continent, and from ten to fifteen fathoms below the ſurface of the water ; the haunts of myriads of uſeful fiſh.

TIDES.

The tides off the *Naze*, and moſt of the coaſts of *Norway*, are very inconſiderable. At the *North Cape*, the ſpring tides have been obſerved to riſe to the height of eight feet one inch ; the neap to ſix feet eight inches ‖. Mr. *William Ferguſon*, an able pilot, who had often the conduct of our fleets in the North ſea, informed me, that on the *Naze*, and many other parts of *Norway*, the tides were hardly perceptible, except with ſtrong weſterly winds, when they roſe two or three feet, and fell with the eaſterly winds.

RIVERS.

Into the ends of moſt of the *Dybrendes* ruſh the furious rivers, or rather torrents, of the mountains ; uſeleſs for navigation, but moſt ſingularly advantageous

* *Pontoppidan*, i.   † The ſame, i. 68.   ‡ *Le Bruyn's Voyages*, i. tab. 1.   § *Olaus Magnus, Gent. Septentr.* lib. ii. c. xi.   ‖ Mr. *Bayley*, in *Phil. Tranſ.* lix. 270.

for

for the conveyance of the great article of commerce, the masts and timber of the country, from the otherwise inaccessible forests. The trees are cut down, and at present conveyed from some distance to the rivers, down which they are precipitated over rocks and stupendous cataracts, until they arrive at the *Lentzes* or booms *, placed obliquely in the stream in fit places. To them the owners of the timber resort; and, on paying a certain rate to the proprietors, receive their pieces, which are all marked before they are committed to the water; but numbers are injured or destroyed in the rough passage.

The species which is of such great value to *Norway*, is the *Fyr* or *Fure*, our *Scotch* Pine, and the *Pinus Sylvestris* of *Linnæus*. It grows in the driest places, and attains the vast age of four hundred years †; and is of universal use in the northern world. Such trees as are not destined for masts are squared, and arrive in *England* under the name of *Balk :* the rest are sawed on the spot, in hundreds of mills, turned by the torrents, and reach us in form of planks. An immense quantity of tar is made from the trees, and even from the roots, very long after they have been divided from the trunk. The *Gran, Pinus Abies*, or what we call *Norway* Fir, is in little esteem. Thousands are cut down annually by the peasants, who feed their cattle with the tender shoots. It is the tallest of *European* trees, growing to the height of a hundred and sixty feet. In winter, the branches are depressed to the ground with snow, and form beneath them the dens of wild beasts.

I must here mention the adventitious fruits, such as nuts and other vegetable productions, which are brought by the waves to these shores, those of *Feroe*, and the *Orknies*, from *Jamaica* and other neighboring parts ‡. We must have recourse to a cause very remote from this place. Their vehicle is the gulph-stream from the gulph of *Mexico*. The trade-winds force the great body of the ocean from the westward through the *Antilles* into that gulph, when it is forced backward along the shore from the mouth of the *Mississipi* to Cape *Florida*; doubles that cape in the narrow sea between it and *Cuba*, and from Cape *Florida* to Cape *Cannaveral* runs nearly north, at the distance of from five to seven leagues from shore, and extends in breadth from fifteen to eighteen leagues. There are regular soundings from the land to the edge of the stream, where the depth is generally seventy fathoms; after that no bottom can be found. The soundings off Cape *Cannaveral* are very steep and uncertain, as the water shallows so quick, that from forty fathoms it will immediately lessen to fifteen, and from that to four, or less; so that, without great care, a ship may be in a few minutes on shore. It must be observed, that, notwithstanding the gulph-stream in general

LENTZES.

EXOTIC FRUITS FOUND ON THE SHORES.

GULPH-STREAM.

* *Pontoppidan*, i. 93. tab. vii.　　† *Amæn. Acad.* iv.　　‡ *Voy. Hebrides.*

is

is faid to begin where foundings end, yet its influence extends feveral leagues within the foundings; and veffels often find a confiderable current fetting to the northward all along the coaft, till they get into eight or ten fathom water, even where the foundings ftretch to twenty leagues from the fhore; but their current is generally augmented or leffened by the prevaling winds, the force of which, however, can but little affect the grand unfathomable ftream. From Cape *Cannaveral* to Cape *Hatteras* the foundings begin to widen in the extent of their run from the fhore to the inner edge of the ftream, the diftance being generally near twenty leagues, and the foundings very regular to about feventy fathoms near the edge of the ftream, where no bottom can be afterwards found. Abreaft of *Savannah* river, the current fets nearly north; after which, as if from a bay, it ftretches north-eaft to Cape *Hatteras*; and from thence it fets eaft-north-eaft, till it has loft its force. As Cape *Hatteras* runs a great way into the fea, the edge of the ftream is only from five to feven leagues diftant from the cape; and the force and rapidity of the main ftream has fuch influence, within that diftance, over fhips bound to the fouthward, that in very high foul winds, or in calms, they have frequently been hurried back to the northward, which has often occafioned great difappointment both to merchant fhips and to men of war, as was often experienced in the late war. In *December* 1754, an exceeding good failing fhip, bound from *Philadelphia* to *Charleftown*, got abreaft of Cape *Hatteras* every day during thirteen days, fometimes even with the tide, and in a middle diftance between the cape and the inner edge of the ftream; yet the fhip was forced back regularly, and could only recover its loft way with the morning breeze, till the fourteenth day, when a brifk gale helped it to ftem the current, and get to the fouthward of the Cape. This fhews the impoffibility of any thing which has fallen into the ftream returning, or ftopping in its courfe.

On the outfide of the ftream is a ftrong eddy or contrary current towards the ocean; and on the infide, next to *America*, a ftrong tide fets againft it. When it fets off from Cape *Hatteras*, it takes a current nearly north-eaft; but in its courfe meets a great current that fets from the north, and probably comes from *Hudfon's Bay*, along the coaft of *Labrador*, till the ifland of *Newfoundland* divides it; part fetting along the coaft through the ftreights of *Belleifle*, and fweeping paft Cape *Breton*, runs obliquely againft the gulph-ftream, and gives it a more eaftern direction: the other part of the northern current is thought to join it on the eaftern fide of *Newfoundland*. The influence of thefe joint currents muft be far felt; yet poffibly its force is not fo great, nor contracted in fuch a pointed and circumfcribed direction as before they encountered. The prevaling winds all over this part of the ocean are the weft and north-weft, and con-

fequently

sequently the whole body of the western ocean seems, from their influence, to have what the mariners call a *set* to the eastward, or to the north-east by east. Thus the productions of *Jamaica*, and other places bordering on the gulph of *Mexico*, may be first brought by the stream out of the gulph, inveloped in the *sargasso* or alga of the gulph round Cape *Florida*, and hurried by the current either along the *American* shore, or sent into the ocean in the course along the stream, and then by the set of the stream, and the prevaling winds, which generally blow two-thirds of the year, wafted to the shores of *Europe*, where they are found *.

The mast of the *Tilbury* man of war, burnt at *Jamaica*, was thus conveyed to the western side of *Scotland*; and among the amazing quantity of drift-wood, or timber, annually flung on the coasts of *Iceland*, are some species which grow in *Virginia* and *Carolina* †. All the great rivers of those countries contribute their share; the *Alatamaha*, *Santee*, and *Roanok*, and all the rivers which flow into the *Chesapeak*, send down in floods numberless trees ‡; but *Iceland* is also obliged to *Europe* for much of its drift-wood; for the common pine, fir, lime, and willows, are among those enumerated by Mr. *Troille*; all which, probably, were wafted from *Norway*.

The mountains of *Norway* might prove a boundless subject of speculation to the traveller. Their extent is prodigious, and the variety of plants, animals, and fishes of the lakes, are funds of constant amusement. The silver mines, wrought ever since 1623, are sources of wealth to the kingdom, and afford the finest specimens of the native kinds yet known. Gold was found in a considerable quantity in 1697. *Christian* V. caused ducats to be coined with it; the inscription was the words of *Job*, VON MITTERNACHT KOMT GOLD, *out of the north comes* GOLD ‖. Copper and iron are found in abundance; lead in less quantities: tin does not extend to this northern region. It is difficult to say which is the beginning of this enormous chain. In *Scandinavia* it begins in the great *Koelen* rock at the extremity of *Finmark*. It enters *Norway* in the diocese of *Drontheim*, bends westward towards the sea, and terminates at a vast precipice, I think, the *Heirefoss*, about three *Norwegian* miles from *Lister*. Another branch of this mountain divides *Norway* from *Sweden*, fills *Lapland*, and rises into

MOUNTAINS.

METALS.

---

* For this curious account, I am indebted to Doctor *Garden*, who, by his long residence in *Charlestown*, is extremely well acquainted with the subject.

† *Troille's Voy. to Iceland*, 47. ‡ Doctor *Garden*.

‖ *Pontoppidan*, i. 179. *Museum Regium Havniæ*, pars ii. sect. v. tab. xx. N° 18.—With more truth, perhaps, our version has it, *out of the north cometh* COLD.

the

the diſtinguiſhed ſummits of *Horrikalero*, *Avaſaxa*, and *Kittis*, and ends in ſcattered maſſes of granite, in the low province of *Finland*. It incloſes *Scandinavia* in form of a horſe-ſhoe, and divides it from the vaſt plains of *Ruſſia*. The antient name of this chain was *Sevo mons*, to this day retained in the modern name *Seveberg*. *Pliny* compares it to the *Riphæan* hills, and truly ſays, it forms an immenſe bay, even to the *Cimbrian* promontory *.

ROMANTIC
VIEWS.

The mountains and iſlands break into very groteſque forms, and would furniſh admirable ſubjects for the pencil. Among the deſiderata of theſe days, is a tour into thoſe parts by a man of fortune, properly qualified, and properly attended by artiſts, to ſearch into the great variety of matter which this northern region would furniſh, and which would give great light into the hiſtory of a race, to which half *Europe* owes its population. Among the views, the mountains of the *Seven Siſters* in *Helgeland* †, and the amazing rock of *Torg-hatten* ‡, riſing majeſtically out of the ſea, with its pervious cavern, three thouſand ells ‖ long, and a hundred and fifty high, with the ſun at times radiating through it, are the moſt capital. Not to mention the tops of many, broken into imaginary forms of towers and Gothic edifices, forts, and caſtles, with regular walls and baſtions.

HEIGHTS OF
MOUNTAINS.

I agree with the Comte *De Buffon*, in thinking that the heights of the *Scandinavian* mountains, given by Biſhop *Pontoppidan*, and Mr. *Browallius*, are extremely exaggerated §. They are by no means to be compared with thoſe of the *Helvetian* Alps, and leſs ſo with many near the equator. The ſober accounts I have received from my northern friends, ſerve to confirm the opinion, that there is an increaſe of height of mountains from the north towards the equatorial countries. M. *Aſcanius*, profeſſor of mineralogy at *Drontheim*, aſſures me, that from ſome late ſurveys, the higheſt in that dioceſe are not above ſix hundred fathoms above the ſurface of the ſea; that the mountains fall to the weſtern ſide from the diſtance of eight or ten *Norwegian* miles ¶; but to the eaſtern, from that of forty. The higheſt is *Dovre-fiæl* in *Drontheim*, and *Tille* in *Bergen*. They riſe ſlowly, and do not ſtrike the eye like *Romſdale-horn*, and *Hornalen*, which ſoar majeſtically from the ſea. In *Sweden*, only one mountain has been properly meaſured to the ſea. Profeſſor *Ritzius* of *Lund*, acquaints me, that *Kinnekulle* in *Weſtro-Gothia* is only eight hundred and fifteen *Engliſh* feet

---

* *Sevo* mons ibi immenſus, nec *Riphæis* jugis minor, immanem ad *Cimbrorum* uſque promontorium efficit ſinum, qui *Codanus* vocatur. Lib. iv. c. 13.

† *Pontoppidan*, i. 46. tab. iii.   ‡ The ſame, i. 47. tab. iii.   ‖ Of two *Daniſh* feet each.
§ *Epoques de la Nature*, Suppl. tom. vi. p. 136. edit. *Amſterdam*.   ¶ Of 18,000 feet each.

above

above the lake *Wenern*, or nine hundred and thirty-one above the fea. He adds, the following have been only meafured to their bafes, or to the next adjacent waters: *Aorfkata*, a folitary mountain of *Jæmtland*, about four or five *Swedifh* miles from the higheft *Alps*, which feparate *Norway* and *Sweden*, is faid to be fix thoufand one hundred and fixty-two *Englifh* feet above the neareft rivers: *Swuckuſtol*, within the borders of *Norway*, four thoufand fix hundred and fifty-eight above lake *Famund*; and that lake is thought to be two or three thoufand above the fea: and finally, *Sylfiællen*, on the borders of *Jæmtland*, is three thoufand one hundred and thirty-two feet perpendicular, from the height to the bafe. *Pontoppidan* gives the mountains of *Norway* the height of three thoufand fathoms: *Browallius* thofe of *Sweden* two thoufand three hundred and thirty-three, which makes them nearly equal to the higheft *Alps* of *Savoy*, or the ftill higher fummits of the *Peruvian Andes.*

In *Finmark*, the mountains in fome places run into the fea: in others recede far, and leave extenfive plains between their bafes and the water. Their extreme height is on the *Fiæll-ryggen, dorfum Alpium*, or *back of the Alps*, a name given to the higheft courfe of the whole chain: the fummits of which are clad with eternal fnow. Thefe are fkirted by lower mountains, compofed of hard fandy earth, deftitute of every vegetable, except where it is mixed with fragments of rock, on which appear the *Saxi-frages* of feveral kinds; *Diapenfia Lapponica*, Fl. Lapp. N° 88; *Azalea Procumbens*, N° 90; the *Andromeda Cærulea*, N° 164; and *Hypnoides*, N° 165, thinly fcatter-ed. Lower down are vaft woods of Birch, N° 341, a tree of equal ufe to the *Laplanders*, and the northern *Indians* of *America*. On the lower *Alps* abound the *Rein-deer Lichen*, N° 437, the fupport of their only cattle; the *Dwarf Birch*, N° 342, the feeds of which are the food of the White Grous beneath the fnow, during the long and rigorous winter; the *Arbutus Alpina*, N° 161; and *Arbutus Uva Urfa*, N° 162; and, finally, the *Empetrum Nigrum*, or *Black Heath Berries*, ufed by the *Laplanders* in their ambrofial difh the *Kappifiàlmas* *.

The *Scotch* Pine, N° 346, and *Norway* Fir, N° 347, form the immenfe forefts of *Lapland*, affociated with the Birch: the Pine affects the dry, the Fir the wet places, and grow to a vaft fize; but, being inacceffible, are loft to the great ufes of man-kind. On their northern fides they are almoft naked, and deprived of boughs by the piercing winds; the wandering *Laplander* remarks this, and ufes it as a compafs to fteer by, amidft thefe wilds of wood. Whole tracts are oft-times fired by light-ning; then proftrated by the next ftorm. The natives make, of the under part of the wood (which acquires vaft hardnefs by length of time) their fnow-fhoes; and

---

* *Fl. Lapp.* p. 108.

k   form

form their bows for shooting the squirrel with pieces united with glue, made from the skin of the perch. Their fragile boats are formed of the thinnest boards: their ropes of the fibrous roots: and finally, the inner bark, pulverized and baked, is the substitute for bread to a people destined to this rigorous climate. These three trees, the *Dwarf Birch*, N° 341, the *Alder*, N° 340, and not less than twenty-three species of Willows, form the whole of the trees of *Lapland*. Every other *Swedish* tree vanishes on approaching that country.

There is a great analogy between the plants of these northern *Alps*, and those of the *Scottish* Highlands. A botanist is never surprized with meeting similar plants on hills of the same height,- be their distance ever so great. It may be remarked, that out of the three hundred and seventy-nine perfect plants which grow in *Lapland*, two hundred and ninety-one are found in *Scotland*; and of the hundred and fifty cryptogamous, ninety-seven are to be met with in *North Britain*.

QUADRUPEDS OF SCANDINAVIA.

The *Alps*, the woods, and marshes of the vast region of *Scandinavia* (for I will consider it in the great) give shelter to numbers of quadrupeds unknown to *Britain*. Those which brave the severity of the extreme north of this country are distinguished by the addition of the *Lapland* name. The Elk, N° 3 of this Work, is found in many parts: the Rein, *Godde*, N° 4, is confined to the chilliest places: the Wolf, *Kumpi*, N° 9, is a pest to the whole: the Arctic Fox, *Njal*, N° 10, skirts the shores of all the northern regions: the Cross Fox, *Raude*, N° 11. β, and the Black Fox, N° 11. α. is scattered every where: the Lynx, *Albos* *, N° 15, inhabits the thickest woods: the Bear, *Guouzhia*, N° 20, and Glutton, *Gjeed'k*, N° 21, have the same haunts: the Sable, N° 30, which continued in *Lapland* till the middle of the last century, is now extinct: the Lesser Otter, or Mænk, of the *Swedes*, is confined to *Finland*: the Beaver, *Majæg*, N° 90, is still found in an unsociable state in several parts: the Flying Squirrel, p. 124, the *Orava* of the *Finlanders*, is found in their forests †, and those of *Lapland*: the Lemmus, *Lumenik*, p. 136, is at seasons the pest of *Norway*, issuing like a torrent from the *Koelen* chain: The Walrus, *Morsh*, N° 71, is sometimes found in the *Finmark* seas: the Harp Seal, *Dælja*, N° 77, the Rough Seal, N° 74, the Hooded, *Oanide?* N° 76, and the Little Seal, *Hist. Quad.* ii. N° 386, omitted by me in this Work, inhabit

---

* I have no proof of this but the name. The *Lynx* inhabits *Norway* and *Sweden*, and all the woody parts of *Sibiria*; a circumstance I omitted in p. 50, of this Work. I scarcely know whether I should apologize for the omission of the Fitchet, *Hist. Quad.* i. N° 195; the *Mustela Putorius*, N° 16, *Faun. Suec.* LINNÆUS speaks with uncertainty of its being found in *Scania*, and that is a latitude rather too far south for my plan.

† See Mr. *Gabriel Bonsdorff's* account of the animals of *Finland*, p. 24.

the

the fame place *. The laſt, ſays Biſhop *Gunner*, is eaten ſalted, not only by the *Laplanders*, but by the better ſort of people in *Finmark*.

Of animals found in *Britain*, the Fox, *Ruopſok*, N° 11; Pine Martin, *Nætte*, N° 27; Ermine, *Boaaid* †, N° 26; Weeſel, *Seibuſh*, N° 25; Otter, *Zhieonares*, N° 34; Varying Hare, *Njaumel*, N° 37; Common Squirrel, *Orre*, p. 122. A; Mouſe, N° 60; Field Mouſe, N° 61; Water Rat, N° 59; and the Shrew, *Vandes* and *Ziebak*, N° 67, are ſeen as high as *Finmark:* the Common Seal, *Nuorroſh*, N° 72, and the Great Seal, N° 73, alſo frequent the ſhores. All the other quadrupeds, common to *Scandinavia*, ceaſe in *Norway*, and ſome even in *Sweden*. *Scandinavia* received its animals from the eaſt; but their farther progreſs was prevented by the intervention of the North ſea between that region and *Britain*. Our extinct ſpecies, the Bear, the Wolf, and the Beaver, came into this iſland, out of *Gaul*, before our ſeparation from the continent. Some of the northern animals never reached us: neither did the north ever receive the Fallow Deer, *Br. Zool*. N° 7; the Harveſt Mouſe, N° 29; the Water Shrew, N° 33; nor yet the Brown Rat, N° 57, of this Work; notwithſtanding it familiarly goes under the name of the *Norway* ‡.

This great tract has very few birds which are not found in *Britain*. We may except the.Collared Falcon, p. 222. G; the *Scandinavian* Owl, p. 237; Rock Crow, p. 252. F; Roller, p. 253; Black Woodpecker, p. 276; Grey-headed, p. 277; Three-toed, N° 168; the Rehuſak Grous, p. 316. B; and the Hazel Grous, p. 316. F. The Ortolan, p. 367. D; the Arctic Finch, p. 379. A; and the Lulean F. p. 380. B. The Grey Redſtart Warbler, p. 417. C; the Blue Throat W. p. 417. E; Boguruſh W. p. 419. I; Fig-eater, 419. K; and Kruka W. p. 422. U. All the cloven-footed water-fowl, except the Spoon-bill, p. 441. A; the Crane, p. 453. A; White and Black Storks, p. 455, 456. C. D; *Finmark* Snipe, p. 471. D; Striated Sandpiper, N° 383; Selninger, p. 480. C; Waved, p. 481. E; Shore, p. 481. F; Wood, p. 482. G; *Alwargrim* Plover, N° 398; and *Alexandrine*, p. 488. B. And all the web-footed kinds, except the Harlequin Duck, N° 490, and *Lapmark*, p. 576. M. are common to both countries; but during ſummer, Fieldfares, Redwings, Woodcocks, and moſt of the water-fowl, retire from *Britain* into *Scandinavia*, to breed in ſecurity: and numbers of both land and water-fowl quit this frozen country during winter, compelled, for want of food, to ſeek a milder climate.

BIRDS.

---

* Conſult *Leems Lapm.* 214, 215, 216. Alſo for the Mouſe, &c. which want the *Lapland* names.
† *Leems*, 220.
‡ It is a native of the *Eaſt Indies*. See *Hiſt. Quad.* ii. N° 44.

The fifhes of this extenfive coaft amount to only one hundred and eleven, and are inferior in number to thofe of *Britain* by twenty eight. The fpecies of the North Sea which differ from the *Britifh*, are not numerous. The depth of water, and the forefts of marine plants which cover the bottom of the *Norwegian* feas, are affuredly the caufe of the preference of certain kinds, in their refidence in them. Infinite numbers of rare Vermes, Shells, Lithophytes, and Zoophytes, are found there, feveral of which, before their difcovery by Bifhop *Pontoppidan*, were the fuppofed inhabitants of only the more remote feas *. Among the fifhes which have hitherto fhunned our fhores, are the Raia Clavata, *Muller*, N° 309; *Squalus Spinax*, 312†; *Sq. Centrina*, 313, which extends to the *Mediterranean*; *Chimera Monftrofa*, 320, a moft fingular fifh; *Sygnathus Typhle*, and *Æquoreus*, 324, 328; the *Regalecus Glefve*, 335, Afcan. Icon. tab. xi.; *Gadus Brofme*, 341; *G. Dypterygius*, or *Byrke-lange*, 346; *Blennius Raninus*, & *Fufcus*, 359, 360; *Echeneis Remora*, 361; *Coryphæna Novacula*, & *Rupeftris*, 362, 363; *Gobius Jozo*, 365; *Plearonectes Cynogloffus, Limanda*, & *Linguatula*, 372, 375, 377; *Sparus Erythrinus*, 380; *Labrus Suillus*, 381; *Perca Norvegica*, and *Lucio-perca*, 390, 391; *Scomber Pelagicus*, 398; *Silurus Afotus*, 404; *Clupea Villofa*, 425.

Thefe are not the fifhes of general ufe. Providence hath, in thefe parts, beftowed with munificence the fpecies which contribute to the fupport of mankind; and made thereby the kingdom of *Norway* a coaft of hardy fifhermen. The chain of iflands, and the fhores, are the populous parts. It is the fea which yields them a harveft; and near to it ftand all the capital towns: the ftaples of the produce of the ocean on one hand, and of the more thinly inhabited mountains on the other. The farther you advance inland, the lefs numerous is the race of man.

The Herring, the Cod, the Ling, and the Salmon, are the maritime wealth of this country. The Herring has two emigrations into this fea: the firft is from *Chriftmas* to *Candlemas*, when a large fpecies arrives, preceded by two fpecies of Whales, who, by inftinct, wait its coming. The fifhermen poft themfelves on fome high cliff, impatiently waiting for the cetaceous fifh, the harbingers of the others. They look for them at the moon *Torre*, or the firft new one after *Chriftmas*, and the moon *Gio*, which immediately follows.

Thefe Herrings frequent the great fand-banks, where they depofit their fpawn. They are followed by the Spring Herrings, a leffer fifh, which approach much nearer to the fhore; after which arrive the Summer Herring, which almoft literally fill every creek: the whole fifhery is of immenfe profit. From *January* to *October*,

CURIOUS FISHES.

THOSE OF USE.

HERRINGS.

---

* See the Plates in *Pontoppidan's Hift. Norway*.

† In the *Britifh Zoology*, iii. N° 40. the trivial *Spinax* is inferted inftead of *Acanthias*.

1751, were exported, from *Bergen* alone, eleven thousand and thirteen lasts; and it was expected that as many more would be shipped off before the expiration of the year. The Herrings which visit this coast are only part of the vast northern army which annually quits the great deeps, and gives wealth and food to numbers of *European* nations.

THE Cod yields another fishery of great profit. They first arrive immediately after the earliest Herrings, and grow so pampered with their fry, that they reject a bait; and are taken in vast nets, which are set down in fifty or seventy fathom water, and taken up every twenty-four hours, with four or five hundred great fish entangled in them. As the Herrings retire, the Cod grows hungry; and after that is taken with hook and line, baited with Herring. In more advanced season, other varieties of Cod arrive, and are taken, in common with Turbot and other fish, with long lines, to which two hundred short lines with hooks are fastened: the whole is sunk to the bottom; its place is marked by a buoy fastened to it by another line of fit length. The extent of the Cod-fishery may be judged of on hearing that 40,000 *tonder*, of four bushels each, of *French* and *Spanish* salt, are annually imported into *Bergen* for that purpose only.

The Ling is taken on the great sand-bank during summer, by hook and line, and, being a fish noted for being capable of long preservation, is much sought after for distant voyages.

The Salmon, a most universal northern fish, arrive in the *Norwegian* rivers, and vast quanties are sent, smoke-dried or pickled, into various countries.

The præfecture of *Nordland*, is the farthest part of the kingdom of *Norway*. In it is the district of *Helgeland*, remarkable for that uncommon genius, *Octher*, or *Ohthere*, who, in a frozen climate, and so early as the ninth century, did shew a passion for discovery, equal perhaps with that of the present. His country was at that time the last in the north which had the left tincture of humanity. In the year 890 he was attracted by the fame of our renowned ALFRED. He visited his court, and related to him his voyages. He told the monarch that he was determined to prove if there was any land beyond the deserts which bounded his country. It appears that he sailed due north, and left, on his starboard side, a waste, the present *Finmark*, occasionally frequented by the *Finnas*, or wandering *Laplanders*, for the sake of fishing and fowling. He went as far as the *Whale-fishers* usually ventured: a proof that the men of *Norway* practised that fishery many centuries before the *English*. He doubled the *North Cape*, and entered the *Cwen Sea*, or *White Sea*, and even anchored in the mouth of the *Dwina*. He was to these parts what *Columbus* was to *America*: but the knowlege of this country was lost for centuries after the days of *Octher*. He mentioned the *Seride Finnas*, who lived to the north-west of

the

CoD

LING.

SALMON.

NORDLAND.

OF OCTHER.

the *Cwen Sea*, and who wore fnow-fhoes. The country about the *Dwina* was well inhabited by a people called *Beormas*, far more civilized than the *Finnas*. The map attending ALFRED's *Orofius* places them in the country of the *Samoieds*, a race at prefent as uncultivated as mankind can be: we therefore muft fuppofe thofe *Beormas* to have been *Ruffians*. *Octher* fays, that in this fea he met with Horfe-Whales (*Walrufes*) and produced to the prince fpecimens of their great teeth, and of thong-ropes made of their fkins; a mark of his attention to every thing curious which occurred to him *.

NORWEGIANS A FINE RACE.

I muft not leave *Norway* without notice of its chief of animals, Man. *Scandinavia*, in the courfe of population, received its inhabitants by colonies of hardy *Scythians*, who, under the name of *Sarmatians*, extended themfelves to the coafts of the *Baltic*. In after-times their virtue was exalted by the arrival of their countryman, *Odin*, and the heroes he fettled in every part of the country. The feverity of the climate has not checked the growth, or diftorted the human form. MAN here is tall, robuft, of juft fymmetry in limbs, and fhews ftrongly the human face divine. Their hair is light: their eyes light grey. The male peafants of the mountains are hairy on their breafts as Bears, and not lefs hardy: active in body:

LONGEVITY.

clear and intelligent in their minds. Theirs certainly is length of days; for out of fix thoufand nine hundred and twenty-nine, who died in 1761, in the diocefe of *Chriftiana*, three hundred and ninety-four lived to the age of nintey; fixty-three to that of a hundred; and feven to that of a hundred and one †. The *Norwegians* juftly hold themfelves of high value; and flightingly call their fellow-fubjects, the the *Danes*, *Jutes* ‡. The *Danes* tacitly acknowlege the fuperiority, by compofing almoft their whole army out of thefe defcendants of the all-conquering *Normans*.

I fhall here fupply an omiffion in my account of the *Scandinavian* antiquities, p. xxxvi. by mentioning the famous tomb, about feven *Swedifh* yards long and two broad, found at *Kivike*, a parifh of *Schonen* in *Sweden*, in the centre of a vaft tumulus of round ftones. It was oblong, and confifted of feveral flat ftones, the infide of which is carved with figures of men and animals, and the weapons of the age, axes and fpears-heads. A figure is placed in a triumphal car; cornets feem founding: captives with their hands bound behind, guarded by armed men; and figures, fuppofed to be female, form part of the conquered people. It is fuppofed that the *Roman* fleet made an accidental defcent here, had a fuccefsful fkirmifh with the natives, might have loft their leader, and left this mark of their victory amidft the

---

* *The Tranflation of Orofius, by the Hon. Daines Barrington, p. 9, &c. and Hackluyt, i. 4.*
† *Phil. Tranf. vol. lix. 119.* ‡ *Lord Molefworth's Account of Denmark, 25.*

barbarous

barbarous conquered. The tomb had been broken open by the country people, and whatsoever it might have contained was stolen away and lost *.

Within the *Arctic* circle, begins *Finmark*, a narrow tract, which winds about the shores eastwards, and bends into the *White Sea*: a country divided between *Norway* and *Russia*. The view from the sea is a flat, bounded, a little inland, by a chain of lofty mountains covered with snow. The depth of water off the shore is from a hundred to a hundred and fifty fathoms †. The inhabitants quit their hovels in winter, and return to them in the summer: and, in the middle of that season, even the *Alpine Laplanders* visit these parts for the sake of fishing; and, like the antient *Scythians*, remove with their tents, their herds, and furniture, and return to their mountains in autumn ‡. Some of them, from living near the sea, have long been called *Siæ Finni*, and *Soe Lappernes*.

In this country begins instantly a new race of men. Their stature is from four to four feet and a half: their hair short, black, and coarse: eyes transversely narrow: irides black: their heads great: cheek-bones high: mouth wide: lips thick: their chests broad: waists slender: skin swarthy: shanks spindle ‖. From use, they run up rocks like goats, and swarm trees like squirrels: are so strong in their arms that they can draw a bow which a stout *Norwegian* can hardly bend; yet lazy even to torpidity, when not incited by necessity; and pusillanimous and nervous to an hysterical degree. With a few variations, and very few exceptions, are the inhabitants of all the *Arctic* coasts of *Europe*, *Asia*, and *America*. They are nearly a distinct species in minds and bodies, and not to be derived from the adjacent nations, or any of their better-proportioned neighbors.

The seas and rivers of *Finmark* abound with fish. The *Alten* of *West Finmark*, after a gentle course through mountains and forests, forms a noble cataract, which tumbles down an immense rock into a fine bason, the receptacle of numbers of vessels which resort here to fish or traffic for Salmon §. The *Tana*, and the *Kola* of the extreme north swarm with them. In the *Alten* they are taken by the natives in weirs built after the *Norwegian* model; and form, with the merchants of *Bergen*, a great article of commerce. These fisheries are far from recent: that on the *Kola* was noted above two centuries ago for the vast concourse of *English* and *Dutch*, for the sake of the fish-oil and Salmon ¶.

The most northern fortress in the world, and of unknown antiquity ††, is *Ward-*

---

* See Mr. *Forssenius*'s curious dissertation on this antiquity, printed at *Lund*, 1780.
† *Anth. Jenkinson's Voy.* in *Hackluyt*, i. 311.   ‡ *Leems*, 169.   ‖ *Scheffer*, 12, and *Lin. Faun. Suec.* 1.   § *Leems*, 342.   ¶ *Hackluyt*, i. 416.   †† *Torfæi Hist. Norwegiæ*, i. 96.

huys,

buys, situated in a good harbour, in the isle of *Wardoe*, at the extremity of *Finmark*; probably built for the protection of the fishing trade, the only object it could have in this remote place.

Sir Hugh Willoughby.

A little farther eastward, in *Muscovitish Finmark*, is *Arzina*, noted for the sad fate of that gallant gentleman, Sir *Hugh Willoughby*, who, in 1553, commanded the first voyage on the *discovery by sea* of *Muscovia*, by the north-east; a country at that time scarcely known to the rest of *Europe*. He unfortunately lost his passage, was driven by tempests into this port, where he and all his crew were found the following year frozen to death. His more fortunate consort, *Richard Chancellor* captain and pilot major, pursued his voyage, and renewed the discovery of the *White Sea*, or *Bay of St. Nicholas*; a place totally forgotten since the days of *Octher*. The circumstances attending his arrival, exactly resemble those of the first discoverers of *America*. He admired the barbarity of the *Russian* inhabitants: they in return were in amaze at the size of his ship: they fell down and would have kissed his feet; and when they left him spread abroad the arrival of ' a strange nation, of singular gentlenesse and courtesie *.' He visited in sledges the court of *Basilowitz* II. then at *Moscow*, and layed the foundation of immense commerce to this country for a series of years, even to the remote and unthought-of *Persia*.

I shall take my departure from the extreme north of the continent of *Europe*, or rather from its shattered fragments, the isle of *Maggeroe*, and other islands, which lie off the coast, in lat. 71. 33.

North Cape.

At the remote end of *Maggeroe* is the *North Cape*, high and flat at top, or what the sailors call *Table-land* †. These are but the continuation of the great chain of mountains which divides *Scandinavia*, and sinks and rises through the ocean, in different places, to the *Seven Sisters*, in about lat. 80. 30, the nearest land to the pole which we are acquainted with.

Cherie Island.

Its first appearance above water, from this group, is at *Cherie Island*, in lat. 74. 30. a most solitary spot, rather more than midway between the *North Cape* and *Spitzbergen*, or about a hundred and fifty miles from the latter. Its figure is nearly round: its surface rises into lofty mountanous summits, craggy, and covered with perpetual snow: one of them is truly called *Mount Misery*. The horror of this isle to the first discoverers must have been unspeakable. The prospect dreary, black, where not hid with snow, and broken into a thousand precipices. No sounds but of the dashing of the waves, the crashing collision of floating ice, the discordant notes of myriads of sea-fowl, the yelping of *Arctic* Foxes, the snorting of the Walruses, or the roaring of the Polar Bears.

---

* *Hacklujt*, i. 246.   † See a view of these islands in *Phil. Transf.* vol. lix. tab. xiv.

This

This island was probably difcovered by *Stephen Bennet* in 1603 *, employed by Alderman *Cherie*, in honor of whom the place was named. The anchorage near it is twenty and thirty fathoms. He found there the tooth of a *Walrus*, but faw none of the animals, their feafon here being paft : this was the 17th of *Auguft*. Encouraged by the hopes of profit, *Bennet* made a fecond voyage the next year, and arrived at the ifland the 9th of *July* ; when he found the Walrufes lying huddled on one another, a thoufand in a heap. For want of experience, he killed only a few ; but in fucceeding voyages the adventurers killed, in 1606, in fix hours time, feven or eight hundred ; in 1608, nine hundred or a thoufand in feven hours ; and in 1610, above feven hundred. The profit, in the teeth, oil, and fkins, was very confiderable † ; but the flaughter made among the animals frightened the furvivors away, fo that the benefit of the bufinefs was loft, and the ifland no more frequented. But from this deficiency originated the commencement of the Whale-fifhery by the *Englifh*.

WALRUSES.

It is remarkable that this ifland produces excellent coals ‡ ; yet none are known nearer than the diocefe of *Aggerhuys*, in the fouth of *Norway*, and there in very fmall quantities. Lead ore is alfo found, both in *Cherie Ifland* and a little one adjacent, called *Gull Ifland* ‖.

COALS.

LEAD.

About a hundred and fifty miles almoft due north, is *South Cape*, north lat. 76. 30, the extreme fouthern point of *Spitzbergen*, the largeft of the group of frozen iflands which go under that name, or *New Groenland*. From this to *Ver-legan-hook*, north lat. 80. 7, the northern extremity, is above three hundred miles ; and the greateft breadth of the group is from *Hackluyt's Headland* to the extreme eaft point of *North Eaftland*, comprizing from 9. to near 24. eaft longitude. The fhores are ragged and indented. A very deep bay runs into the eaft fide from fouth to north ; and a large trifurcated one from north to fouth. *Stat's Forland* is a large ifland rent from the fouthern part of the eaft fide. *North Eaftland* is divided from the north-eaft fide by the *Waygat* and *Hinlopen* ftraits, ufually blocked up with ice, and fo fhallow as to be, in one part, only three fathoms deep §. The long ifle of King *Charles* lies parallel to the weft fide. At the fouthern end is *Black Point* ; the coaft high, black, and inacceffible ; in parts feeming foaring above the clouds ; and the interjacent vallies filled with ice and fnow. *Fair Foreland*, or *Vogel-hook*, is the northern headland, made by failors. And due north of it, at the weftern point of *Spitzbergen*, is the fmall lofty ifle of *Hackluyt's Headland*, another object of the mariners fearch.

SPITZBERGEN.

To the north of the great group is *Moffen's Ifle*, in lat. 80, oppofite to the mouth of *Leifde* bay. This ifland is very low, and fufpected to be a new creation,

MOFFEN'S ISLE.

---

* *Purchas*, iii. 566.   † The fame, pp. 560. 565.   ‡ The fame, 564.
‖ The fame, 558. 564.   § *Barrington's Mifcel*. 35.

1   by

by the meeting of the ftreams from the great ocean, rufhing along the weft fide of *Spitzbergen*, and through the *Waygat*, and forcing up the gravelly bottom of this fhallow part, where the lead touches the bottom at from two to five fathoms water, at half a mile from its weftern fide *.

Low Island. Basaltic.

To the eaftward of this is another low ifland, almoft oppqfite to the mouth of the *Waygat* : it is remarkable for being part of the *Bafaltic* chain, which appears in fo many places in the northern hemifphere. The columns were from eighteen to thirty inches in diameter, moftly hexagonal, and formed a moft convenient pavement.

Plants. Animals.

The middle of the ifle was covered with vegetables, Moffes, Sorel, Scurvy Grafs, and Ranunculufes in bloom on *July* 30th. Of. quadrupeds, the Reindeer fattened here into excellent venifon; the *Arctic* Fox; and a fmall animal larger than a Weefel, with fhort ears, long tail, and fpotted with black and white, were feen.

Birds.

Small Snipes, like Jack Snipes; Ducks, then hatching; and Wild Geefe feeding, helped to animate this dreary fcene †.

Drift-wood.

The beach was formed of an antient aggregate of fand, whale-bones, and old timber, or drift-wood. Fir-trees feventy feet long, fome torn up by the roots, others frefh from the axe, and marked with it into twelve feet lengths, lay confufedly fixteen or eighteen feet above the level of the fea, intermixed with pipeftaves, and wood fafhioned for ufe; all brought into this elevated fituation by the fwell of the furious furges.

The appearance of drift-wood is very frequent in many parts of thefe high latitudes : in the feas of *Greenland*, in *Davis*'s ftreights, and in thofe of *Hudfon*; and again on the coafts of *Nova Zemlja*. I have only two places from whence I can derive the quantity of floating timber which appears on the coaft of *Nova Zemlja* and thefe iflands : the firft is from the banks of the *Oby*, and perhaps other great rivers, which pour out their waters into the Frozen ocean. In the fpring, at the breaking up of the ice, vaft inundations fpread over the land, and fweep away whole forefts, with the aid of the vaft fragments of ice; thefe are carried off, rooted up, and appear entire in various places. Such as are found marked into lengths, together with pipe-ftaves, and other fafhioned woods, are fwept by the *Norwegian* floods out of the rivers, on the breaking of a *lentze* ‡, a misfortune which fometimes happens, to the bankruptcy of multitudes of timber-merchants. At fuch times not only the trees which are floating down the torrents, but the faw-mills, and all other places in which bufinefs is carried on, undergo the fame calamity; and the timber, in whatfoever form it happens to be, is forced into the ocean, and conveyed by tides or tempefts to the moft diftant parts of the north.

* *Phips*, 54.   † The fame, 58.   ‡ *Purchas*, iii. 527.

Let

Let no one be staggered at the remoteness of the voyage: I have before shewn instances, but from a contrary course, from west to east. Part of the masts of the *Tilbury*, burnt at *Jamaica*, was taken up on the western coast of *Scotland*; and multitudes of seed or fruits of the same island, and other hot parts of *America*, are annually driven on shore, not only on the western side of *Scotland* †, but even on those of more distant *Norway* ‡, and *Iceland*.

The islands of the *Seven Sisters*, last of known land, lie due north from *North-East-land:* the extreme point of the most remote is in lat. 80. 42. They are all high primæval isles: from a high mountain on the farthest, the hardy navigators of 1773 had a sight of ten or twelve leagues of smooth unbroken ice to the east and north-east, bounded only by the horizon; and to the south-east certain land laid down in the *Dutch* maps. Midway between these islands and *North-Eastland*, Lord *Mulgrave*, after every effort which the most finished seaman could make to accomplish the end of his voyage, was caught in the ice, and was near experiencing the unhappy fate of the gallant *Englishman*, Sir *Hugh Willoughby*, who was frozen in 1553, with all his crew, in his unhappy expedition.

The scene, divested of the horror from the eventful expectation of change, was the most beautiful and picturesque :—Two large ships becalmed in a vast bason, surrounded on all sides by islands of various forms: the weather clear: the sun gilding the circumambient ice, which was low, smooth, and even; covered with snow, excepting where the pools of water on part of the surface appeared crystalline with the young ice ‖: the small space of sea they were confined in perfectly smooth. After fruitless attempts to force a way through the fields of ice, their limits were perpetually contracted by its closing; till at length it beset each vessel till they became immoveably fixed §. The smooth extent of surface was soon lost: the pressure of the pieces of ice, by the violence of the swell, caused them to *pack*; fragment rose upon fragment, till they were in many places higher than the main-yard. The movements of the ships were tremendous and involuntary, in conjunction with the surrounding ice, actuated by the currents. The water shoaled to fourteen fathoms. The grounding of the ice or of the ships would have been equally fatal: the force of the ice might have crushed them to atoms, or have lifted them out of the water and overset them, or have left them suspended on the summits of the pieces of ice at a tremendous height, exposed to the fury of the winds, or to the risque of being dashed to pieces by the failure of their frozen dock ¶. An

VOYAGE BY
LORD MULGRAVE,
IN 1773.

* P. 21. of this Work.   † *Voy. to the Hebrides.*   ‡ *Amæn. Acad.* vii. *Rariore Norvegiæ,* 477.   ‖ *Phips Voy.* tab. iv.   § Same, tab. iii.

¶ See these distressful situations in tab. B. of *Fr. Marten's Voyage,* and *Gerard le Ver, Voy. au Nord,* p. 19, edition 1606.

l 2   attempt

attempt was made to cut a paſſage through the ice; after a perſeverance worthy of *Britons*, it proved fruitleſs. The commander, at all times maſter of himſelf, directed the boats to be made ready to be hauled over the ice, till they arrived at navigable water (a taſk alone of ſeven days) and in them to make their voyage to *England*. The boats were drawn progreſſively three whole days *. At length a wind ſprung up, the ice ſeparated ſufficiently to yield to the preſſure of the full-ſailed ſhips, which, after laboring againſt the reſiſting fields of ice †, arrived on the 10th of *Auguſt* in the harbor of *Smeeringberg*, at the weſt end of *Spitzbergen*, between it and *Hackluyt's Headland*.

It was the hard fortune of Lord *Mulgrave*, at this ſeaſon, to meet with one of thoſe amazing ſhoals of ice which cover, at times, theſe ſeas, for multitudes of leagues. He made the fulleſt trial, from long. 2 to 21 eaſt, and from about lat. 80. 40, as low as about 78. 30, oppoſed by a face of ice without the leaſt opening, and with all the appearance of a ſolid wall. It is well known, that the coaſts of *Sibiria* are, after a northern tempeſt, rendered inacceſſible for a vaſt extent, by the polar ice being ſet in motion. It is as well known, that a ſtrong ſouthern wind will again drive them to their former ſeats, and make the ſhores of the Frozen ocean as clear as the equatorial ſeas. A farther diſcovery on this ſide was denied to the noble navigator. His misfortune will for ever redound to his honor, as it proved his ſpirit, his perſeverance, and a ſoul fertile in expedients among the greateſt difficulties!

That navigators have gone into higher latitudes I cannot deny : the authenticated inſtances only ſhew their accidental good fortune, in having the ice driven towards the pole, and in making a retreat before they were enveloped in the returning ice. The *Ruſſians*, under vice-admiral *Tſhitſhaghef*, within theſe very few years, made an attempt to ſail to the pole by the eaſtern ſide of *Spitzbergen*; but after ſuffering great hardſhips, returned without effecting any diſcovery. Curioſity has been amply ſatisfied : and I believe we may reſt fully content with the common paſſage to *India*, on the conviction of this tract being totally impracticable.

ICE.

The forms aſſumed by the ice in this chilling climate, are extremely pleaſing to even the moſt incurious eye. The ſurface of that which is congealed from the ſea-water (for I muſt allow it two origins) is flat and even, hard, opake, reſembling white ſugar, and incapable of being ſlid on, like the *Britiſh* ice ‡. The greater pieces, or fields, are many leagues in length: the leſſer, are the meadows

* *Phips Voy.* tab. v.
† Same, tab. vi.
‡ *Crantz.* i. 31.

of

of the Seals, on which thofe animals at times frolic by hundreds. The motion of the leffer pieces is as rapid as the currents : the greater, which are fometimes two hundred leagues long, and fixty or eighty broad *, move flow and majeftically; often fix for a time, immoveable by the power of the ocean, and then produce near the horizon that bright white appearance, called by mariners the *blink of the ice* †. The approximation of two great fields produces a moft fingular phænomenon ; it forces the leffer (if the term can be applied to pieces of feveral acres fquare) out of the water, and adds them to their furface : a fecond, and often a third fucceeds ; fo that the whole forms an aggregate of a tremendous height. Thefe float in the fea like fo many rugged mountains, and are fometimes five or fix hundred yards thick ‡; but the far greater part is concealed beneath the water. Thefe are continually encreafed in height by the freezing of the fpray of the fea, or of the melting of the fnow, which falls on them. Thofe which remain in this frozen climate, receive continual growth ; others are gradually wafted by the northern winds into fouthern latitudes, and melt by degrees, by the heat of the fun, till they wafte away, or difappear in the boundlefs element.

The collifion of the great fields of ice, in high latitudes, is often attended with a noife that for a time takes away the fenfe of hearing any thing elfe ; and the leffer with a grinding of unfpeakable horror.

The water which dafhes againft the mountanous ice freezes into an infinite variety of forms ; and gives the voyager ideal towns, ftreets, churches, fteeples, and every fhape which imagination can frame ‖.

The *Icebergs*, or *Glacieres* of the north-eaft of *Spitzbergen*, are among the capital wonders of the country ; they are feven in number, but at confiderable diftances from each other : each fills the vallies for tracts unknown, in a region totally inacceffible in the internal parts. The *glacieres* of *Switzerland* feem contemptible to thefe ; but prefent often a fimilar front into fome lower valley. The laft exhibits over the fea a front three hundred feet high, emulating the emerald in color: cataracts of melted fnow precipitate down various parts, and black fpiring mountains, ftreaked with white, bound the fides, and rife crag above crag, as far as eye can reach in the back ground §.

At times immenfe fragments break off, and tumble into the water, with a moft alarming dafhing. A piece of this vivid green fubftance has fallen, and grounded in twenty-four fathoms water, and fpired above the furface fifty feet **. Simi-

[ICEBERGS.

---

* *Crantz*, i. 31.   † *Phips*, 72.   ‡ *Ellis's Voy.* 127.   ‖ *Marten*, 37.   *Crantz*, i. 31.   § See the beautiful plate in *Phips's Voy.* tab. vii.   ** *Phips*, p. 70.

lar

lar *icebergs* are frequent in all the *Arctic* regions ; and to their lapses is owing the solid mountanous ice which infests those seas.

Frost sports also with these *icebergs*, and gives them majestic as well as other most singular forms. Masses have been seen, assuming the shape of a Gothic church, with arched windows and doors, and all the rich tracery of that style, composed of what an *Arabian* tale would scarcely dare to relate, of crystal of the richest sapphirine blue : tables with one or more feet ; and often immense flat-roofed temples, like those of *Luxxor* on the *Nile*, supported by round transparent columns of cærulean hue, float by the astonished spectator *.

These *icebergs* are the creation of ages, and receive annually additional height by the falling of snows and of rain, which often instantly freezes, and more than repairs the loss by the influence of the melting sun †.

SNOW.   The snow of these high latitudes is as singular as the ice. It is first small and hard as the finest sand ‡ ; changes its form to that of an hexagonal shield, into the shape of needles, crosses, cinquefoils, and stars, plain and with serrated rays. Their forms depend on the disposition of the atmosphere ; and in calm weather it coalesces, and falls in clusters §.

SEASONS.   Thunder and lightning are unknown here. The air in summer is generally clear ; but the sky loaden with hard white clouds. The one night of this dreadful country begins about *October* 20th, O. S. ; the sun then sets, and never appears till about the 3d of *February* ‖ : a glimmering indeed continues some weeks after its setting : then succeed clouds and thick darkness, broken by the light of the moon, which is luminous as that in *England*, and shines without intermission during the long night ¶. Such also is the case in *Nova Zemlja* ** The cold, according to the *English* proverb, strengthens with the new year ; and the sun is ushered in with unusual severity of frost. The splendor of that luminary on the snowy summits of the mountains was the most glorious of sights to the single party who survived to relate the account. The Bears stalk forth at the same time from their dens, attended by their young cubs. By the beginning of *March*, the chearful light grows strong : the *Arctic* Foxes leave their holes, and the sea-fowls resort in great multitudes to their breeding-places ††.

BEARS.

FOXES.
FOWLS.

---

* *Marten*, 43.   † The same,   ‡ The same.   § The same, 51.   ‖ Relation of Eight *Englishmen*, &c. *Churchill's Coll.* iv. 818.—Relation of Seven *Dutchmen*, &c. *Churchill*, ii. 430.   ¶ Narrative of Four *Russian* sailors, 94.   ** *De Ver, trois Voy. au Nord.* 22, b.   †† Relation of Eight *Englishmen*, &c. 817, 818, 819.

The

The fun, in the height of fummer, has at times heat enough to melt the tar on the decks of fhips. It fets no more after the third of *May*, O. S. Diftinction of day and night is loft; unlefs it be fact what *Fr. Marten* alleges, that during the fummer night of thefe countries, the fun appears with all the faintnefs of the moon *. This is denied by Lord *Mulgrave* †. From *Auguft* the power of the fun declines, it fets faft; in *September* day is hardly diftinguifhable; and by the middle of *October* takes a long leave of this country; the bays become frozen; and winter reigns triumphant.

Nature, in the formation of thefe iflands, preferves the fame rule which fhe does in other places: the higheft mountains are on the weftern fide; and they gradually lower to the eaft. The altitude of the moft lofty which has been taken by Lord *Mulgrave*, feems to have been one a little to the north of *Black Point*, which was found by the megameter to be fifteen hundred and three yards ‡: that of a hill on the little ifle, the *Norways*, a fmall diftance to the north-eaft of *Spitzbergen*, was two thoufand four hundred feet: one on *Vogel Sang*, fixteen hundred and fifty; another, on the ifle near *Cloven Cliff*, in about lat. 80, eight hundred and fixty-five; a third on that near *Cook's Hole*, feven hundred and eleven; and one on *Hackluyt's Ifland*, only three hundred and twenty-one §. Thefe are the moft northern lands which ever were meafured; and the experiments favor the fyftem of the decreafe of the heights of the mountains toward the poles.

Earth and foil are denied to thofe dreadful regions: their compofition is ftone, formed by the fublime hand of Almighty Power; not frittered into fegments by fiffures, tranfverfe or perpendicular, but at once caft into one immenfe and folid mafs; a mountain is but a fingle ftone throughout, deftitute of fiffures, except in places cracked by the refiftlefs power of froft, which often caufes lapfes, attended with a noife like thunder, fcattering over their bafes rude and extenfive ruins. The ftone is granite, moftly grey and black; fome red, white, and yellow. I ftrongly fufpect, that veins of iron are intermixed; for the meltings of the fnow tinge the rocks frequently with a ferruginous ochre. A potter's clay and a gypfum are to be met with on the eaftern part of the iflands ‖.

The vallies, or rather glens, of this country, are filled with eternal ice or fnow; are totally inacceffible, and known only by the divided courfe of the mountains, or where they terminate in the fea in form of a *glaciere*. No ftreams water thefe dreary bottoms; even fprings are denied; and it is to the periodical

---

* *Marten*, 48.   † *Voy*. 71.   ‡ *Phips Voy*. 33.   § The fame, on tab. viii.
‖ Narrative of Four *Ruffian* failors, 78, 89.

cataracts of melted snow of the short summer, or to the pools in the middle of the fields of ice, to which the mariners are indebted for fresh water.

HARBOURS.

The harbours on the west side are frequent; penetrate deep into the island of *Spitzbergen*; and are the only channels by which the slight knowlege of the interior parts is attained. *North Harbour* is a scene of picturesque horror, bounded by black craggy *Alps*, streaked with snow; the narrow entrance divided by an island; and at seasons affording a land-locked shelter to multitudes of ships.

TIDE AND SEA.

The tide at the *Vogel Sang* flows only four feet, and the flood appears to come from the south. The depth of the sea is very irregular: near the shore it is generally shallow: off *Low Island* only from ten to twenty fathoms; yet suddenly deepens to a hundred and seventeen: off *Cloven Cliff* from fourteen to twenty-eight, and deepens to two hundred. The shallows are usually on rock; the great depths on soft mud: the former I look on as submarine islands; but, from the small number of fish, the bottoms must be universally barren.

SOIL!

The grit worn from the mountains by the power of the winds, or attrition of cataracts of melted snow, is the only thing which resembles soil, and is the bed for the few vegetables found here. This indeed is assisted by the putrefied *lichens* of the rocks, and the dung of birds, brought down by the same means.

PLANTS.

Even here *Flora* deigns to make a short visit, and scatter over the bases of the hills a scanty stock. Her efforts never rise beyond a few humble herbs, which shoot, flower, and seed, in the short warmth of *June* and *July*; then wither into rest till the succeeding year.—Let me here weave a slender garland from the lap of the goddess, of such, and perhaps all, which she hath bestowed on a country so repugnant to her bounty. Let the salubrious Scurvy Grass, the resource of distempered seamen, be remarked as providentially most abundant in the composition.

Let me first mention its only tree, the *Salix Herbacea*, or Dwarf Willow, described by *Marten*, p. 65, *Phips*, 202, which seldom exceeds two inches in height, yet has a just title to the name. The plants are, a new species of Grass, now named *Agrostis Algida : Tillæa Aquatica*, Sp. Pl. 186. Fl. Suec. 156: *Juncus Campestris*, Fl. Sc. i. 186 : *Sibbaldia Procumbens?* Fl. Lap. 111.; Marten's Spitz. tab. H. fig. b : *Polygonum Viviparum*, Fl. Lap. 152; Marten's Spitz. tab. I. fig. a: *Saxifraga Oppositafolia*, Fl. Lap. 179, 222 : *Sax. Cernua*, Sp. Pl. i. 577 ; Fl. Lap. 172 : *Sax. Rivularis*, Sp. Pl. 577 ; Fl. Lap. 174 : *Sax. Cæspitosa*, Sp. Pl. 578; Fl. Suec. 376 : *Sedum Annuum?* Sp. Pl. 620; Marten's Spitz. tab. F. fig. c: *Cerastium Alpinum*, Sp. Pl. 628; Fl. Lap. 192 : *Ranunculus Sulphureus*, Phips Voy. 202; Mart. Spitz. 58 : *R. Lapponicus*, Fl. Lap. 461, 503 : *R. Nivalis?* 232; Mart. Spitz. tab. F. fig. a: *Cochlearia Danica*, Sp. Pl. 903; Fl. Suec. 578, 579:

Cochl.

Cochl. Grœnlandica, Sp. Pl. 904 : *Polytrichum commune,* Fl. Lap. 395 : *Bryum Hypnoides,* Fl. Lap. 396 : *Bryum Trichoides?* Dill. 391 ; Muſc. tab. 50, fig. 61 : *Bryum Hypnoides?* Dill. Muſc. 394, tab. 50, fig. 64, C : *Hypnum Aduncum,* Sp. Pl. 1592 ; Fl. Suec. 879, 1025 : *Jungermannia Julacea,* Sp. Pl. 1601 : *Jung.* like the *Lichenaſtrum Ramoſius, fol. trif.* Dill. Muſc. 489, tab. 70, fig. 15 : *Lichen Ericetorum,* Fl. Lap. 936, 1068 : *L. Iſlandicus,* 959, 1085 : *L. Nivalis,* 446 : *L. Caninus,* 441 : *L. Polyrhizos,* Sp. Pl. 1618 ; Fl. Suec. 1108 : *L. Pyxidatus,* Fl. Lap. 428 : *L. Cornutus,* 434 : *L. Rangiferinus,* 437 : *L. Globiferus,* Lin. Mantiſſ. 133 : *L. Paſchalis,* Fl. Lap. 439 : *L. Chalybeiformis,* Sp. Pl. 1623 ; Fl. Suec. 988, 1127 : and the *Fucus Saccharinus?* Fl. Lap. 460 ; Mart. Spitz. tab. F. fig. 6.

It is matter of curioſity to trace the decreaſe of vegetables from our own iſland to this ſpot, where ſo few are to be found. They decreaſe with the numbers of herbivorous animals, and the wants of mankind. The following catalogue may not be quite juſt, but is probably pretty near the truth :

		Perfect.		Imperfect.		Total.
England has	—	1,124	—	590	—	1,714
Scotland	—	804	—	428	—	1,232
The *Orknies*		354	—	144	—	498
Sweden	—	933	—	366	—	1,299
Lapland	—	379	—	155	—	534
Iceland	—	309	—	233	—	542

Thoſe of *Spitzbergen* are given above.

The three terreſtrial quadrupeds of theſe iſlands are confined here without poſſibility of migration. The Polar Bears paſs the greateſt part of the winter in a torpid ſtate : appear in numbers at the firſt return of the ſun, when, probably, they take to the ice, in queſt of their prey, Seals, or dead Whales.

It is difficult to account for the means which the Foxes find for ſupport, as the iſland is deſtitute of birds during the whole winter ; and, the bays being totally frozen up, they can find no ſubſiſtence from the ſea. Perhaps they lay up proviſion for winter, on which they ſubſiſt till the arrival of the birds in *March* ; at which ſeaſon they have been obſerved firſt to quit their holes, and appear in multitudes *. The Rein Deer have at all times their favorite *lichen,* which they can readily get at, by help of their palmated horns.

WALRUSES and Seals are found in great abundance ; the latter are often the object of chace, for the ſake both of oil and ſkins : the *Ruſſians* make voyages on

* *Churchill,* iv. 819.

m   purpoſe.

purpose. In 1743, four unhappy mariners of that nation were accidentally left on shore on *North Eastland*, called by the *Russians Maloy Broun*. Here three (the fourth died in the last year) lived till *August* 15th 1749 ; when they were providentially relieved by the arrival of a ship, after passing six years, realizing in ingenious contrivances the celebrated *English* fable of *Robinson Crusoe* *.

In the year 1633 seven *Dutch* sailors were left voluntarily on the western part of *Spitzbergen*, to pass the winter, and form their remarks. They were furnished with medicines, and every requisite to preserve life ; but every one perished by the effects of the scurvy. In the next year, seven other unhappy men devoted themselves, and died in the same manner. Of the first set, it appeared by his journal, that the last was alive the 30th of *April* 1634 ; of the second, the life of the last survivor did not continue far beyond the 28th of *February* 1635 †. Yet eight *Englishmen*, left in 1630 in the same country, by accident, and unprovided with every thing, framed themselves a hut from some old materials, and were found by the returning ships, on *May* 28th 1631, in good health ‡. Thus *Russian* hardiness and *British* spirit braved a climate, which the phlegmatic constitution of a *Dutchman* could not resist.

BIRDS.

To meet with the Snow Bunting, N° 222, a bird whose bill, in common with the rest of that genus, is calculated for granivorous life, is a kind of miracle. The country has a very scanty provision of seeds ; the earth yields no worms, the air no insects ; yet these birds are seen in flocks innumerable, and that chiefly on the ice around *Spitzbergen* : as it breeds early, possibly the old and young may have quitted the land, and collected on the ice at the time of the arrival of the ships.

Of cloven-footed water-fowl, the Purre, N° 390, alone is seen here.

Of web-footed, the Puffin Auk, N° 427 ; the Razor Bill, N° 425 ; the Little Auk, N° 429 ; the Foolish Guillemot, N° 436 ; the Black Guillemot, N° 437 ; the Northern Diver, N° 439 ; the Ivory Gull, N° 457 ; the Herring Gull, N° 452 ; the Arctic Gull, N° 459 ; the Kittiwake, N° 456 ; and the Greater Tern, N° 448 : these, with the Eider Duck, N° 480, complete the short list of the feathered tribe of *Spitzbergen*. All these breed in the frost-rent cracks of the mountains, and appear even in these regions before the 16th of *March* §.

FISH.

The Whale is lord paramount of these seas ; and, like a monstrous tyrant, seems to have terrified almost every other species of fish away. A few Coal Fish, *Br. Zool.* iii. N° 78, and two of the unctuous Suckers, N° 58, were the whole which were taken by Lord *Mulgrave*, after several trials by hook and by net.

* See the curious Narrative.   † *Churchill's Coll.* ii. 415, 427.   ‡ The same, iv. 808.
§ The same, p. 818.

I can

I can never imagine that the ſhallow, barren, and turbulent ſhores of the polar regions receive, as is popularly thought, the immenſe ſhoals of Herrings and Cod which annually repair to other more ſouthern ſeas. Their retreat muſt be in the great depths before deſcribed *, where they are ſecure from the greateſt ſtorms, and probably enjoy a bottom luxuriant in plants and *vermes*.

The Whale which inhabits theſe ſeas, and occaſions the great reſort of ſhipping, is the common ſpecies, *Br. Zool.* iii. N° 16. I have in that Work given its hiſtory; therefore ſhall add no more, than that during ſpring theſe animals keep near *Greenland* and the iſland of *John Mayen*; and towards ſummer they appear in the ſeas of *Spitzbergen*. The Fin Fiſh, *Br. Zool.* iii. N° 18, is another ſpecies: on their appearance, the Common Whale makes its retreat. The *Beluga* or White Whale, p. 183 of this Work, is ſeen here in ſummer, and prognoſticates a good fiſhery.

The inſects, *vermes*, and ſhells, of *Spitzbergen*, are very few. The Prawn, *Br. Zool.* iv. N° 28, and Sea Flea, N° 33, are found there. The *Cancer Boreas*, *Ampulla*, and *Nugax*, are three new ſpecies †, added to the genus by the noble navigator.

Of the known ſpecies of *vermes*, the *Aſcidia Gelatinoſa*, Lin. Syſt. 1087: the *Aſcidia Ruſtica*, 1087, 5: the *Lernea Branchialis*, 1092: and the *Clio Helicina*, the ſmall Slime Fiſh of *Marten*, p. 141, tab. Q. fig. e: and the *Clio Limacina*, the Sea May Fly of the ſame, p. 169, tab. P. fig. 5: the *Sipunculus Lendix*, a new ſpecies, *Phips*, 194, tab. xiii. are found here: the two laſt, the ſuppoſed food of the Common Whale, are met with in vaſt abundance ‡: the *Meduſa Capillata*, the *Aſterias Pappoſa*, Lin. Syſt. 1098: *Aſt. Rubens*, 1099; *Aſt. Pectinata*, 1101; *Br. Zool.* iv. N° 70: *Aſt. Ophiura*, 1100; *Br. Zool.* iv. N° 62: and *Aſt. Caput Meduſæ*, Lin. Syſt. 1101; *Br. Zool.* iv. N° 73. And of Shells, the *Chiton Ruber*, 1107; *Lapes Tintinnabulum*, 1168: the *Mya Truncata*, 1112; *Br. Zool.* N° 14: and *Mytilus Rugoſus*, 1156; *Br. Zool.* iv. N° 72: the *Buccinum Carinotum*, a new ſpecies, *Phips*, 197, tab. xiii: *Turbo Helicinus* of the ſame, 198: the *Serpula Spirorbis*, Lin. Syſt. 1265; *Br. Zool.* iv. N° 155: *Serpula Triquetra*, 1265; *Br. Zool.* iv. N° 156: and the *Sabella Fruſtuloſa*, Phips, 198, complete the liſt of this claſs. Among the Zoophytes is the *Millepora Polymorpha*, Lin. Syſt. 1285; and *Millep.* 1286; and a moſt curious new genus, diſcovered in the voyage, named the *Synoicum Turgens*, 199, tab. xiii: the *Fluſtra Piloſa*, Lin. Syſt. and *Fl. Membranacea*, 1301, 3, 5: and, to conclude, that very curious *Zoophyte*, the foundation of the foſſil *Encrini*,

---

* See p.   † *Phips Voy.* 190, &c. tab. xii.   ‡ The ſame, p. 194, 195.

the

the *Vorticella Encrinus*, Lin. Syſt. N° 1317, engraven in our Tranſactions, vol. xlviii. p. 305, and taken in lat. 79, off this coaſt : two of them being drawn up with the founding-line, in 236 fathom water.

DISCOVERY OF SPITZBERGEN.

The priority of diſcovery of theſe iſlands has been a great matter of controverſy between the *Engliſh* and the *Dutch*. We clame it from the fight which Sir *Hugh Willoughby* is pretended to have had of it in his unfortunate voyage; but if what he ſaw, in lat. 72, was not a fog-bank, we muſt ſuppoſe it to have been either *John Mayen*'s iſle, or part of *Eaſt Greenland*. The abſurd zeal of the *Engliſh* compilers makes *Stephen Boroughs* the ſecond diſcoverer of this country, in 1556; but it is very certain, that he never got higher than lat. 70. 42, nor ever meant any diſcovery but a paſſage to the river *Ob* *. It doubt-leſsly was firſt diſcovered by the *Dutch Barentz*; who, in his third voyage, in 1596, for the finding out the north-eaſt paſſage, met with a land in lat. 79 ½, and anchored in a good road, in eighteen fathom water. He afterwards ſailed as high as 80, and found two of the iſlands of which *Spitzbergen* is compoſed †. Embar-raſſed with ice, he took a ſouthern courſe, and was ſoon after wrecked on the coaſt of *Nova Zemlja* : but the *Engliſh* and *Dutch* purſued the hint; and the Whale-fiſhery, which before was chiefly carried on by the *Biſcayeners* in the bay of *St. Laurence*, was commenced here with great ſucceſs. So active were we, that our ſhips frequented the place within two years after its diſcovery.

I now return to the *North Cape* on the coaſt of *Finmark*; and after paſſing by the ſeveral places mentioned in pages lxxix. and lxxx. enter a ſtreight, bounded by *Muſcovitiſh Finmark*, conſiſting of low hills, and the flat province of *Meſen*, on the eaſt. This leads into the *Bioele Mari*, or *White Sea*, or, more pro-perly, gulph; for its waters are ſhallow, its bottom full of mud, brought by the great rivers which diſcharge themſelves into it, which almoſt deprive it of ſaltneſs. This was the *Cwen* ſea of *Octher*; but had been forgotten fince his time. The *Dwina*, or *Double River*, is the greateſt, which takes its name from being formed by the *Suchona* and the *Yug*, very remote from its mouth. It is navigable to a great diſtance, and brings the commodities of the interior parts of the empire to *Archangel*, a city ſeated on its banks, about ſix miles from the ſea. It roſe from a caſtle built there by *Baſilowitz* II. to protect the in-

WHITE SEA.

ARCHANGEL.

* *Hackluyt*, i. 274, 280.
† *Trois Voyages au Nord*, &c. par *Girard de Ver*, p. 14, 15.

creaſing

creafing trade brought here on the difcovery of the *White Sea* by the *Englifh*; for fhips of all nations reforted to this port, even as far as from *Venice*. Its exports, in 1655, amounted to thrée hundred and thirty thoufand pounds *. *Peter* the Great, intent on aggrandizing his creation, *Peterfburg*, prohibited all trade to *Archangel*, except from the neighboring provinces. Still its exports of tar were confiderable : in 1730, to the amount of forty thoufand lafts, of eleven barrels each †. It fends, during winter, great quantities of the *Nawaga*, a fmall fpecies of three-finned Cod ‡, to *Peterfburg*, frozen, as *Kola* does Herrings in the fame ftate.

The *White Sea* is every winter filled with ice from the Frozen ocean, which brings with it the Harp Seal, N° 77; and the Leporine, N° 75, frequent it during fummer. Whoever furveys the maps of the provinces between this fea and the gulphs of *Bothnia* and *Finland*, will obferve them to be more occupied by lakes than land, and be at once fatisfied of the probability of the once-infulated ftate of *Scandinavia*. As foon as thefe ftreights were clofed, the *White Sea* loft its depth, and is at prefent kept open only by the force of its great rivers.

On the eaftern fide of the entrance into the ftreight is the ifle of *Kandinos*, often fpoken of by our early navigators in their way to the *Waygatz*, in their fearch for a north-eaft paffage. Between it and the main land is a very narrow channel. After doubling the cape of *Kandinos*, the fea forms two great bays. A confiderable part of the fhore to the eaft confifts of low fandy hills ‖. Into the moft remote bay flows, in lat. 68. 30, by many mouths, the vaft river *Peczora*, a place of great trade before the time of *Peter* I. Thoufands of *Samoieds* and other favages reforted to the town, with feathers of White Grous, and other birds; Sables, and the moft valuable furs; fkins of Elks and other deer; the oil from the Walrus, N° 71, from the *Beluga*, p. 182; and different fort of fifh §. Here was, in 1611, a great fifhery of *Beluga*: above fifty boats, with three men each, were employed to harpoon them ¶. The entrance into the river is dangerous, by reafon of a fandy fhoal. The tide rifes there only four feet.

The coafts eaft of *Archangel*, even as far as the river *Ob*, are inhabited by the *Samoieds*; a race as fhort as the *Laplanders*, more ugly, and infinitely more brutalized; their food being the carcaffes of horfes, or any other animals. They ufe the Rein Deer to draw their fledges, but are not civilized enough to

SAMOIEDS.

---

* *Anderfon's Dict.* i. 97.   † The fame, 328.   ‡ *Nov. Com. Petrop.* xiv. 484. tab. xii. Its length does not exceed eleven inches.   ‖ *Hackluyt*, i. 277.   § *Purchas*, i. 546. ¶ The fame, 549.

make

make it the substitute for the Cow. These are in fact the *Hottentots* of the north.

To the east of the *Peczora* commences the continent of

A S I A,

Which has most natural and strongly-marked limits. Here appear the *Werchoturian* mountains, or famous *Urallian* chain, which begins distinctly (for it may be traced interruptedly farther south) near the town of *Kungur*, in the government of *Kasan*, in lat. 57. 20, runs north, and ends opposite to the *Waygatz* streight, and rises again in the isle of *Nova Zemlja*. The *Russians* also call this range *Semennoi Poias*, or the *Girdle of the World*, from a supposition that it encircled the universe. These were the *Riphæi montes : Pars mundi damnata a natura rerum, et densa mersa Caligine* *, of which only the southern part was known to the antients, and that so little as to give rise to numberless fables. Beyond these were placed the happy *Hyperborei*, a fiction most beautifully related by *Pomponius Mela* †. Moderns have not been behind-hand in exaggerating several circumstances relative to these noted hills. *Ysbrand Ides*, who crossed them in his embassy to *China*, asserts that they are five thousand toises or fathoms high : others, that they are covered with eternal snow. The last may be true in their more northern parts ; but in the usual passages over them, they are free from it three or four months.

The heights of part of this chain have been taken by *M. l' Abbè d' Auteroche*; who, with many assurances of his accuracy, says, that the height of the mountain *Kyria*, near *Solikamskaia*, in lat. 60, does not exceed four hundred and seventy-one toises from the level of the sea, or two hundred and eighty-six from the ground on which it stands ‡. But, according to *M. Gmelin*, the mountain *Pauda* is much higher, being seven hundred and fifty-two toises above the sea ‖. From *Petersburg* to this chain

---

* *Plinii Hist. Nat.* lib. iv. c. 12.

† In *Asiatico* littore primi *Hyperborei*, super aquilonem *Riphæosque* montes, sub ipso siderum cardine jacent ; ubi sol non quotidiè, ut nobis, sed primùm verno Æquinoctio exortus, autumnali demum occidit ; et ideò sex mensibus dies, & totidem aliis nox usque continua est. Terra augusta, aprica, per se fertilis. Cultores justissimi, et diutiùs quam ulli mortalium & beatiùs vivunt. Quippe festo semper otio læti, non bella novêre, non jurgia ; sacris operati, maximè *Apollinis*; quorum primitias *Delon* misisse, initio per virgines suas, deinde per populos subinde tradentes ulterioribus ; moremque eum diu, & donec vitio gentium temeratus est, servasse referuntur. Habitant lucos sylvasque ; et ubi eos vivendi satietas magis quam tædium cepit, hilares, redimiti sertis, semet ipsi in pelagus ex certa rupe præcipiti dant. Id eis funus eximium est. Lib. iii. c. 5.

‡ *Voyage de la Siberie*, ii. 605.　　　　‖ Preface to *Flor. Sibir.* i. 54.

is

:s a vaſt plain, mixed with certain elevations or platforms, like iſlands in the midſt of an ocean. The eaſtern ſide deſcends gradually to a great diſtance into the wooded and moraſſy *Sibiria*, which forms an immenſe inclined plane to the *Icy Sea*. This is evident from all the great rivers taking their riſe on that ſide, ſome at the amazing diſtance of lat. 46; and, after a courſe of above twenty-ſeven degrees, falling into the Frozen ocean in lat. 73. 30. The *Yaik* alone, which riſes near the ſouthern part of the eaſtern ſide, takes a ſouthern direction, and drops into the *Caſpian* ſea. The *Dwina*, the *Peczora*, and a few other rivers in *European Ruſſia*, ſhew the inclined plane of that part: all of them run to the northern ſea; but their courſe is comparatively ſhort. Another inclination directs the *Dnieper* and the *Don* into the *Euxine*, and the vaſt *Wolga* into the *Caſpian Sea*.

The *Altaic Chain*, its ſouthern boundary, which begins at the vaſt mountain *Bogdo*, paſſes above the head of the *Irtiſch*, and then takes a courſe rugged, precipitous, cloathed with ſnow, and rich in minerals, between the *Irtiſch* and *Ob*; then proceeds by the lake *Telezkoi*, the riſe of the *Ob*; after which it retires, in order to comprehend the great rivers which form the *Jeneſei*, and are locked up in theſe high mountains; finally, under the name of the *Sainnes*, is uninterruptedly continued to the lake of *Baikal* *. A branch inſinuates itſelf between the ſources of the rivers *Onon* and *Ingoda*, and thoſe of *Ichikoi*, accompanied with very high mountains, running without interruption to the north-eaſt, and dividing the river of *Amur*, which diſcharges itſelf into the eaſt, in the *Chineſe* dominions, from the river *Lena* and lake *Baikal*. Another branch ſtretches along the *Olecma*, croſſes the *Lena* below *Jakoutſk*, and is continued between the two rivers *Tongouſka* to the *Jeneſei*, where it is loſt in wooded and moraſſy plains. The principal chain, rugged with ſharp-pointed rocks, approaches and keeps near the ſhores of the ſea of *Ockhozt*, and paſſing by the ſources of the rivers *Outh*, *Aldan*, and *Maia*, is diſtributed in ſmall branches, which range between the eaſtern rivers which fall into the *Icy Sea*; beſides two principal branches, one of which, turning ſouth, runs through all *Kamtſchatka*, and is broken, from the cape *Lopatka*, into the numerous *Kurile* iſles, and to the eaſt forms another marine chain, in the iſlands which range from *Kamtſchatka* to *America*; moſt of them, as well as *Kamtſchatka* itſelf, diſtinguiſhed by fierce vulcanoes, or the traces of vulcanic fires. The laſt chain forms chiefly the great cape *Tſchutſki*, with its promontories and rocky broken ſhores.———I have ſo far pillaged the labors of my friend †, to trace the boundaries of the vaſt region which has ſo amply furniſhed my *Zoologic* part.—To that, and the Table of Quadrupeds, I refer the ſeveral peculiarities of their ſituations.

ALTAIC CHAIN.

HOW DISTRIBUTED.

---

* *Obſervations ſur la Formation des Montagnes*, par P. S. PALLAS, p. 18.   † Docter PALLAS.

At the northern end of the great *Urallian* chain, is the *Waygatz* ſtreight, which cuts them from *Nowyia Zemlja*, *Nova Zembla*, or the *New Land*. The paſſage is narrow, obſtructed by iſlands, and very frequently by ice. The flux and reflux is here uncertain, by reaſon of the winds; but the tide has been obſerved to riſe only four feet *: the depth from ten to fourteen fathoms. It was diſcovered by *Stephen Boroughs*, in 1556; and the navigation was often attempted by the *Dutch*, in hopes of a paſſage that way to *China*. Continual obſtructions from the floating ice baffled their deſigns, and obliged them to return.

Nova Zemlja conſiſts of five iſlands; but the channels between them are always filled with ice †. It is quite uninhabited, but is occaſionally frequented by the people of *Meſen*, who go there to kill Seals, Walruſes, *Arctic* Foxes, and White Bears, the ſole animals of the place, excepting a few Rein Deer. Attempts have been made to find a way to the *Eaſt Indies* to the north of it; but with equal bad ſucceſs as through the *Waygatz*. *Barentz* juſt doubled the eaſtern end in 1596; ſuffered ſhipwreck there with his crew; and paſſed there a moſt miſerable winter, continually beſieged by the Polar Bears: ſeveral of the crew died of the ſcurvy or exceſs of cold; the ſurvivors made a veſſel of the remains of their ſhip, and arrived ſafe in *Europe* the following year; but their great pilot ſunk under the fatigue ‡.

The ſouthern coaſts of theſe iſlands are in a manner unknown. Between them and the continent is the *Kara* ſea, which forms a deep bay to the ſouth, in which the tide has been obſerved to flow two feet nine inches. Fiſhing people annually come here from the *Peczora* through the *Waygatz*, for the ſake of a ſmuggling trade in furs with the *Samoieds* of the government of *Tobolſki* ||. In the reign of the Empreſs *Anne* attempts were made to double the great cape *Jalmal*, between the gulph of *Kara* and that of the *Ob*; one of which (in 1738) only ſucceeded, and that after encountering the greateſt difficulties §. Had the diſcovery of *Sibiria* depended on its approach by ſea, it might have ſtill remained unknown.

THE RIVER OB. The mouth of the *Ob* lies in a deep bay, which opens into the *Icy Sea*, in lat. 73. 30. This is the firſt and greateſt of the *Sibirian* rivers: it riſes from a large lake in lat. 52, has a gentle courſe through eight hundred leagués of country, navigable almoſt to its ſource ¶: is augmented by the vaſt river *Irtiſch*, in lat. 61, which again receives on each bank a multitude of vaſt rivers in its extenſive progreſs. *Tobolſki*, capital of *Sibiria*, lies on the forks, where it takes in the *Tobol*. The

---

* *Hackluyt.* i. 282. † Doctor PALLAS. ‡ See this curious voyage, as related by *De Veer.* || PALLAS. § *Coxe's Ruſſian Diſcoveries*, 306. ¶ *Gmelin Introd. Fl. Sib.* vii. xxx. By *Leuca* he ſeems to mean a *Verſt*, of which 104½ make a degree. See cxxiii. and Mr. *Coxe's Ruſſian Diſcoveries, Introd.* xiii.

banks

banks of the *Irtifch* and *Ob*, and other *Sibirian* rivers, are, in many places, covered with immenfe forefts, growing on a foft foil; which being torn up by the refiftlefs force of the vaft fragments of ice brought down by the torrents occafioned by the melting of the fnows, are conveyed into the *Icy* and other feas, and form the drift-wood I have before fpoken of. The channel of the *Ob*, from its fource to the *Ket*, is ftony: from that river to the mouth it runs through a fat land. After it has been frozen fome time, the water grows foul and fetid. This is owing to the vaft moraffes it in fome places goes through, to the flownefs of the current, and to the *earth-falt* (*erdfaltz*) with which fome of the rivers which run into it are impregnated. The fifh therefore fhun the waters of the *Ob*, and refort in vaft fhoals to the mouths of thofe rivers which rufh into it from ftony countries, and in fuch places are taken in great abundance. This ftench continues till the river is purified in the fpring by the melting of the fnow. The *Taz*, another river which empties itfelf into the eaft of the gulph of *Ob*, is liable to the fame impurity.

The *Jenefei* next fucceeds. Mr. *Gmelin*, as a naturalift, would confider this as the boundary between *Europe* and *Afia*. From its eaftern banks every thing puts on a new appearance: a certain new and unufual vigour reigns in every thing. The mountains, which to the weftward, as far as the *Urallian* chain, appeared only fcattered, now take full poffeffion, and are interfperfed with moft beautiful vallies. New animals, fuch as the *Argali*, p. 12, and *Mufk*, p. 34, and feveral others, begin to fhew themfelves. Many *European* plants difappear, and others peculiar to *Afia*, gradually mark the alteration *. This river is fcarcely inferior to the *Ob*. It rifes from the two rivers *Ulu-kem* and *Bei-kem*, in north lat. 51. 30, long. 111, and runs due north into the *Icy Sea*, forming a mouth filled with multitudes of iflands: its channel for the moft part ftony or gravelly: its courfe fwift: its fifhes moft delicate: its banks, efpecially the eaftern, mountanous and rocky; but from the fort of *Saiaenes* to the river *Dubtches*, rich, black, and cultivated. It is fed by numbers of rivers. The *Tungufca*, and the lower *Tungufca*, are the moft noted. The firft rufhes, near *Irkutz*, out of the great lake *Baikal*, under the name of the *Angara*, between two vaft rocks, natural, but with all the appearance of being cut through by art, and tumbling over huge ftones in a bed a mile wide, and for a fpace nearly the fame †. The collifion of the waters againft the ftones is attended with a moft dreadful noife, which, with the magnificence of the fcenery, forms the moft awful approach imaginable to this facred water. A deity prefided over the lake; and no one dared call it by that degrading name, for fear of incurring the penalty of the difrefpect. Inftead of *lake*, the borderers ftyle it the *Holy Sea*; and its vaft mountains, the *Holy Mountains*. St. *Nicholas* prefides over them, and has

* Pref. *Fl. Sibir.* xliv.   † *Bell's Travels,* 8vo. ed. i. 279.

here

here his chapel. The mountains are cloathed with forefts : of large trees on the lower parts ; with fewer and leffer as they gain the heights. Thefe are the retreat of the Wild Boar, and variety of game. Its depth of water is very great : its clearnefs perfect : free from iflands, except the *Olchon* and *Saetchia* : navigable in all parts : and in ftorms, the waves like thofe of the fea. Its length is a hundred and twenty-five common leagues : its breadth from four to feven *. The Com-

SEALS. mon Seal abounds in this lake. It is a fmall variety, but fo fat as to appear almoft fhapelefs. Thefe animals muft have been here aboriginally ; for, befides the vaft diftance from the fea, their paffage muft have been entirely obftructed by the cata- racts which intervene. I am got eight degrees beyond my plan ; but I could not refift the defcription of this prince of lakes.

TOWN OF MAN-GAZEA. The *Angara* runs nearly due north for a great way ; then affumes the name of *Tungufca,* turns weftward, and joins the *Jenefei* in lat. 58. The lower *Tungufca* rifes far to the fouth-weft, approaches very near to the *Lena,* and falls into the *Jenefei* in lat. 65. 40. Above its junction ftands the town of *Mangazea,* cele- brated for its great fair of furs of every kind, brought there by the furrounding pagans, who pafs the long winter in the chace. Many *Ruffians* have alfo migrated, and fettled here for the fame purpofe, and draw great profit from the fpoils of the animals. This neighborhood is, during fummer, the great refort of multitudes of fpecies of water-fowl. About the feaft of St. *Peter,* here *Flora* begins to difclofe her beauties : the country is covered with the moft beautiful *Sibirian* flowers ; many of which enliven the gardens of our more fouthern climate. The fowls now exult, and unite in emitting their various notes ; none particularly melodious in them- felves, but together form a concert far from difagreeable † ; perhaps from the hear- er being confcious that they are the notes of happinefs, at the enjoyment of the reviving rays of the fun.

In antient times, *Mangazea,* or, as it was then called, *Mongozey,* and *Mongolmy,* was feated near the mouth of the *Taz* ‡ ; but was removed by the inhabitants into a milder climate, *i. e.* juft to the fouth fide of the *Arctic* circle. Before that period it was a place of great trade, and was eagerly vifited from *Archangel,* through a complication of difficulties, by fea, by rivers, by land, by rein-drawn fledges, and by drawing the veffels from river to river over frequent carrying-places ‖. Thefe tracts were certainly *Le pais prefque inacceffible à caufe de boües, & de glaces,* and, *Le pais de tenebres,* fpoken of by *Marco Polo* §, as the regions from whence the *Chams* of *Tartary* procured the richeft furs.

CAPE TAIMURA. From the mouth of the *Jenefei,* the immenfe promontory *Taimura* ftretches

---

* *Voyage en Siberie,* i. 213.   † Same, ii. 56.   ‡ Same, 57.   ‖ Same, and *Purchas,* iii. 539.   § In *Bergeron's Collection,* 160, 161.

3

farthe ft

fartheſt north of all this region into the *Icy Sea*, nearly into lat. 78. To the eaſt of it the *Chatunga*, *Anabara*, and *Olenek*, rivers little known, fall into the ſea, and have before the mouth of each a conſiderable bay. Remarks have been made on the tide which flows into the *Katanga*, that at the full and new moon it riſes two feet ; at other times is much leſs *. We may conclude, that if it flows no higher in this contracted place, and that of the gulph of *Kara*, its encreaſe muſt be very ſmall on the open ſhores of the *Icy Sea*. The coaſts are in general ſhallow, which has proved a ſafety to the few ſmall veſſels which have navigated this ſea ; for the ſhoalneſs of the water preſerves them from the montanous ice, which grounds before it can reach them.

BEYOND the *Olenek*, the vaſt *Lena*, which riſes near lake *Baikal*, after a gentle and free courſe over a ſandy or gravelly bottom, diſcharges itſelf by five great mouths, the eaſtern and weſtern moſt remote from each other. The middle, or moſt northerly, is in lat. 73. 20. To form an idea of the ſize of this river, I muſt remark, that at *Iakutſk*, in lat. 61, twelve degrees from its diſcharge, the breadth is near three leagues †. Beyond this river the land contracts itſelf, and is bounded to the ſouth by the gulph of *Ochotz*. The rivers *Jana*, *Indigirſka*, and *Kolyma* or *Kowyma*, have a comparatively ſhort courſe. The laſt is the moſt eaſterly of the great rivers which fall into the *Icy Sea*. Beyond it is a woodleſs tract, which cuts off the Beaver, the Squirrels, and many other animals to whom trees are eſſential in their œconomy. No foreſts can exiſt farther north than lat. 68 ; and at 70, bruſh-wood will ſcarcely grow. All within lat. 68, form the *Arctic Flats*, the ſummer haunts of water-fowl ; a bare heath or moor, mixed with rocky mountains : and beyond the river *Anadyr*, which in lat. 65. falls into the *Kamtſchatkan Sea*, the remainder of the tract between it and the *Icy Sea* has not a ſingle tree ‡.

RIVER LENA.

ARCTIC FLATS WOODLESS.

I ſhall now take a review of the vaſt extent of ſhore which borders on the *Icy Sea*. The *Jouratzkaine* coaſt, which lies between the *Ob* and the *Jeneſei*, is high but not mountanous, and almoſt entirely compoſed of gravel or ſand ; but in many places there are low tracts. Not only on theſe, but on more elevated ſituations, are found great fragments of wood, and often entire trees, all of the ſame ſpecies ; Fir, Larch, and Pine, green and freſh ; in other places, elevated beyond the reach of the ſea, are alſo great quantities of *floated* wood, antient, dried, and rotting §. This is not the only proof of the loſs of water in the *Icy* as well as other ſeas ; for in theſe places is ſeen a ſpecies of clay, called by the *Ruſſians*, *Il*, which is exactly like the kinds uſually depoſited by the water : and of this there is, in theſe parts, a bed about eight inches thick, which univerſally forms the upper ſtratum ‖. Still farther to

* *Voy. en Siberie*, ii. 30.   † Poſſibly *Verſts.* See *Voy. en Siberie*, i. 407.   ‡ Doctor PALLAS.
§ *Voy. en Siberie*, ii. 27, 28.   ‖ Same, ii. 362.

         the

the eaſt, it grows mountanous, covered with ſtones, and full of coal. On the ſummit of the chain, to the eaſt of *Simovie Retchinoïe*, is an amazing bed of ſmall Muſſels, of a ſpecies not obſerved in the ſubjacent ſea. I think them brought there by ſea-fowl, to eat at leiſure; for it is not wonderful that numbers of objects of natural hiſtory ſhould eſcape the eye in ſuch a ſea as this. Many parts again are low; but in moſt places the ſea near the ſhore is rugged with pointed rocks. The coaſt about the bay of cape *Tſchutſki*, the moſt eaſtern extremity of *Aſia*, is in ſome places rocky, in others ſloping and verdant; but within land riſing into a double ridge of high mountains.

FREEZING OF THE ICY SEA.

About the end of *Auguſt*, there is not a day in which this ſea might not be frozen; but in general it never eſcapes later than the firſt of *October*. The thaw commences about the twelfth of *June*, at the ſame time with that of the mouth of the *Jeneſei* . From the great headlands, there is at all times a fixed, rugged, and mountanous ice, which projects far into the ſea. No ſea is of ſo uncertain and dangerous navigation : it is, in one part or other, always abundant in floating ice. During ſummer, the wind never blows hard twenty-four hours from the north, but every part of the ſhore is filled for a vaſt diſtance with ice; even the ſtreights of *Bering* are obſtructed with it †. On the reverſe, a ſtrong ſouth wind drives it towards the pole, and leaves the coaſt free from all except the fixed ice. During winter, the ſea is covered, to the diſtance of at leſt ſix degrees from land. *Markoff*, a hardy *Coſſac*, on *March* 15th, O. S. in the year 1715, attempted, with nine other perſons, a journey from the mouth of the *Jana*, in 71 north lat. to the north, over the ice, on ſledges drawn by dogs. He went on ſucceſsfully ſome days, till he had reached lat. 77. or 78: he was then impeded by moſt mountanous ice. He climbed to the ſummit of one of the *Icebergs*; and ſeeing nothing but ice as far as his eye could reach, returned on *April* 3d, with the utmoſt difficulty: ſeveral of his dogs died, and ſerved as food for the reſt ‡.

I ſhall juſt mention ſome of the attempts made to paſs through the *Icy Sea* to that of *Kamtſchatka*. The firſt was in 1636, from the ſettlement of *Yakutzk*. The rivers from the *Jana* to the *Kolyma* were in conſequence diſcovered. In 1646 a company of *Ruſſian* adventurers, called *Promyſchleni*, or Sable-hunters, made a voyage from the *Kolyma* to the country of the *Tſchutſki*, and traded with thoſe people for the teeth of the *Walrus*. A ſecond, but unſucceſsful voyage was made in the next year; but in 1648 one *Deſchnew*, on the 20th of *June*, began his memorable voyage, was fortunate in a ſeaſon free from ice, doubled the *Tſchutſki-noſs*, arrived near the river *Olutora*, ſouth of the river *Anadyr*, where he ſuffered ſhip-

---

* *Voy. en Siberie*, ii. 29.   † PALLAS : Alſo Narrative of four *Ruſſian* ſailors caſt away on *Eaſt Spitzbergen*, 55.   ‡ *Forſter's Obſ.* 81.

wreck,

wreck, but efcaped to enjoy the honor of his difcovery. Many other attempts were made, but the moft which the adventurers have done was to get from the mouth of one great river to another in the courfe of a fummer. I find very few names, except of rivers, in a tract fo vaft as it is, on account of its being fo little frequented. To the eaft of the promontory *Taimura*, that of *St. Transfigurationis* bounds the eaft fide of the bay of *Chatanga*, in lat. 74. 40, long. from *Ferro* 125. *Swaitoi-nofs*, or the *Holy Cape*, in lat 73. 15, is a far-projecting headland, and, with the ifles of the *Lena*, and another intervening headland, forms two vaft bays. Out of the moft eaftern, into which the river *Yana* difcharges itfelf, one *Schalourof*, a broken *Ruffian* merchant, took his departure for an eaftern difcovery. He began his voyage in *July* 1760 from the *Lena*, but was fo obftructed with ice that he was forced into the *Yana*, where he was detained the whole winter, by the fame caufe, till *July* 29th, 1761. He doubled the *Swaitoi-nofs September* the 6th; according to fome, faw to the north a montanous land, poffibly an ifland. He was eight days in getting through the paffage between the continent and the ifle of *St. Diomede*, which lies a little to the fouth eaft of the *Nofs*. He paffed with a favorable wind the mouths of the *Indigirka* and *Alazeia*, and getting entangled among the ice between the *Medviedkie Oftrova*, or *Bear Iflands*, was obliged to lay up his veffel in one of the mouths of the *Kolyma* during winter, where he fubfifted on rein-deer, which frequented thofe parts in great herds during the fevere feafon; and on various fpecies of falmon and trout, which were pufhing their way up the river before it was frozen. After this he made two other attempts. In the year 1763 he paffed the *Pefzcanoi-nofs*, and got into a deep bay, called *Tfchaoún Skaja Goúba*, with the ifle of *Sabedei* at its mouth; the great *Schalatfkoi-nofs* to the eaft; and at its bottom the little river *Tfchaoún*, which difcharges itfelf here out of the land of the *Tfchutfki*, fome of whom he faw on the fhore, but they fled on his appearance. He found no means of fubfifting in this bay, therefore was obliged to return to the *Lena*, and was greatly affifted in his paffage by the ftrength of the current, which uniformly fet from the eaft. In 1764 he made his laft attempt, and was, as is conjectured, flain by the *Tfchutfki*; but whether he doubled the famous cape of that name, is left uncertain. A MS. map, which Doctor PALLAS favored me with, places the montanous ifle before mentioned in lat. 75, oppofite to the cape *Schalatfkoi* *. Thus clofes all the accounts I can collect of the voyages along

* This was fuppofed to have been part of the continent of *America*; but in 1768, M. *Tchitfcherin*, governor of *Sibiria*, put the matter out of doubt; for he fent there three young officers in the winter, on the ice. They found fome fmall defart ifles, without the left appearance of land on the north; but on one they met with a fort of defence, formed of floating wood, on the fide of a precipice, but by whom formed, or againft what enemy, is hard to guefs. PALLAS. MS.

this

this diftant coaft.   Part is taken from Mr. *Coxe*'s *Ruffian Difcoveries* *, and part from a manufcript for which I am indebted to the learned Profeffor before mentioned.

The wind which paffes over the ice of this polar fea, has rendered *Sibiria* the coldeft of inhabited countries : its effects may perhaps extend much farther.   At *Chamnanning*, in *Thibet*, in lat. 30. 44. (according to Major *Rennel*'s claffical map) Mr. *Bogle* found, during winter, the thermometer in his room at 29° below the freezing point.   In the middle of *April* the ftanding waters were all frozen, and heavy fnows perpetually fell †.   I have heard of ice even at *Patna*, in lat. 25. 35 ; and of the *Seapoys* who had flept on the ground being found in the morning torpid.   Near the fort of *Argun*, not higher than lat. 52, the ground feldom thaws deeper than a yard and a half ‡.   At *Iakutfk*, in lat. 62, the foil is eternally frozen even in fummer, from the depth of three feet below the furface. An inhabitant, who by the labor of two fummers funk a well to the depth of ninety-one feet, loft his labor, and found his fartheft fearches frozen §.   Birds fall down, overcome with the cold ; and even the wild beafts fometimes perifh. The very air is frozen, and exhibits a moft melancholy gloom ‖.

AURORA BOREA-
LIS.

The *Aurora Borealis* is as common here as in *Europe*, and ufually exhibits fimilar variations : one fpecies regularly appears between the north-eaft and eaft, like a luminous rainbow, with numbers of columns of light radiating from it : beneath the arch is a darknefs, through which the ftars appear with fome brilliancy.   This fpecies is thought by the natives to be a forerunner of ftorms.   There is another kind, which begins with certain infulated rays from the north, and others from the north-eaft.   They augment little by little, till they fill the whole fky, and form a fplendor of colors rich as gold, rubies, and emeralds : but the attendant phænomena ftrike the beholders with horror, for they crackle, fparkle, hifs, make a whiftling found, and a noife even equal to artificial fire-works.   The idea of an electrical caufe is fo ftrongly impreffed by this defcription, that there can remain no doubt of the origin of thefe appearances.   The inhabitants fay, on this occafion, it is a troop of men furioufly mad which are paffing by.   Every animal is ftruck with terror ; even the dogs of the hunters are feized with fuch dread, that they will fall on the ground and become immoveable till the caufe is over ¶.

FISH.

I am flightly acquainted with the fifh of the *Icy* fea, except the anadromous kinds, or thofe which afcend from it into the *Sibirian* rivers.   The *Ob*, and other

---

* P. 323 to 329.   † *Ph. Tranf.* lxvii. 471.   ‡ Pref. *Flora Sib.* 78.   § *Forfter*'s *Obf.* 85. quoted from *Gmelin*.   ‖ Pref. *Flora Sib.* 73.   ¶ *Voy. en Siberie*, ii. 31, 52.

Sibirian rivers, are vifited by the *Beluga* Whale, the common Sturgeon, and the Sterlet or *Acipenfer Ruthenus*, Lin. Syft. 403; but I am informed by Doctor *Pallas*, that they have neither Carps, Bream, Barbels, nor others of that genus, nor yet Eels, *Silurus Glanis*, Lin. Syft. 501; *Perca Lucioperca*, 481; or common Trout: all which are found in the *Amur*, and other rivers which run into the eaftern ocean: in the latter, our common Cray-fifh is found. In return, the *Sibirian* rivers abound in vaft variety of the Salmon kind, and many unknown to us in *Europe*, which delight in the chilly waters of thefe regions. The common Salmon, *Br. Zool.* iii. N° 143, is one of the fcarcer kinds: the *Salmo Nelma*, Pallas Itin. ii. 716, or *Salmon Leucichthys* of *Guldenftaedt*, Nov. Com. Petrop. xiv. 531, is a large fpecies, growing to the length of three feet: the head greatly protracted: the lower jaw much the longeft: the body of a filvery white: fcales oblong: tail bifid. P. D. Rad. 14. The *Salmo Taïmen*, or *Hucho*; Pallas, ii. 716, grows to the weight of ten or fifteen pounds, and the length of a yard and a half: the color of the back is dufky; towards the fides filvery: the belly white: fpotted with dufky on the back: anul fin of a deep red: tail bifurcated: flefh white: *Salmo Lavaretus*, iii. 705, or *Gwiniad*, Br. Zool. iii. N° 152: *Salmo Albula*, Lin. Syft. 512: *Salmo Schokur*, Pallas Itin. iii. 705; a fpecies about two feet long, not unlike the *Gwiniad*: the *Salmo Pidfchian*, Pallas Itin. iii. 705; about two fpans long, broader than the *Gwiniad*, and with a gibbous back: *Salmo Wimba*, Lin Syft. 512: and *Salmo Nafus*, Pallas Itin. iii. 705 *, are extremely common in the *Ob*. Others fhun that ftill river, and feek the *Jenefei*, and other rapid ftreams with ftony bottoms. Such are the *Salmo Lenok*, Pallas Itin. ii. 716 †: *Salmo Oxyrhynchus*, Lin. Syft. 512: and *Salmo Autumnalis*, or *Omul*, Pallas Itin. iii.705; which annually force their way from the fea, from lat. 73. to lat. 51. 40, into lake *Baikal*, a diftance of more than twenty-one degrees, or near thirteen hundred miles. The *Omul* even croffes the lake, and afcends in *Auguft* the river *Selinga*, where it is taken by the inhabitants in great quantities, and is preferved for the provifion of the whole year. After dropping its fpawn in the ftony beds of the river, it again returns to the fea. The *Salmo Arcticus*, Pallas Itin. iii. 206; and *S. Thymallus*, or Grayling, Br. Zool. iii. N° 150; may be added to the fifh of the *Sibirian* rivers. The *Salmo Cylindraceus*, or *Walok* of the *Ruffians*, is a fifh very flender, and almoft cylindrical, with a very fmall mouth, large filvery fcales, and the under fins reddifh. This is found only in the *Lena*, the *Kowyma*,

SURPRIZING MIGRATIONS OF FISH.

* The *Schokur* and *Nafus* are two fpecies of *Coregoni*, or Salmons, with very fmall teeth.
† *Voy. en Siberie*, i. 237. It alfo afcends through the *Jenefei* and the *Tuba* to the *Madfbar*, a lake an amazing diftance in the mountains.

and

and *Indigirſka*. M. *Gmelin* and the *Abbé D'Auterocbe* aſſure us, that Pikes, Perch, Ruffs, Carp, Bream, Tench, Crucians, Roach, Bleaks, and Gudgeons, are alſo met with in the *Ob*, and different rivers of this country *. I cannot reconcile this to the former account given me by ſo able a naturaliſt, to whom I owe this hiſtory of the *Arctic* fiſh. The *Salmo Kundſha*, Pallas Itin. iii. 706, abounds in the gulphs of the *Icy* ſea, but does not aſcend the rivers; and the *Pleuronectes-Glacialis*, Pallas Itin. iii. 706, is frequent on the ſandy ſhores.

To review the inhabitants of the *Arctic* coaſts, I ſhall return as far as *Finmark*. I refer the reader to p. LXXIX. for what I have ſaid of the *Laplanders*. The *Samoieds* line the coaſts from the eaſt ſide of the *White* ſea, as far (according to the *Ruſſian* maps) as the river *Ob*, and even the *Anabara*, which falls into the *Icy* ſea in lat. 73. 30; and poſſeſs the wildeſt of countries inland, as low as lat. 65. After them ſucceeds, to the eaſt, a race of middle ſize; and, extraordinary to ſay, inſtead of degeneracy, a fine race of men is found in the *Tſchutſki*, in a climate equally ſevere, and in a country equally unproductive of the ſupports of life, as any part of theſe inhoſpitable regions. The manners of all are brutal, ſavage, and nearly animal; their loves the ſame; their living ſqualid and filthy beyond conception: yet on the ſite of ſome of theſe nations *Mela* hath placed the elegant *Hyperborei*: and our poet, *Prior*, giving free looſe to his imagination, paints the manners of theſe *Arctic* people in the following beautiful fiction, after deſcribing the condition of the natives of the torrid zone.

And may not thoſe, whoſe diſtant lot is caſt
North beyond *Tartary*'s extended Waſte;
Where, thro' the plains of one continual day,
Six ſhining months purſue their even way,
And ſix ſucceeding urge their duſky flight,
Obſcur'd with vapors, and o'erwhelm'd in night;
May not, I aſk, the natives of theſe climes
(As annals may inform ſucceeding times)
To our quotidian change of heaven prefer
Their own viciſſitude, and equal ſhare
Of day and night, diſparted thro' the year? }
May they not ſcorn our ſun's repeated race,
To narrow bounds preſcrib'd, and little ſpace,

Haſt'ning from morn, and headlong driven from
    noon,
Half of our daily toil yet ſcarcely done?
May they not juſtly to our climes upbraid
Shortneſs of night, and penury of ſhade?
That, ere our weary'd limbs are juſtly bleſt
With wholeſome ſleep, and neceſſary reſt,
Another ſun demands return of care,
The remnant toil of yeſterday to bear?
Whilſt, when the ſolar beams ſalute their ſight,
Bold and ſecure in half a year of light,
Uninterrupted voyages they take
To the remoteſt wood, and fartheſt lake;

* *Voy en Siberie*, par *Gmelin*, i. 84, 89, 241. ii. 167, 170, 219.—*Voy. en Siberie*, par *l'Abbé D'Auteroche*, i. 200. *Engl. Ed.* 231.

Manage

Manage the fishing, and pursue the course
With more extended nerves, and more continued
    force ?
And when declining day forsakes their sky;
When gathering clouds speak gloomy Winter nigh,
With plenty for the coming season blest,
Six solid months (an age) they live releas'd
From all the labor, process, clamor, woe,
Which our sad scenes of daily action know:

They light the shining lamp, prepare the feast,
And with full mirth receive the welcome guest:
Or tell their tender loves (the only care
Which now they suffer) to the list'ning Fair;
And rais'd in pleasure, or repos'd in ease,
(Grateful alternates of substantial peace)
They bless the long nocturnal influence shed
On the crown'd goblet, and the genial bed.

With greater reality speaks that just observer of nature, the naturalist's poet, of the inhabitants of this very country, as a true contrast to the foregoing lines:

Hard by these shores, where scarce his freezing
    stream
Rolls the wild *Oby*, live the last of men;
And half enliven'd by the distant sun,
That rears and ripens man as well as plants,
Here human nature wears its rudest form.
Deep from the piercing season, sunk in caves,
Here, by dull fires, and with unjoyous chear,

They waste the tedious gloom.   Immers'd in
    furs,
Doze the gross race. Nor sprightly jest, nor song,
Nor tenderness they know; nor aught of life,
Beyond the kindred bears that stalk without.
Till morn appears, her roses dropping all,
Sheds a long twilight bright'ning o'er the fields,
And calls the quiver'd savage to the chace.
                 THOMSON.

This amazing extent of the *Asiatic Russian* dominions remained undiscovered to a very late period. The *Czars*, immersed in sensuality, or engaged in wars, had neither taste or leisure to explore new countries. A plundering excursion was made into it in the reign of *Basilovitz* I; a second was made under his successor: but a stranger, the celebrated *Cossac, Yermac*, driven from his country on the shores of the *Caspian* sea, pushed his way with a resolute band as far as *Orel*, near the head of the *Kama*, on the western side of the *Urallian* chain. There he met with one *Strogonoff*, a *Russian* merchant, recently settled in those parts for the sake of the traffic of furs. He continued in that neighborhood the whole winter, and was supplied by the *Russians* with all necessaries. In the spring he turned his arms against *Kutchum Chan*, one of the most powerful of the petty princes of the country which now forms part of the government of *Tobolski*. In 1581, he fought a decisive battle with the Chan, overthrew him, and seated himself on the throne. Finding his situation precarious, he ceded his conquests to *Basilovitz*, who seized on the opportunity of adding this country to his dominions. He sent *Yermac* a supply of men. But at length his good fortune forsook him. He was surprized by the Chan; and, after performing all that a hero could do, perished in attempting to escape.

                                                    The

The *Ruffians*, on the death of their ally, retired out of *Sibiria* ; but they foon returned, recovered the conquefts made by *Yermac*, and, before the middle of the following century, added to their antient poffeffions a territory fourteen hundred and feventy leagues in length, and near feven hundred in breadth (without including the *Ruffian* colonies on the ifland of *Oonalafhka*, on the coaft of *America**) yet is fo thinly peopled, and with fuch barbarians, as to add no ftrength to the empire by any fupplies to the army or navy. They are almoft torpid with inaction; lazy to the higheft degree, from their neceffary confinement to their ftoves during the long winter of the country. In that feafon, the ground is clad with deep fnow, and the froft moft tremendoufly fevere. The fpring, if fo it may be called, is diftinguifhed by the muddied torrents of melting fnows, which rufh from the mountains, and give a fea-like appearance to the plains. Mifts, and rain, and fnow, are the variations of that feafon, and they continue even to the fourth of *June*. The fhort fummer is hot, and favorable to vegetation. Corn may be feen a foot high by the 22d of *June*; and the grafs is moft luxuriant. Culinary plants will fcarcely grow about *Tobolfki*. Fruits of every kind, except a currant, are unknown. A fingle crab-like apple, raifed in a hot-houfe, was once produced there, fliced in a large difh, at a great entertainment, and ferved up with as much oftentation as we would in *England* a pine-apple.

The animals of *Sibiria*, the furs of which were the original object of its conqueft, are now fo reduced, that the *Ruffians* are obliged to have recourfe to *England* for a fupply from *North America*, which they add to their own ftock of furs exported into *China*. Metals feem the ftaple trade of the country. Thofe of iron and copper are abundant and excellent. Gold and filver are found in feveral places, and in fuch abundance, as to form a moft important article in the revenues of *Ruffia*. The copper mines of *Kolyvan*, from which thofe pretious metals are extracted, employ above forty thoufand people, moftly colonifts. The filver mines of *Nertfchinfk*, beyond lake *Baikal*, above fourteen thoufand. The whole revenue arifing from the mines of different metals, is not lefs than £. 679,182. 13 *s*. †

PLANTS.   Next to the difcovery of the new world, no place has added more to the entertainment of naturalifts than *Sibiria*. As has been before obferved, nature there affumes a new appearance in the animal world : it does the fame in the vegetable ; at leaft, very few trees are found common to *Europe* and *Afia*. Let me juft mention the nobler kinds : the Oak, frequent as it is in *Ruffia* and in *Cafan*, is not to be feen in this vaft region nearer than the banks of the *Argun*

---

* *D'Auteroche, Voy. en Siberie*, i. 83.   † *Coxe's Travels.*

and

and *Amur*, in the *Chinese* dominions. The White Poplar, *Populus alba*; and the Aspen, *Populus tremula*, are extremely common. The Black Poplar, *Populus nigra*; the Common Sallow, *Salix caprea*; Sweet Willow, *Salix pentandra*; White Willow, *Salix alba*, are very frequent. The Hazel, *Corylus Avellana*, is circumstanced like the Oak. The Common Birch, *Betula alba*, is most abundant; and, as in all northern nations, of universal use. The Dwarf Birch, *Betula nana*, is confined to the neighborhood of lake *Baikal*. The Alder, *Betula Alnus*, is very frequent. The Pinaster, *Pinus Pinea*; the Pine with edible seeds, or *Pinus Cembra*; and Larch, *Pinus Larix*; all trees of the first use, medicinal or œconomical, cover many parts of the country. The *Norway* Fir, *Pinus Abies*, and the Silver Fir, *Pinus Picea*, form, in most parts of the country, great forests: the first grows in this country not farther north than lat. 60; the last not higher than lat. 58; yet the former flourishes in *Europe*, and composes in *Lapmark*, far beyond the *Arctic* circle, woods of great extent: a proof of the superior rigour of cold in the *Asiatic* north. These form the sum of *European* trees growing in *Sibiria*. Of other plants, common to both continents, M. *Gmelin* gives the reader, in p. xciv. of his Preface, a slender list of such which fell under his observation.

The trees or shrubs peculiar to *Sibiria* and *Tartary*, are the *Acer Tartaricum*, Sp. Pl. ii. 1495: the *Ulmus pumila*, 327: *Prunus Sibirica*, Amman. *Ruth.* 272, tab. 29: *Pyrus baccata*, 274: *Robinia Caragana, frutescens*, and *pygmæa*, Sp. Pl. ii. 1044. I may also observe, that the *Taccamahacca*, or *Populus balsamifera*, 1463, common also to *North America*, abounds about the upper part of the *Lena*, the *Angara*, and *Jenesei*, and between the *Onon* and *Aga*. An infusion of its buds is used by the natives as an excellent remedy for an infamous disorder, frequent in this great country.

EUROPE is obliged to *Sibiria* for that excellent species of Oat, the *Avena Sibirica*, Fl. Sib. i. 113. tab. 22. Lin. Sp. Pl. i. 117; and our gardens are in a most peculiar manner enlivened with the gay and brilliant flowers introduced from that distant and severe climate. I shall only select a few out of the multitude*. *Veronica Sibirica, Iris Sibirica*, Fl. Sib. i. 28. *Eryngium planum*, i. 185. *Lilium bulbiferum*, i. 41. *L. pomponium*, i. 42. *L. Martagon*, i. 44. *Delphinium grandiflorum*, Sp. Pl. i. 749. *Erythronium Dens canis*, i. 39. tab. 7. *Hemerocallis flava*, i. 37. *Saxifraga crassifolia*, Sp. Pl. i. 573. *Lychnis chalcedonica*, Sp. Pl. i. 625. *Pyrus baccata, Lythrum virgatum*, Sp. Pl. 642. *Amyg-*

* This list was communicated to me by an able botanist; but I think some of the plants are also found in *Europe*.

dalus nana, Sp. Pl. 677. *Pæonia tenuifolia,* Sp. Pl. i. 748. *Clematis integrifolia,* Sp. Pl. i. 767. *Adonis vernalis,* Sp. Pl. i. 771. *Aftragalus alopecuroides,* Sp. Pl. ii. 1064. *Hypericum Afcyron,* Sp. Pl. ii. 1102. *Echinops Ritro,* Fl. Sib. ii. 100. *Veratrum nigrum,* Fl. Sib. i. 76.

Tschutski.

    After the conqueſt of *Sibiria,* the *Tſchutſki* were the firſt people diſcovered by the *Ruſſians,* who were indebted to the adventure of *Deſchnew* for the knowlege of them. They are a free and brave race, and in ſize and figure ſuperior to every neighboring nation; tall, ſtout, and finely made, and with long and agreeable countenances; a race inſulated ſtrangely by a leſſer variety of men. They wore no beards. Their hair was black, and cut ſhort, and covered either with a cloſe cap, or hood large enough to cover the ſhoulders. Some hung beads in their ears, but none had the barbariſm to bore either noſes or lips. They wore a ſhort and cloſe frock, breeches, and ſhort boots · ſome had trowſers. The materials of their cloathing was leather admirably dreſſed, either with or without the hair *. It is ſaid that at times they wear jackets made of the inteſtines of whales †, like the *Eſkimaux;* probably when they go to ſea, for they excel their neighbors in fiſhing, and uſe open boats covered with ſkins ‡, and like the women's boats of the *Greenlanders.* They have alſo the leſſer or *kajak.* They make uſe of ſledges, and have large fox-like dogs of different colors, with long ſoft woolly hair, which are probably deſigned for the draught. Some ſay that they uſe rein-deer, of which they have vaſt abundance, but neither milk them nor kill them for food, preferring the fleſh of ſea animals, except one dies by chance, or is killed by the wolves. They are a brave and warlike people; are armed with bows and arrows; the laſt pointed with ſtone or bone. They had ſpontoons headed with ſteel, procured by traffic from the *Ruſſians;* theſe they uſually ſlung over their right ſhoulder; and a leathern quiver of moſt elegant workmanſhip hung over the left §. The *Ruſſians* have often gained dear-bought victories over this brave people, but never were able to effect their conqueſt. They retained an high ſenſe of liberty, and conſtantly refuſed to pay tribute; and the ambitious *European* miſcalled them rebels. They will not on any conſideration part with their weapons: poſſibly a *Tſchutſki* may think a diſarmed man diſhonored. Captain Cook, in his three hours viſit to them, found their attachment to their arms, notwithſtanding they willingly parted with any thing elſe, and even without the proſpect of exchange. They treated him with great civility, but prudent caution: ſaluted him by bow-

---

* *Voyage,* ii. 450, tab. 51.   † *Hiſt. Kamtſchatka,* Fr.   ‡ *Voyage,* ii. 452.
§ See tab. 51 of the *Voyage.*

ing and pulling off their caps, possibly a piece of politeness they learned from the *Russians*. They treated him with a song and dance, and parted friends ; but not without a most remarkable and consequential event :—A year after the interview between Captain Cook and the *Tschutski*, a party of these people came to the frontier post of the *Russians*, and voluntarily offered friendship and tribute. These generous people, whom fear could not influence, were overcome by the civility and good conduct of our illustrious commander : they mistook him and his people for *Russians*, and, imagining that a change of behaviour had taken place, tendered to their invaders a lasting league *. Possibly the munificent empress may blush at the obligation conferred by means of *British* subjects, in procuring to her empire a generous ally, at the instant her armed neutrality contributed to deprive us of millions of lawful subjects.

From the shortness of the interview little knowlege could be gained of their customs. I shall only observe, that they bury their dead under heaps of stones, or carnedds : several were seen here with the rib of a whale on the top instead of a pillar † ; a proof of the universality of these memorials of the dead.

TUMULI.

The country of the *Tschutski* forms the most north-easterly part of *Asia*. It is a peninsula, bounded by the bay of *Tchaoŭn*, by the *Icy Sea*, the streights of BERING, and the gulph and river of *Anadir*, which open into the sea of *Kamtschatka*. It is a mountanous tract, totally destitute of wood, and consequently of animals which require the shelter of forests. The promontory *Schalotskoi*, before mentioned, is the most westerly part. Whether it extends so far north as lat. 74, as the *Russians* place it, is very doubtful : there is the opinion of our great navigator against it. From his own reasonings he supposed that the tract from the *Indigirska*, eastward, is laid down in the maps two degrees to the northward of its true position ‡. From a map he had in his possession, and from information he received from the *Russians*, he places the mouth of the *Kowyma*, in lat. 68, instead of lat. 71. 20, as the *Petersburg* map makes it. It is therefore probable, that no part of *Asia* in this neighborhood extends further than lat. 70, in which we must place the *Schalotskoi Nofs* ; and after the example of Mr. *Campbell*, who formed his map of this country chiefly from the papers of Captain BERING §, give the land which lies to the east of that promontory a very southern trend. As Captain Cook had cause to imagine that the former charts erred in longitude as well as latitude, it is probable that he reached within sixty miles of the *Schalotskoi Nofs* ‖. There we find him on *August* 29th, 1778, and from this period are enabled, from his remarks, to proceed securely accurate.

CORRECTIONS IN GEOGRAPHY BY CAPT. COOK.

* *Voy.* iii. 217.   † *Ellis's Narrative*, i. 332.   ‡ *Voyage* iii. 268.   § In
Harris's Voy. ii. 1016.   ‖ *Voyage* iii. 270.

After

After croſſing the *Icy Sea* from the moſt extreme part of the coaſt of *America* which he could attain, he fell in with land. It appeared low near the ſea, and high inland; and between both lay a great lake. To a ſteep and rocky point, nearly in lat. 68. 56, and long. 180. 51, his *ne plus ultra* on the *Aſiatic* ſide, he gave the name of *Cape North*; beyond which he could not ſee any land, notwith-ſtanding the weather was pretty clear. The ſea, at three miles diſtance from the ſhore, was only eight fathoms deep: this, with a riſing wind, approaching fog, and apprehenſion of the coming down of the ice, obliging him to deſiſt from farther attempts in theſe parts, he proceeded as near to the coaſt as he could with prudence, towards the ſouth-eaſt, and found it retain the ſame appearance. In lat. 67. 45, he diſcovered a ſmall iſle, about three leagues from the main, with ſteep and rocky ſhores, on which he beſtowed the name of *Burney*, in honor of one of his officers; gratefully immortalizing the companions of his voyage, in this and other inſtances. After paſſing the iſland, the continent inland roſe into mountains of conſiderable height, the termination of the great chain I before deſcribed.

In lat. 67. 3, long. 188. 11, he fell in with *Serdze Kamen* *, a lofty promontory, faced towards the ſea with a ſteep rocky cliff. To the eaſtward the coaſt continues high and bold, towards the *North Cape* low, being a continuation of the *Arctic* flats. This was the northern limit of the voyage of another illuſtrious navigator, Captain VITUS BERING, a *Dane* by birth, and employed on the ſame plan of diſcovery in theſe parts as our great countryman was in the late voyage. He was in the ſervice of PETER the GREAT; who, by the ſtrength of an extenſive genius, conceiving an opinion of the vicinity of *America* to his *Aſiatic* dominions, laid down a plan of diſcovery worthy of ſo extraordinary a monarch, but died before the attempt was begun; but his ſpirit ſurvived in his ſucceſſor. BERING, after a tedious and fatiguing journey through the wilds of *Sibiria*, arrived in *Kamt-ſchatka*, attended with the ſcanty materials for his voyage, the greateſt part of which he was obliged to bring with him through a thouſand difficulties. Several of the circumſtances of his adventures will be occaſionally mentioned †. I ſhall only ſay here, that he ſailed from the river of *Kamtſchatka* on *July* 15th, 1728; on the 15th of *Auguſt* ſaw *Serdze Kamen*, or the heart-ſhaped rock, a name be-ſtowed on it by the firſt diſcoverer.

From *Serdze Kamen* to a promontory named by Captain COOK *Eaſt Cape* ‡, the land trends ſouth-eaſt. The laſt is a circular peninſula of high cliffs, projecting

CAPE NORTH.

BURNEY'S ISLE.

SERDZE KAMEN.

CAPT. BERING.

EAST CAPE.

* See tab. 84 of the *Voyage*.   † The account of the voyage is extremely worthy of peruſal, and is preſerved by the able Doctor *Campbell*, in *Harris's Collection*, ii. 1018.   ‡ See tab. 84 of the *Voyage*.

far into the fea due eaft, and joined to the land by a long and very narrow ifthmus, in lat. 66. 6. This is the *Tfchutfki Nofs* of our navigators, and forms the beginning of the narrow ftreights or divifion of the old and new world. The diftance between *Afia* and *America* in this place is only thirteen leagues. The country about the cape, and to the north-weft of it, was inhabited. About mid-channel are two fmall iflands, named by the *Ruffians* the ifles of St. *Diomedes*; neither of them above three or four leagues in circuit †. It is extremely extraordinary that BERING fhould have failed through this confined paffage, and yet that the object of his miffion fhould have efcaped him. His misfortune could only be attributed to the foggy weather, which he muft have met with in a region notorious for mifts ‡; for he fays that he faw land neither to the north nor to the eaft §. Our generous commander, determined to give him every honor his merit could clame, has dig-nified thefe with the name of BERING'S STREIGHTS.

The depth of thefe ftreights is from twelve to twenty-nine or thirty fathoms. The greateft depth is in the middle, which has a flimy bottom; the fhalloweft parts are near each fhore, which confifts of fand mixed with bones and fhells. The current or tide very inconfiderable, and what there was came from the weft.

From *Eaft Cape* the land trends fouth by weft. In lat. 65. 36, is the bay in which Captain COOK had the interview with the *Tfchutfki*. Immediately beyond is the bay of St. *Laurence*, about five leagues broad in the entrance, and four deep, bounded at the bottom by high land. A little beyond is a large bay, either bounded by low land at the bottom, or fo extenfive as to have the end invifible. To the fouth of this are two other bays; and in lat. 64. 13, long. 186. 36, is the extreme fouthern point of the land of the *Tfchutfki*. This formerly was called the *Anadirfkoi Nofs*. Near it BERING had converfation with eight men, who came off to him in a *baidar*, or boat covered with the fkins of feals; from which BERING and others have named it the *Tfchutfki Nofs*. A few leagues to the fouth-eaft of this point lies *Clerke*'s ifland, in lat 63. 15, difcovered by Capt. COOK; and immediately beyond a larger, on which BERING beftowed the name of St. *Laurence*: the laft, the refort of the *Tfchutfki* in their fifhing parties ‖. Both of thefe confift of high cliffs, joined by low land. A fmall ifland was feen about nineteen leagues from St. *Laurence*'s, in a north-eaft by eaft half eaft direction; I fufpect it to be that which Capt. COOK named *Anderfon*'s, in memory of his furgeon, who died off it, and from his amiable character feems to have well

---

* See the chart of them, *Voyage*, vol. ii. tab. 53.   † *Voy.* ii. 445. iii. 243.   ‡ *Voyage* ii. 470. and Meteorolog. Tables, iii. App. 512, 513, 520, 521.   § *Harris's Coll.* ii. 1020. ‖ *Muller's Voy. des Ruffes*, i. 148.

    merited

merited this memorial. It lies in lat. 63. 4, long. 192. An anonymous iflet, imperfectly feen, and lying in lat. 64. 24, long. 190. 31, in mid-channel, completes the fum of thofe feen remote from land between the ftreights and the ifle of St. *Laurence*. As to thofe named in the chart given by Lieut. *Synd*, who in 1764 made a voyage from *Kamtfchatka* towards BERING's Streights, they feem to exift only in imagination, notwithftanding the *Ruffian* calendar has been exhaufted to find names for them. St. *Agathon*, St. *Titus*, St. *Myron*, and many others, fill the fpace paffed over by Capt. COOK, and which could not have efcaped the notice of his fucceffor *.

The land from BERING's *Tfchutfki Nofs* trends vaftly to the weft, and bounds on that fide the vaft gulph of *Anadir*, into the bottom of which the river of the fame name empties itfelf; and limits the territory of the *Tfchutfki*.

From thence is a large extent of coaft trending fouth-weft from Cape St. *Thaddeus*, in lat. 62. 50, long. 180, the fouthern boundary of the gulph of *Anadir*, to *Oljutorfkoi Nofs*, beyond which the land retires full weft, and forms in its bofom a gulph of the fame name. Off *Thaddeus Nofs* appeared, on *June* 29th, abundance of walrufes and great feals; and even the wandering *albatrofs* was feen in this high latitude †. Between this and the *Penginfk* gulph, at the end of the fea of *Ochotfk*, is the *ifthmus* which unites the famous peninfula of *Kamtfchatka* to the main land, and is here about a hundred and twenty miles broad, and extends in length from 52 to 61, north lat. The coafts are often low: often faced with cliffs, in many parts of an extraordinary height; and out at fea are rude and fpiring rocks, the haunts of leonine feals, whofe dreadful roarings are frequently the prefervation of mariners, warning them of the danger, in the thick fogs of this climate ‡. The coaft has but few harbours, notwithftanding it juts frequently into great headlands. The moft remarkable are, the *North Head*, with its needle rocks, at the entrance of the bay of *Awatcha* (*Voyage*, vol. iii. tab. 58); *Cheepoonfkoi Nofs*, ftill further north, engraven in vol. ii. tab. 84; and *Kronotfkoi Nofs*, with its lofty cliffs. The peninfula widens greatly in the middle, and leffens almoft to a point at Cape *Lopatka*, which flopes into a low flat, and forms the fouthern extremity of the country. The whole is divided lengthways by a chain of lofty rocky mountains, frequently covered with fnow, and fhooting into conic fummits, often fmoking with vulcanic eruptions. They have broken out in numbers of places: the extinct are marked by the craters, or their broken tops. The vulcano near *Awatcha* §, that of *Tolbatchick*, and that of the mountain of *Kamtfchatka* ‖, are the modern. They burft out fometimes in whirlwinds of flames,

VULCANOS.

---

* *Coxe's Ruffian Difcovery* Map, p. 300.—*Voy*. iii. 503. † *Voyage* iii. 241. ‡ *Defcr. Kamtfch.* 429. § See tab. 85, *Voyage*, vol. iii.; and defcription of its eruption, p. 235. ‖ See *Defcr. Kamtfchatka*, tab. xv. p. 342.

and

and burn up the neighboring forests : clouds of smoke succeed, and darken the whole atmosphere, till dispersed by showers of cinders and ashes, which cover the country for thirty miles round. Earthquakes, thunder, and lightning, join to fill the horror of the scenery at land ; while at sea the waves rise to an uncommon height, and often divide so as to shew the very bottom of the great deep *. By an event of this kind was once exposed to sight the chain of submarine mountains which connected the *Kuril* isles to the end of this great peninsula. I do not learn that they overflow with lava or with water, like the vulcanos of *Europe*. There are in various parts of the country hot springs, not inferior in warmth to those of *Iceland* † : like them they in some places form small *jets d'eaux*, with a great noise, but seldom exceed the height of a foot and a half‡.

HOT SPRINGS.

The climate during winter is uncommonly severe; for so low as *Bolcheretsk*, lat. 52, 30, all intercourse between neighbors is stopped. They dare not stir out for fear of being frost-bitten. Snow lies on the ground from six to eight feet thick as late as *May*; and the storms rage with uncommon impetuosity, owing to the subterraneous fires, the sulphureous exhalations, and general vulcanic dispo-sition of the country. The prevaling winds are from the west, which passing over the frozen wilds of *Sibiria* and *Tartary*, add keenness and rigour to the winters of *Kamtschatka*. Winter continues till the middle of *June :* from that month to the middle of *September* may be called summer, if a season filled with rain, and mists, and ungenial skies, merits that name. Rye, barley, and oats, are committed to the earth, but seldom come to perfection. The subsistence of the *Russians* and *Cossacks* depends therefore on importation from *Sibiria*. In some parts grass grows to a great height, and hay of uncommon nutriment is harvested for the fattening of cattle §. Grain is a luxury for the colonists only : the natives have other resources, the effects of necessity. Excepting in few places, this is a land of in-corrigible barrenness. As soon as the sea otters and other pretious furs are ex-hausted, *Kamtschatka* will be deserted by the *Russians*, unless they should think fit to colonize the continent of *America*, which the furs of that country, or the prospect of mineral wealth, may induce them to attempt.

CLIMATE.

Few ores have as yet been discovered in this peninsula : not that it wants either copper or iron ; but every necessary in those metals is imported at so cheap a rate, that it is not worth while for a people ignorant in mining and smelting to search for them in the almost inaccessible mountains.

ORES.

From the climate and the barren nature of *Kamtschatka*, the reader need not be

PLANTS.

---

* *Descr. Kamtsch. Fr.* 340, 341.   † *Voyage* iii. 206, 332.   ‡ *Descr. Kamtsch. Fr.* 348, and tab. iv. v. in which are given the course of the warm streams.   § *Voy.* iii. 327.

surprized

furprized at the poverty of its *Flora*. It muft not be fuppofed that the fcanty enu-
meration of its plants arifes from a neglect of fearch, or the want of a botanift to
explore its vegetable kingdom. STELLER, a firft-rate naturalift of *Germany*, who
attended BERING in his laft voyage, refided here a confiderable time after his
efcape from that unfortunate expedition, exprefsly to complete his remarks in
natural hiftory. The refult of his botanical refearches was communicated to
Doctor *Gmelin*, another gentleman fent by the *Ruffian* government to examine into
the natural hiftory of its dominions. *Europe* has from time to time been ranfacked
for men of abilities to perform this meritorious miffion, and the fruits of their
labors have been liberally communicated to a public thirfting for knowlege.
The names of MULLER, GMELIN, STELLER, DE L'ISLE, KRASHANINICOFF,
GUILDENSTAEDT, LEPECHIN, and PALLAS, will ever be held in refpect, for adding
to the ftock of natural knowlege. But how much is it to be lamented that *England*
wants a patron to encourage the tranflation of their works, locked up at prefent
in *Ruffian* or *German*, concealed from the generality of readers, to the great fup-
preffion of knowlege!

I here give a lift of the plants of *Kamtfchatka* in fyftematic order; and from it
annex an account of the ufes made of them by the natives of the peninfula. I
muft not omit my thanks to the Rev. Mr. *Lightfoot*, and the Rev. Mr. *Hugh
Davies* of *Beaumaris*, for the great affiftance I received from them. Let me
premife, that the plants marked *A.* are common to *America* and *Kamtfchatka*;
with *B.* to BERING's Ifle; with *E.* to *England* or *Scotland*; and with *Virg.* thofe
which extend to *Virginia*, or the eaftern fide of *North America* *. It is remarkable,
that the *European* plants, which had deferted *Sibiria* about the *Jenefei*, appear here
in great abundance.

Veronica. *Gmel. Sib.* iii. 219. N° 33.	Sanguiforba canadenfis. *A.*
V. incana.	Cornus fuecica.
V. ferpyllifolia. *E.*	Pulmonaria virginica. *A. Am. Acad.* ii.
Iris fibirica.	310.
Iris. *Gm. Sib.* i. 30. N° 28.	Cerinthe major. *A.*
Dactylis. *Gm. Sib.* i. 130. N° 68.	Cortufa Gmelini. *Am. Acad.* ii. 313.
Bromus criftatus. *Amœn. Acad.* ii. 312.	Anagallis. *Gm. Sib.* iv. 87, 37.
Triticum. *Gm. Sib.* i. 119. N° 56.	Azalea procumbens. *E.*
Plantago major. *A. E. Virg.*	Phlox fibirica. *Am. Acad.* ii. 314.
Pl. afiatica.	Convolvulus perficus. *Ibid.*

* Taken from Doctor *Forfter's* FLORA AMERICÆ SEPTENTRIONALIS. It is highly probable that
many, not noted as fuch, may be common to both fides of the continent, notwithftanding they efcaped
the notice of *Steller* or our navigators.

Polemonium

Polemonium cæruleum. *A. E.*

Lonicera Xylosteum. *A ?*

L. cærulea.

Ribes alpinum. *A. E.*

R. rubrum. *Virg.*

R. grossularia. *A. Virg.*

Claytonia virginica. *A. Am. Acad.* ii. 310.

Salsola prostrata - - 318.

Anabasis aphylla. - - 319.

Heuchera americana. - - 310.

Swertsia dichotoma. - - 317.

Sw. corniculata. - - *ibid.*

Gentiana amarella. *E.*

G. aquatica. *Am. Acad.* ii. 316.

Heracleum panaces. *A.*

Angelica archangelica.

Ang. Sylvestris. *E. Virg.*

Cicuta virosa.

Chærophyllum Sylvestre.

Chær. aureum ?

Sambucus racemosa.

Tradescantia. *Virg ?*

Allium ursinum. *E. Virg.*

Allium triquetrum.

Lilium martegon.

L. Camschatcense. *A. Virg. Am. Acad.* ii. 320.

Uvularia perfoliata. - - 310.

Convallaria bifolia.

Juncus filiformis. *E. Virg.*

J. campestris. *E.*

Rumex acetosa. *Virg.*

Melanthium sibiricum. *Am. Acad.* ii. 320.

Trillium erectum. - - ii. 310.

Alisma plantago aquatica. *E.*

Alsinanthemos. *Gm. Sib.* iv. 116. N° 86.

Epilobium latifolium.

Vaccinium myrtillus. *A. E.*

Vaccinium uliginosum. *E.*

Vac. vitis idæa. *A. E.*

Vac. oxycoccos. *E. Virg.*

Erica. *Gm. Sib.* iv. 130. N° 21. *B.*

Er. *Gm. Sib.* iv. 131. N° 22. *A.*

Bryanthus. *Gm. Sib.* iv. 133. N° 23.

Polygonum bistorta. *E.*

Pol. viviparum. *E.*

Adoxa moschatellina. *A. E.*

Sophora Lupinoides. *Am. Acad.* ii. 321.

Ledum palustre.

Andromeda. *Gm. Sib.* iv. 121. N° 9.

Chamærhododendros. *Gm. Sib.* iv. 126. N° 13. *B.*

Arbutus uva ursi. *E. Virg.*

Pyrola rotundifolia. *E. Virg.*

Tiarella trifoliata. *Am. Acad.* ii. 322.

Sedum verticillatum. ii. 323.

Prunus padus. *E.*

Sorbus aucuparia. *E. Virg.*

Cratægus oxyacantha. *Voyage,* iii. 334.

Spiræa hypericifolia. *Am. Acad.* ii. 310.

Sp. Sorbifolia. - - 324.

Spiræa. *Gm. Sib.* iii. 192. N° 55.

Spiræa. - - 192. N° 56.

Sp. aruncus.

Rosa alpina.

Rubus Idæus. *A. E. Virg.*

R. Cæsius. *E.*

R. fruticosus. *E. Virg.*

R. arcticus. *Virg.*

R. chamæmorus. *E.*

Fragaria vesca. *A. E.*

Potentilla fruticosa. *E.*

Dryas pentapetala.

Actæa cimicifuga. *Am. Acad.* ii. 325.

Papaver nudicaule.

Aconitum napellus.

Anemone

Anemone narciffifolia.

Anem. ranunculoides.

Anem. Dichotoma. *Am. Acad.* ii. 310.

Thaliſtrum flavum. *E.*

Ranunculus.

Troillius europeus. *E.*

Helleborus trifolius. *Am. Acad.* ii. 327.

Bartfia pallida. - - - *ibid.*

Pedicularis verticillata.

Linnæa borealis. *Virg.*

Myagrum fativum. *E.*

Thlafpi burfa paſtoris. *E. Virg.*

Arabis grandiflora.

Turritis hirfuta. *E.*

Geranium pratenfe. *E.*

Lathyrus. *Gm. Sib.* iv. 85.

Aftragalus alopecuroides. *Am. Acad.* ii. 330.

Aftr. alpinus.

Aftr. *Gm. Sib.* iv. 44. N° 58.

Aftr. phyfodes. *Am. Acad.* ii. 329.

Hypericum. *Gm. Sib.* iv. 279. N° 3.

Picris hieraciodes. *E.*

Sonchus. *Gm. Sib.* ii. 13. N° 13.

Prenanthes repens. *Am. Acad.* ii. 331.

Serratula noveboracenfis. *Virg.*

Circium. *Gm. Sib.* ii. 69. N° 49.

Cacalia fuaveolens. *Am. Acad.* ii. 310.

Artemifia vulgaris. *A. E.*

Gnaphalium margaritaceum. *E. Virg.*

Erigeron acre. *A. E.*

Tuffilago. *B. Gm. Sib.* ii. 145. N° 125.

Senecio. *B.* - - 136. N° 118.

After. *A. B. Gm. Sib.* ii. 175. N° 145.

After. - - 186. N° 152.

Solidago virga aurea. *A. B E.*

Solidago. *Gm. Sib.* ii. 170. N° 190.

Cineraria fibirica.

Pyrethrum. *A. B. Gm. Sib.* ii. 203. N° 170.

Orchis bifolia. *E. Virg.*

Orchis latifolia. *E.*

Ophrys Camtfcatca. *Am. Acad.* ii. 332.

Drachontium Camtfcatcenfe. *Am. Acad.* ii. 332.

Carex panicea. *E. Virg.*

Carex. *Gm. Sib.* i. 139. N° 77.

Betula alba. *E.*

Betula nana. *E. Virg.*

Betula alnus. *A. E. Virg.*

Urtica dioica. *E.*

Sagittaria latifolia. *E.*

Pinus cembra.

Pinus Larix. *A. Virg.*

Pinus picea.

Salix retufa.

Salix viminalis. *E.*

Empetrum nigrum *A. E. Virg.*

Populus alba. *E.*

Juniperus communis. *E.*

Equifetum hyemale. *E. Virg.*

Afplenium Rhyzophyllum. *Am. Acad.* ii. 311. *Virg.*

Lycopodium rupeſtre. *Virg.* *ibid.*

Lycop. Sanguinolentum. ii. 333.

Uses.   The *Kamtſchatkans* boaſt of their ſkill in the knowlege of the application of the vegetable kingdom to the ufes of mankind. The *Sibirians* cure the venereal difeafe by a decoſtion of the root of the *Iris Sibirica*, which aſts by purging and vomiting. They keep the patient eight days in a ſtove, and place him in a bed of the leaves

of

of the *Arctium Lappa*, or common Burdock, which they frequently change till the cure is effected.

The *Heracleum Panaces*, or *Sweet grafs*, was a plant of the firft ufe with the *Kamtf-chatkans*, and formerly made a principal ingredient in all their difhes ; but fo powerful does the love of hot liquors fway with the *Ruffians*, that, fince their arrival, it is entirely applied to diftillation. The beginning of *July* the more fucculent ftalks and leaves are gathered ; after the down is fcraped off with fhells, they are layed to ferment ; when they grow dry, they are placed in bags, and in a few days are covered with a faccharine powder : only a quarter of a pound of powder is collected from a pood, or thirty-fix pounds of the plant, which taftes like liquorice. They draw the fpirit from it by fteeping bundles of it in hot water ; then promote the fermentation in a fmall veffel, by adding the berries of the *Lonicera Xylofteum*, Sp. Pl. i. 248, and *Vaccinium uliginofum*, 499. They continue the procefs by pouring on more water, after drawing off the firft: they then place the plants and liquor in a copper ftill, and draw off, in the common manner, a fpirit equal in ftrength to brandy *. Accident difcovered this liquor. One year, the natives happening to collect a greater quantity of berries of feveral kinds, for winter provifion, than ufual, found in the fpring that a great quantity had fermented, and become ufelefs as a food. They refolved to try them as a drink, and mixed the juice with water. Others determined to experience it pure ; and found, on trial, the *Arctic* beatitude, drunkenefs †. The *Ruffians* caught at the hint, introduced diftillation, and thus are enabled to enjoy ebriety with the production of the country.

The *Moucho-more* of the *Ruffians*, the *Agaricus mufcarius*, Sp. Pl. 1640, is another inftrument of intoxication. It is a fpecies of Toadftool, which the *Kamtf-chadales* and *Koriaks* fometimes eat dry, fometimes immerfed in a fermented liquor made with the *Epilobium*, which they drink notwithftanding the dreadful effects. They are firft feized with convulfions in all their limbs, then with a raving fuch as attends a burning fever ; a thoufand phantoms, gay or gloomy (according to their conftitutions) prefent themfelves to their imaginations : fome dance ; others are feized with unfpeakable horrors. They perfonify this mufhroom ; and, if its effects urge them to fuicide, or any dreadful crime, they fay they obey its commands. To fit themfelves for premeditated affaffinations, they take the *Moucho-more*. Such is the fafcination of drunkenefs in this country, that nothing can induce the natives to forbear this dreadful potion ‡ !

---

* *Voyage*, iii. 337. † *Gmelin, Fl. Sib.* i. 217. ‡ *Hift. Kamtfchatka*, 99, 100.

As a food, the *Saranne*, or *Lilium Kamtfchatcenfe*, is among the principal. Its roots are gathered by the women in *Auguft*, dried in the fun, and layed up for ufe : they are the beft bread of the country ; and after being baked are reduced to powder, and. ferve inftead of flour in foups and feveral difhes. They are fometimes wafhed, and eaten as potatoes ; are extremely nourifhing, and have a pleafant bitter tafte. Our navigators boiled and eat them with their meat. The natives often parboil, and beat it up with feveral forts of berries, fo as to form of it a very agreeable confection. Providentially it is an univerfal plant here, and all the grounds bloom with its flower during the feafon *. Another happinefs re-marked here is, that while fifh are fcarce, the *Saranne* is plentiful ; and when there is a dearth of this, the rivers pour in their provifions in redoubled profufion. It is not to the labors of the females alone that the *Kamtfchatkans* are indebted for thefe roots. The *œconomic Moufe*, p. 134. A. faves them a great deal of trouble. The *Saranne* forms part of the winter provifions of that little animal: they not only gather them in the proper feafon, and lay them up in their ma-gazines, but at times have the inftinct of bringing them out, in funny weather, to dry them, leaft they fhould decay †. The natives fearch for their hoards ; but with prudent tendernefs leave part for the owners, being unwilling to fuffer fuch ufeful caterers to perifh.

Let me add, that STELLER enumerates other fpecies of the Lilly genus, which I believe are edible. Every fpecies of fruit, except berries, is denied to this un-kind climate ; but the inhabitants ufe various forts of them as wholefome fubfti-tutes, which they eat frefh, or make into palatable jams, or drefs with their fifh, either frefh or when preferved for winter ufe : fuch are thofe of the *Lonicera Xylo-fteum* or *Gimoloft*, a fort of Honeyfuckle : the *Rubus Chamæmorus*, *Morochka*, or Cloudberries : the *Vaccinium Myrtillus*, *Uliginofum*, *Vitis Idæa*, and *Oxycoccos*, or Bilberries, Marfh Bilberries, Red Bilberries, and Cranberries : the *Empetrum Nigrum*, or Heathberries : the *Prunus Padus*, or Bird Cherry : *Cratægus Oxyacan-tha*, or White Thorn with red and with black berries : the *Juniperus Communis*, or Common Juniper : and finally, of thofe of the *Sorbus Aucuparia*, or Common Service.

Of the *Epilobium Latifolium*, Sp. Pl. 494, or *Kipri*, is brewed a common beve-rage ; and, with the affiftance of the Sweet Plant, is made an excellent vinegar : the leaves are ufed as a tea, and the pith is mixed with many of the difhes, and ferved up green as a defert. When the infufion of it is mixed with the Sweet Herb in the diftillation, much more brandy is procured than if water alone is ufed ‡.

---

* *Defc. Kamtfch.* 363. † PALLAS, *Nov. Sp. Mur.* 230. ‡ *Defc. Kamtfch.* 368.

The

The *Polygonum Biſtorta*, Snake-weed, or *Jikoum*, is eaten freſh or dried, and often pounded with the *Caviar*. The *Chærophyllum Sylveſtre*, Wild Chervil, or Cow-weed, the *Morkavai* of the natives, is eaten green in the ſpring, or made into four krout. The *Solidago Itſchitſchu*, Fl. Sib. ii. 170, is dried and boiled with fiſh; and the broth from it taſtes as if the fleſh of the *Argali* or wild ſheep had been ſeethed in it. The root of *Kotkonnia*, a ſpecies of *Tradeſcantia*, is eaten either freſh, or uſed with the roes of fiſh: the berries have an agreable acidity, like an unripe apple, but will not keep, therefore they muſt be eaten as ſoon as they are gathered. *Allium Urſinum*, *Tcheremcha*, our Wild Garlic, is very common, and uſeful in medicine as well as food; both *Ruſſians* and natives gather it in great quantities for winter ſervice: they ſteep it in water, then mix it with cabbage, onions, and other ingredients, and form out of them a ragout, which they eat cold. It is alſo the principal remedy for the ſcurvy. As ſoon as this plant appears above the ſnow, they ſeem to put this dreadful diſorder at defiance, and find a cure almoſt in its worſt ſtages. The *Potentilla fruticoſa*, Sp. Pl. i. 709, or Shrubby Cinquefoil, is very efficacious in the dyſentery, or in freſh wounds. The *Dryas pentapetala*, Sp. Pl. i. 717, or *Ichagban*, is employed in ſwellings or pains of the limbs. That dreadful poiſon the *Cicuta viroſa*, Sp. Pl. i. 366, Water Hemlock, the *Omeg*, is applied to uſe, by the bold practitioners of this country, in caſes of pains in the back. They ſweat the patient profuſely, and then rub his back with the plant, avoiding to touch the loins, which, they ſay, would bring on immediate death.

The trees of uſe are a dwarf ſpecies of *Pinus Cembra*, or Pine with edible kernels; it grows in great quantities on both the mountains and plains, covered with moſs. It never grows upright, but creeps on the ground, and is therefore called by the *Ruſſians*, *Slanetz*. The natives eat the kernels, with even the cones, which brings on a teneſmus; but the chief uſe of the tree is as a ſovereign medicine in the ſcurvy. BERING taught the *Kamtſchatkans* to make a decoction of it: but they have neglected his inſtructions, notwithſtanding they ſaw numbers of his people reſtored to health in a ſhort time, and ſnatched, as it were, from the jaws of death *. Even at this time the *Ruſſian* coloniſts periſh miſerably with the diſorder, notwithſtanding the remedy is before their eyes.

TREES.

The *Pinus Larix*, or Larch-tree, grows only on the river of *Kamtſchatka*, and the ſtreams which run into it. This tree is of the firſt uſe in the mechanical ſervices of the country: with it they build their houſes, their fortifications, and boats. They make uſe of the *Populus alba*, or White Poplar, for the ſame pur-

* *Voyage*, iii. 332.—*Gm. Fl. Sib.* i. 181.——Reſpecting the trees, conſult *Voyage*, iii. 332. *Deſc. Kamtſchatka*, 359, and the preceding catalogue.

poſes.

poſes. Of the *Betula alba*, or Common Birch, a tree ſo uſeful to theſe northern nations, they make their ſledges and canoes; and cut the freſh bark into ſmall ſlices like vermicelli, and eat it with their dried caviar : they alſo tap the trees, and drink the liquor without any preparation. With the bark of the alder they dye their leather ; but that, and every tree they have near the coaſt, is ſtunted, ſo that they are obliged to go far inland for timber of proper ſize.

I muſt add, as a vegetable of uſe in œconomics, the *Triticum*, Gm. Sib. i. 119, Nº 56, which grows in great quantities along the ſhores, which they mow, and work into mats, which ſerve for bed clothes and curtains; into mantles, ſmooth on one ſide, and with a pile on the other, which is water-proof. They alſo make with it ſacks, and very elegant baſkets ; theſe, as well as the mats, they ornament with ſplit whale-bones, and work into variety of figures *. The *Urtica dioica*, or Common Nettle, is another plant of great uſe : this they pluck in *Auguſt* or *September*, tie in bundles, and dry on their huts : they tear it to pieces, beat, and clean it ; then ſpin it between their hands, and twiſt the thread round a ſpindle. It is the only material they have to make their nets ; which, for want of ſkill in the preparation, will rot, and laſt no longer than one ſeaſon †.

QUADRUPEDS.   In reſpect to the quadrupeds of this country, I have reaſon to think, from the great aſſiſtance I have received from the *Ruſſian* academiſts, or their labors, that my account of them, in my zoological part of this Work, can receive little addition. I requeſt that the *Brown Bear*, Nº 20, may be ſubſtituted inſtead of the Black, Nº 19, as the native of *Kamtſchatka*. I was led into the miſtake by the ſuſpicions of a moſt able naturaliſt. I am ſince informed, by the beſt authority (that of Captain KING ‡) that it is the brown ſpecies which is found there ; that they are carnivorous §, and prey at times on the *Argali* or wild ſheep ; but do not attack man, except urged by extreme hunger, or provoked by wounds, or by the ſlaughter of their young ; when nothing but their death can ſecure the ſafety of the perſons who fall in their way. In the firſt caſe, they will hunt mankind by the ſcent, and ſacrifice them to their want of food, which uſually is fiſh or berries.—The *Kamtſchatkans* never read *Pope*, but obſerve his advice :

Learn from the Beaſts the phyſic of the field.

The Bear is their great maſter ; and they owe all their knowlege in medicine and ſurgery, and the polite arts, to this animal. They obſerve the herbs to which he has recourſe when he is ill, or when he is wounded, and the ſame ſimples prove

---

* *Hiſt. Kamtſchatka*, 373.   † Same, 375.   ‡ See *Voy.* iii. 304 to 308, where Mr. *King* gives a full account of the preſent method of hunting.   § The reader is requeſted, at p. 58, l. 26, to change the word *carnivorous* into *animal*.

equally

equally reftorative to the two-legged Urfine race. The laft even acknowlege the Bear as their dancing-mafter, and are moft apt fcholars in mimicking his attitudes and graces *. I was informed by one of the gentlemen who was on the voyage, that the *Sea Otter*, N° 36, was feen on the firft arrival on the *American* coaft ; but, as it is not mentioned in that excellent and magnificent work till the arrival of the fhips in *Nootka* found, I will not infift on the accuracy of its latitude.

The *Argali* yields a difh of moft excellent flavor. The natives work the horns into fpoons, fmall cups, and platters ; and have frequently a fmall one hanging at their belts, by way of a drinking horn, in their hunting expeditions †.

ARGALI.

The Dogs are like the *Pomeranian*, but vaftly larger; the hair rather coarfer, and the ufual color light dun, or dirty creme-color. Bitches are never ufed for the draught, but dogs alone ; which are trained to it from their puppy-hood, by being tied with thongs to ftakes, with their food placed at a fmall diftance beyond their reach ; fo that by conftant laboring and ftraining, they acquire both ftrength of limb and habit of drawing ‡.

DOGS.

The leonine and urfine Seals, and the Manati, muft have been on their migrations during the time the navigators vifited this peninfula ; for they faw not one of thofe curious animals. The common Seals, being ftationary, were met with in great numbers. The bottle-nofed Seal, or Sea-Lion of Lord *Anfon*, is totally unknown in thefe feas. I refer the reader, for a view of the quadrupeds and birds of *Kamtfchatka*, to the catalogue which Captain KING honored with a place in the third volume of the Voyage §. I fhall only add, that the clafs of Auks is far the moft numerous of any, and contains fix fpecies unknown to *Europe*; that the only bird which has efcaped me is a fmall *Blue Petrel* ||, feen in numbers in about lat. 59. 48, off the northern part of the peninfula.

SEALS.

Kamtfchatka is deftitute of every fpecies of ferpent and frog. Lizards are very frequent, and are detefted by the natives, who believe them to be fpies fent by the infernal gods to examine their actions, and predict their deaths. If they catch one, they cut it into fmall pieces, to prevent it from giving any account of its miffion: if it efcapes out of their hands, they abandon themfelves to melancholy, and expect every moment their diffolution, which often happens through fear, and ferves to confirm the fuperftition of the country ¶. The air is very unfavorable

REPTILES.

INSECTS.

---

* *Voy.* iii. 308.   † Same, 344.   ‡ Same, 345.

§ By fome typographical miftake, the greater part of the *webbed-footed birds* are, in the firft edition, placed under the divifion of *cloven-footed*. The naturalift reader will eafily fee, that the birds, from CRANE, p. 357, to PIED OYSTER-CATCHER, ought to be placed in the divifion of *cloven-footed*; and from GREAT TERN, p. 356, to RED-FACED CORVORANT, p. 357, fhould be put after RED-THROAT-ED DIVER, p. 358, the *webbed footed*. || *Narrative*, ii. 246. ¶ *Defcr. Kamtfch.* Fr. 509.

to infects, except lice and fleas, which are in all their quarters; and, filthy to relate! are eaten by thefe beaftly people *. Bugs are acquifitions of late years, imported into the bay of *Awatcha*.

FISH.

The fifh of *Kamtfchatka* are with difficulty enumerated. There does not feem to be any great variety of genera; yet the individuals under each fpecies are found in moft aftonifhing abundance. Providence hath been peculiarly attentive to the natives of this peninfula, by furnifhing them in fo ample a manner, who for the greater part muft for ever be deprived of fupport derived from grain and cattle. The vegetables they have are fufficient to correct the putrefcent quality of the dried fifh, and often form an ingredient in the difhes; which are prepared different ways. The *Joukola* is made of the falmon kind, cut into fix pieces, and dried either in the open air or fmoked: the roes are another difh in high efteem with them, either dried in the air, or rolled in the leaves of different plants, and dried before the fire. They can live a long time on a fmall quantity of this food, and eat with it the bark of birch or willow trees, to affift them in fwallowing a food fo very vifcid; but their ambrofial repaft is the *Huigul*, or fifh flung into a pit till it is quite rotten, when it is ferved up in the ftate of carrion, and with a ftench unfupportable to every nofe but that of a *Kamtfchatkan* †.

WHALE.

The Fin Whale, *Br. Zool.* iii. N° 18, is very frequent, and is of fingular ufe to the inhabitants. They eat the flefh; preferve the fat for kitchen ufe and for their lamps; with the corneous laminæ they few the feams of their canoes, and make nets for the larger fort of fifh; they form the fliders of their fledges with the under jaw-bones, and likewife work them into knives; with the blade-bones, worked down to a fharp edge, they form fcythes, and moft fuccefsfully mow the grafs. The *Tfchutfki* verify the relation of *Pliny* ‡, and, like the *Gedrofi* of old, frame their dwellings with the ribs §; with the ligaments they make excellent fnares for different animals; with the inteftines dried, cleaned, and blown, they make bags for their greafe and oil; and with the fkins the foles of their fhoes, and ftraps and thongs for various purpofes. The *Tfchutfki* take thefe animals by harpooning; the *Oloutores*, in nets made of thongs cut out of the fkins of the *Walrus*; and the *Kamtfchatkans*, by fhooting them with darts or arrows, the points of which, having been anointed with the juice of the *Zgate*, a fpecies of *Anemone* and *Ranunculus* ‖, are fo noxious as to bring fpeedy death from the flighteft wound, like the celebrated poifon of the *Paragua Indians*. The vaft animals in queftion,

---

* *Defcr. Kamtfchatka*, Fr. 507.   † *Hift. Kamtfchatka*, Engl. 194. Fr. 46.   ‡ *Hift. Nat.* lib. ix. c. 3.   § *Voyage*, iii. 450.   ‖ I cannot difcover the fpecies. *Gmelin*, in his *Flora Sibirica*, does not give the leaft account of thefe plants.

when

when ſtruck with it, are infeſted with ſuch agonies that they cannot bear the ſea, but ruſh on ſhore, and expire with dreadful groans and bellowing.

The *Kaſatka* or Grampus, *Br. Zool.* iii. Nº 26, is very common in theſe ſeas : they are dreaded by the natives, who even make offerings to them, and entreat their mercy, leaſt they ſhould overſet their boats ; yet, if theſe fiſh are thrown on ſhore, they apply them to the ſame uſes as the Whale *.

The *Motkoïa* or *Akoul*, or White Shark, *Br. Zool.* iii. Nº 42, is among the uſeful fiſh. They eat the fleſh, and form of the inteſtines and bladder, bags to hold their oil. In the chaſe of this fiſh they never call it by its name, for fear of provoking it to burſt its bladder †.

Lampries, *Br. Zool.* iii. Nº 27 ; Eels, — 57 ; Wolf-fiſh, — 65 ; common Cod-fiſh ? — 73 ; Hadock, — 74 ; and Hake, — 81, are found in the *Kamtſchatkan* ſea : and I alſo ſuſpeſt, that the three-bearded Cod, — Nº 87, is alſo met with : it is called there *Morſkie Nalimi* ‡. An elegant ſpecies of Flounder, of excellent flavor, was taken here in abundance by our navigators : the back was ſtudded with prickly tubercles, and marked longitudinally with lines of black on a brown ground. The *Jerchei*, poſſibly our Ruffe, — Nº 127, is among the fiſh of the country ; as is a ſpecies of the *Engliſh* Sticklebacks.

But the fiſh of the firſt importance to the *Kamtſchatkans*, and on which they depend for ſubſiſtence, are the anadromous kinds, or thoſe which at ſtated ſeaſons aſcend the rivers and lakes out of the ſea. Theſe are entirely of the Salmon genus, with exception to the common Herring, which in autumn quits the ſalt water. It is ſayed, that every ſpecies of Salmon is found here. I may with certainty adjoin, that ſeveral of the *Sibirian* ſpecies, with variety peculiar to this country, aſcend the *Kamtſchatkan* rivers in multitudes incredible. The inhabitants dignify ſome of their months by the names of the fiſh. One is called *Kouiche*, or the month of *Red Fiſhes* ; another, *Ajaba*, or that of *Little White Fiſh* ; a third, *Kaiko*, or of the fiſh *Kaiko* ; and a fourth, *Kijou*, or the month of the *Great White Fiſh* §. It is obſervable, that each ſhoal keeps apart from others of different ſpecies, and frequently prefers a ſeparate river, notwithſtanding the mouths may be almoſt contiguous. They often come up in ſuch numbers as to force the water before them, and even to dam up the rivers, and make them overflow their banks ; inſomuch that, on the fall of the water, ſuch multitudes are left on dry ground, as to make a ſtench capable of cauſing a peſtilence, was it not fortunately diſperſed by the violence of the winds ; beſides, the bears and dogs aſſiſt, by preying on them, to leſſen the ill effeſts.

* *Deſcr. Kamtſch.* 462.   † *Same*, 466.   ‡ *Br. Zool.* iii. 261.   § *Hiſt. Kamtſch.* 218.

Every species of Salmon dies in the same river or lake in which it is born, and to which it returns to spawn. In the third year, male and female consort together, and the latter deposits its spawn in a hole formed with its tail and fins in the sand; after which both sexes pine away, and cease to live. A fish of a year's growth continues near the place, guards the spawn, and returns to the sea with the new-born fry in *November**. The Salmons of this country spawn but once in their lives: those of *Sibiria* and *Europe*, the rivers of which are deep, and abound with insect food, are enabled to continue the first great command of nature during the period of their existence. In *Kamtschatka* the rivers are chilly, shallow, rapid, full of rocks, and destitute of nourishment for such multitudes: such therefore which cannot force their way to the neighborhood of the tepid streams, or get back to the sea in time, universally perish; but Providence has given such resources, in the spawners, that no difference in numbers is ever observed between the returning seasons. It is singular, that neither the lakes or rivers have any species of fish but what come from the sea. All the lakes (for this country abounds with them) communicate with the sea; but their entrance, as well as that of many of the rivers, is entirely barred up with sand brought by the tempestuous winds, which confine the fish most part of the winter, till they are released by the storms taking another direction.

TSHAWYTSCHA.

The species which appears first is the *Tshawytscha*. This is by much the largest; it weighs sometimes between fifty and sixty pounds, and its depth is very great in proportion to the length. The jaws are equal, and never hooked: the teeth large, and in several rows: the scales are larger than those of the common Salmon; on the back dusky grey, on the sides silvery: the fins bluish white, and all parts unspotted: the tail is lunated: the flesh, during its residence in the sea, is red; but it becomes white in fresh waters. It is confined, on the eastern side of the peninsula, to the river of *Kamtschatka* and *Awatcha*; and on the western to the *Bolchaia-reka*, and a few others; nor is it ever seen beyond lat. 54. It enters the mouths of the rivers about the middle of *May*, with such impetuosity as to raise the water before it in waves. It goes in far less numbers than the other species; is infinitely more esteemed; and is not used as a common food, but reserved for great entertainments. The natives watch its arrival, which is announced by the rippling of the water; take it in strong nets; and always eat the first they take, under a notion that the omission would be a great crime.

† P. D. 12.
　P. 16.
　V. 10.
　A. 15.

* *Descr. Kamtsch.* 471.　　　　† Numbers of rays in the dorsal, pectoral, ventral, and anal fins.

The

The *Nærka* is another species, called by the *Ruffians, Krafnaya ryba*, from the intenfe purplifh rednefs of the flefh. It is of the form of the common Salmon ; but never exceeds fixteen pounds in weight. When it firft enters the rivers it is of a filvery brightnefs, with a bluifh back and fins : when it leaves the fea the teeth are fmall, and jaws ftrait ; but after it has been fome time in the frefh water, the jaws grow crooked (efpecially in the male) and the teeth large. It begins to afcend the rivers in vaft numbers in *June* ; penetrates to their very fources ; and returns in *September* to the fea, firft refting for fome time in the deep parts of the intervening lakes. It is taken in nets, either in the bays, as it approaches the rivers, or in the rivers, after it has quitted the fea *.

The *Kyfutch*, or *Bjelaya ryba*, or White Fifh of the *Ruffians*, afcends the rivers in *July*, particularly fuch as are difcharged from the inland lakes, and remain till *December*, when all the old fifh perifh, and the fry take to the fea. The upper jaw of the male, in its laft period, becomes crooked. This fpecies has the form of a common Salmon, but never attains three feet in length. It is of a filvery gloffy color, fpotted about the back ; but in the rivers acquires a reddifh caft : the jaws are long and blunt : the teeth large : the flefh is reddifh before it quits the fea ; but in the frefh water grows white. It is reckoned the moft excellent of the light-colored fifh.

The *Keta* or *Kayko*, in form and fize refembles the laft ; but the head is fhorter and more blunt : the tail is lunated : the flefh white : the color of the fcales a filvery white : the back greenifh ; and the whole free from fpots. It afcends the rivers in *July*, and the fifhery continues till *October*. This fpecies is found in great abundance ; and is fo common, that the *Joukola* made with it is called *houfhold bread*.

The *Gorbufcha*, or Hunch-back, arrives at the fame time with the laft. In form it refembles the Grayling : never exceeds a foot and a half in length : is of a filvery color, and unfpotted : the tail forked : the flefh white. After it has been fome time in the frefh water it changes its fhape (the male efpecially) in a moft furprizing manner. The jaws and teeth grow prodigioufly long, efpecially the upper, which at firft is fhorteft, but foon fhoots beyond the under, and grows crooked downwards ; the body becomes emaciated, and the meat bad : but what is moft characteriftic, an enormous bunch rifes juft before the firft dorfal fin, to which it owes its name. Its flefh is bad ; fo that this fifh falls to the fhare of the dogs.

* This fpecies is defcribed (*Voyage*, iii. 351) under the name of *Red Fifh* ; the preceding, in p. 350, under that of *Tchavitfi*.

r   The

MALMA.

P. D. 12.
P. 14.
V. 8.
A. 10.

The *Malma*, or *Golet* of the *Ruffians*, grows to the weight of twenty pounds, and to the length of about twenty-eight inches. It is the moft flender and cylindrical of all the genus. The head refembles that of a trout: the fcales are very fmall: the back and fides bluifh, with fcattered fpots of fcarlet red: the belly white: ventral and anal fins red: tail flightly forked. This and the two following are fporadic, going difperfedly, and not in fhoals. It afcends the rivers with the laft, and attains their very fources. It feeds on the fpawn of the other fpecies, and grows very fat. The natives falt thofe they take in autumn, and preferve frozen thofe which are caught when the frofts commence *.

MILKTSCHITSCH

P. D. 11.
P. 14.
V. 10.
A. 13.

The *Milktfchitfch* is a fcarce fpecies, in form like a young Salmon; but the fcales larger in proportion, and the body more flat: it never exceeds a foot and a half in length: is of a filvery white, with a bluifh back: nofe conical: jaws equal: tail flightly forked.

MYKISS.

P. D. 12.
P. 14.
V. 10.
A. 12.

The *Mykifs*, appears at firft very lean, but grows foon fat: it is very voracious: feeds not only on fifh, but infects and rats, while fwimming over the rivers; and is fo fond of the berries of *vaccinium vitis idæa*, that it will dart out of the water, and fnatch at both leaves and berries, which hang over the banks †. In fhape it refembles a common Salmon: feldom grows above two feet long: has large fcales, blunt nofe, and numerous teeth: the back is dufky, marked with black fpots; and on each fide is a broad band of bright red: the belly white. It is a fpecies of excellent flavor; but is fcarcer than the other kinds. Its time of arrival is not known: M. STELLER therefore fufpects that it afcends the rivers beneath the ice ‡.

KUNSHA.

The *Kunfha*, mentioned in page CIV, frequents the bays of this country, but never advances inland; and grows to the length of two feet: the nofe is fhort and pointed: the back and fides dufky, marked with great yellowifh fpots, fome round, others oblong: the belly white: the lower fins and tail blue: the flefh white, and excellent. It is a fcarce fifh in thefe parts; but near *Ochotfk* afcends the rivers in great fhoals.

I conclude this divifion of the tribe with the common Salmon, which is frequent here, and, like the others, afcends the rivers, equally to the advantage of the natives of the country.

INGHAGHITSH.

P. D. 8, 9
P. 12.
V. 10.
A. 12.

Of the Salmon which LINNÆUS diftinguifhed by the title of *Coregoni* is the *Inghaghitfh*, which has the habit of a fmall carp, with very large fcales: the jaws nearly of equal length: the eyes very great, and filvery: the teeth very minute: the body filvery, bluifh on the back: tail forked: it does not exceed five inches

---

* *Defcr. Kamtfch.* 482.   † Same, 482.   ‡ Same, 482.

in

in length. It arrives in spring and autumn, and in both seasons is full of spawn, and smells like a smelt.

The *Innyagha* is another small kind, about five inches long, and not unlike the *S. Albula* of LINNÆUS. It is a rare species, and found but in few rivers. P. D. 9. P. 11. V. 8. A. 16.

The most singular is the *Ouiki*, or *Salmo Catervarius* of STELLER. It belongs to the *Osmeri* of LINNÆUS. Swims in immense shoals on the eastern coast of *Kamtschatka*, and the new-discovered islands, where it is often thrown up by the sea to the height of some feet, upon a large extent of shore : is excessively unwholesome as a food, and causes fluxes even in dogs. It never exceeds seven inches in length. Just above the side-line is a rough fascia, beset with minute pyramidal scales, standing upright, so as to appear like the pile of shag : their use is most curious— while they are swimming, and even when they are flung on shore, two, three, or even as many as ten, will adhere as if glued together, by means of this pile, insomuch that if one is taken up, all the rest are taken up at the same time.

To conclude this list of *Kamtschatkan* Salmon, I must add the *Salmo Thymallus*, or Grayling ; the *S. Cylindraceus*, before described ; the *Salmo Albula*, Lin. Syst. 512 ; and the *Salmo Eperlanus*, or common Smelt, to those which ascend the rivers.— For this account I am indebted to Doctor PALLAS, who extracted it from the papers of STELLER, for the use of this Work.

The Herring, both the common and the variety, found in the gulph of *Bothnia*, called the *Membras*, and by the *Suedes*, *Stroeming*, Faun. Suec. p. 128, visit these coasts in shoals, perhaps equal to those of *Europe*. There are two seasons, the first about the end of *May*, the second in *October*. The first species are remarkably fine and large * ; they ascend the rivers, and enter the lakes : the autumnal migrants are closed up in them by the shifting of the sand at the mouths of the entrance, and remain confined the whole winter. The natives catch them in summer in nets ; and in winter in most amazing numbers, by breaking holes in the ice, into which they drop their nets, then cover the opening with mats, and leave a small hole for one of their companions to peep through, and observe the coming of the fish ; when they draw up their booty : and string part on packthread for drying ; and from the remainder they press an oil white as the butter of *Finland* †.

The sea, on which these people depend for their very existence, is finely adapted for the retreat and preservation of fish. It does not consist of a level uniform bottom, liable to be ruffled with storms, but of deep vallies and lofty

* *Voyage*, iii. 350.　　　† *Descr. Kamtsch.* 485.

r 2　　　　　mountains,

mountains, such as yield security and tranquillity to the finned inhabitants. We find the soundings to be most unequal: in some places only twenty-two fathoms, in others the lead has not found a bottom with a hundred and sixty fathoms of line. On such places the fish might rest undisturbed during the rage of the tempestuous winters. I do not find the least notice of shells being met with in these seas: either there are none, or they are pelagic, and escape the eyes of the navigators. But nature probably hath made ample provision for the inhabitants of the sea, in the quantity of sea-plants which it yields; STELLER, the great explorer of this region, enumerates the following, many of which are of uncommon elegance:

Fucus peucedanifolius, *Gm. Hift. Fucor.* 76	Fucus rofa marina	-	- 102
Fucus turbinatus - - - 97	Fucus crenatus	-	- 160
Fucus corymbiferus, *E.* - - 124	Fucus fimbriatus	-	- 200
Fucus dulcis, *E.* - - 189	Fucus anguftifolius	-	- 205
Fucus tamarifcifolius *, *E.*	Fucus agarum	-	- 210
Fucus bifidus - - - 201	Fucus quercus marina †		
Fucus polyphyllus - - 206	Fucus veficulofus, *Sp. Pl.* 1626, *E.*		
Fucus clathrus - - 211	Ulva glandiformis	-	- 232
Fucus myrica - - 88	Ulva Priapus -	-	- 231

Of these the *Quercus marina* is used as a remedy in the dysentery; and the females of *Kamtschatka* tinge their cheeks with an infusion of the *Fucus tamarifcifolius* in the oil of Seals.

TIDES.

In the harbours of Sts. *Peter* and *Paul* the greatest rise of the tides was five feet eight inches at full and change of the moon, at thirty-six minutes past four, and they were very regular every twelve hours ‡. The *Russian* philosophers observed here a singular phænomenon in the flux and reflux of the sea twice in the twenty-four hours, in which is one great flood and one small flood; the last of which is called *Manikha.* At certain times nothing but the water of the river is seen within its proper channel; at other times, in the time of ebb, the waters are observed to overflow their banks. In the *Manikha*, after an ebb of six hours, the water sinks about three feet, and the tide returns for three hours, but does not rise above a foot; a seven-hours ebb succeeds, which carries off the sea-water, and leaves the bay dry. Thus it happens three days before and

* *Hift. Kamtfchatka,* 43.   † Same, 124.   ‡ *Voyage,* iii. 323.

after

after the full moon ; after which the great tide diminifhes, and the *Manikha*, or little tide, increafes *.

The rivers of the country rife in the midft of the great chain of mountains, and flow on each fide into the feas of *Ochotfk*, or that of *Kamtfchatka*. They furnifh a ready paffage in boats or canoes (with the intervention of carrying-places) quite acrofs the peninfula. As has been mentioned, the waters yield no fifh of their own, but are the retreat of myriads of migrants from the neighboring feas.

NATIVES.

This peninfula, and the country to the weft, are inhabited by two nations ; the northern parts by the *Koriacs*, who are divided into the Rein-deer or wandering, and the fixed *Koriacs* ; and the fouthern part by the *Kamtfchatkans*, properly fo called : the firft lead an erratic life, in the tract bounded by the *Penfchinfka* fea to the fouth-eaft ; the river *Kowyma* to the weft ; and the river *Anadir* to the north †. They wander from place to place with their Reindeer, in fearch of the mofs, the food of thofe animals, their only wealth ‡. They are fqualid, cruel, and warlike, the terror of the fixed *Koriacs*, as much as the *Tfchutfki* are of them. They never frequent the fea, nor live on fifh. Their habitations are *jourts*, or places half funk in the earth : they never ufe *balagans*, or fummer-houfes elevated on pofts, like the *Kamtfchatkans* : are in their perfons lean, and very fhort : have fmall heads and black hair, which they fhave frequently : their faces are oval : nofe fhort : their eyes fmall : mouth large : beard black and pointed, but often eradicated.

KORIACS.
WANDERING.

The fixed *Koriacs* are likewife fhort, but rather taller than the others, and ftrongly made : they inhabit the north of the peninfula : the *Anadir* is alfo their boundary to the north ; the ocean to the eaft ; and the *Kamtfchatkans* to the fouth. They have few Rein-deer, which they ufe in their fledges ; but neither of the tribes of *Koriacs* are civilized enough to apply them to the purpofes of the dairy. Each fpeak a different dialect of the fame language ; but the fixed in moft things refemble the *Kamtfchatkans* ; and, like them, live almoft entirely on fifh. They are timid to a high degree, and behave to their wandering brethren with the utmoft fubmiffion ; who call them by a name which fignifies *their flaves*. Thefe poor people feem to have no alternative ; for, by reafon of the fcarcity of Rein-deer, they depend on thefe tyrants for the effential article of cloathing. I cannot trace the origin of thefe two nations ; but from the features may pronounce them offspring of *Tartars*, which have fpread to the eaft, and degenerated in fize and ftrength by the rigour of the climate, and often by fcarcity of food.

FIXED.

* *Defcr. Kamtfch.* 510.   † *Hift. Kamtfch.* 136.   ‡ See p. 25 of this Work.

The

KAMTSCHAT-KANS.

The true *Kamtfchatkans* * poffefs the country from the river *Ukoi* to the fouthern extremity, the cape *Lopatka*. They are fuppofed, by M. STELLER, to have been derived from the *Mongalian Chinefe*, not only from a fimilarity in the termination of many of their words, but in the refemblance of their perfons, which are fhort. Their complexion is fwarthy : their beard fmall : their hair black : face broad and flat : eyes fmall and funk : eye-brows thin : belly pendent : legs fmall—circumftances common to them and the *Mongalians*. It is conjectured, that in fome very remote age they fled hither, to efcape the yoke of the eaftern conquerors, notwithftanding they believe themfelves to be aboriginal, created and placed on the fpot by their god *Koutkou*.

RELIGION,

In refpect to their deity, they are perfect minute philofophers. They find fault with his difpenfations ; blafpheme and reproach him with having made too many mountains, precipices, breakers, fhoals, and cataracts ; with forming ftorms and rains ; and when they are defcending, in the winter, from their barren rocks, they load him with imprecations for the fatigue they undergo. In their morals they likewife bear a great fimilitude to numbers among the moft polifhed rank in the *European* nations—they think nothing vitious that may be accomplifhed without danger ; and give full loofe to every crime, provided it comes within the pale of fecurity.

GENII.

They have alfo their leffer deities, or genii. Each of them have their peculiar charge ; to thefe they pay confiderable veneration, and make offerings to them, to divert their anger or enfure their protection. The *Kamouli* prefide over the mountains, particularly the *vulcanic* ; the *Ouchakthou*, over the woods ; *Mitg*, over the fea ; *Gaetch*, over the fubterraneous world ; and *Fouila* is the author of earthquakes. They believe that the world is eternal ; that the foul is immortal; that in the world below it will be reunited to the body, and experience all the pains ufual in its former ftate ; but that it never will fuffer hunger, but have every thing in great abundance : that the rich will become poor, and the poor rich ; a fort of juft difpenfation, and balance of former good and evil †. But almoft all thefe fuperftitions are vanifhed by the attention of the *Ruffians* to their converfion. There are few who have not embraced the Chriftian religion. Churches have been built, and fchools erected, in which they are fuccefsfully taught the language of their conquerors, which has already almoft worn out that of the native people.

NUMBERS OF PEOPLE.

The country was very populous at the arrival of the *Ruffians* ; but, after a dreadful vifitation of the fmall-pox, which in 1767 fwept away twenty thoufand

---

* The moft proper word for the natives of this country is *Kamtfchadales* ; but as I have on many occafions ufed this, I wifh to continue it.   † *Hift. Kamtfch.* 68, 71.

                                             fouls,

fouls *, at prefent there are not above three thoufand who pay tribute, the inhabitants of the *Kuril* ifles included. Here are about four hundred of the military
Ruffians and *Coffacks*, befides a number of *Ruffian* traders and emigrants perpetually pouring in, who intermix with the natives † in marriage, and probably
in time will extinguifh the aboriginal race. The offspring is a great improvement ;
for it is remarked, that the breed is far more active than the pure *Ruffian* or
Coffack. Sunk in lordly indolence, they leave all the work to the *Kamtfchatkans*,
or to their women; and fuffer the penalty of their lazinefs, by the fcurvy in its
moft frightful forms.

DRESS.

The *Kamtfchatkans* feem to retain the antient form of their drefs; but during
fummer it is compofed of foreign materials; in the warm feafon both fexes ufe
nankeen, linen, and filk; in winter, the fkins of animals well dreffed : the
drefs of men and women refembles a carter's frock with long fleeves, furred at
the wrifts, the bottom, and about the neck. On their head is a hood of fur, fometimes of the fhaggy fkin of a dog, and often of the elegant fkin of the earlefs
Marmot. Troufers, boots, and furred mittens, compofe the reft. The habit of
ceremony of a *Toion* or chieftain is very magnificent, and will coft a hundred
and twenty rubels : in antient times it was hung over with the tails of animals,
and his furred hood flowed over each fhoulder, with the refpectability of a full-
bottomed perriwig in the days of *Charles* II. The figure given in the *Hiftory
of Kamtfchatka*, tranflated into *French*, exhibits a great man in all his pride or
drefs ‡ ; but fo rapidly has the prefent race of natives copied the *Ruffians*, that
poffibly in fo fhort a fpace as half a century, this habit, as well as numbers of other
articles and cuftoms, may be ranked among the antiquities of the country.

ARMS.

Bows and arrows are now quite difufed. Formerly they ufed bows made of
larch-wood, covered with the bark of the birch. The arrows were headed with
ftone or bone, and their lances with the fame materials. Their armour was
either mats, or formed of thongs cut out of the fkins of Seals, and fewed together, fo as to make a pliable cuirafs; which they fixed on their left fide; a board
defended their breaft, and a high one on their back defended both that and the
head.

HOSPITALITY.

Their favage and beaftly hofpitality is among the obfolete cuftoms. Formerly, as a mark of refpect to a gueft, the hoft fet before him as much food as
would ferve ten people. Both were ftripped naked : the hoft politely touched nothing, but compelled his friend to devour what was fet before him, till he was

* *Voyage*, iii. 366. † Same, 367. ‡ See *Hift. Kamtfchatka*, tab. vi.—It differs
much from the habit of ceremony defcribed by Captain KING, iii. 377.

quite

quite gorged ; and at the same time heated the place, by inceſſantly pouring water on hot ſtones, till it became unſupportable. When the gueſt was crammed up to the throat, the generous landlord, on his knees, ſtuffed into his mouth a great ſlice of whale's fat, cut off what hung out, and cried, in a ſurly tone, *Tana*, or *There!* by which he fully diſcharged his duty ; and, between heat and cramming, obliged the poor gueſt to cry for mercy, and a releaſe from the heat, and the danger of being choaked with the noble welcome : oftentimes he was obliged to purchaſe his diſ-miſſion with moſt coſtly preſents ; but was ſure to retaliate on the firſt oppor-tunity *.

DWELLINGS.

From the birds they learned the art of building their *balagans* or ſummer-houſes. They ſeem like neſts of a conic form, perched on high poles inſtead of trees; with a hole on one ſide, like that of the magpie, for the entrance. Their *jourts*, or winter reſidences, are copied from the *œconomic Mouſe*, p. 134; but with leſs art, and leſs cleanlineſs. It is partly ſunk under ground ; the ſides and top ſupported by beams, and wattled, and the whole covered with turf. In this they live gregariouſly, to the number of ſix families in each; in a ſtate in-tolerable to an *European*, by reaſon of ſmoke, heat, and ſtench, from their ſtore of dried or putrid fiſh, and from their lazineſs, in never going out to perform their offerings to *Cloacina* †.

Inſtigated by avarice, the *Ruſſians* made a conqueſt of this ſavage country; and found their account in it, from the great value of its furry productions. They have added to their dominions this extremity of *Aſia*, diſtant at leaſt four

ROADS TO KAMTS-CHATKA.

thouſand miles from their capital. The journey to it is ſtill attended with great difficulties, through wild and barren regions, over dreadful mountains ; and poſſibly impracticable, but for the multitude of *Sibirian* rivers, which, with ſhort intervals of land, facilitate the paſſage. Travellers uſually take their de-parture out of *Sibiria* from *Jakutz*, on the river *Lena*, in lat. 62 : they go either by water along the river, to its conflux with the *Aldun*, along the *Aldun* to the *Mai*, and from that river up the *Judoma* ; and from near the head of that river to *Ochotſk*, the port from whence they embark, and croſs the ſea of *Ochotſk* to *Bolſchaia-reka*, the port of the weſtern ſide of *Kamtſchatka*. The whole journey uſually takes up the ſhort ſummer : that over the hills to *Ochotſk* (and which is moſt convenient) was performed by STELLER in thirty-four days, excluding ſeven of reſt ‡.

KURIL ISLES.

The *Kuril* or *Kurilſki* iſles, which probably once lengthened the peninſula of *Kamtſchatka*, before they were convulſed from it, are a ſeries of iſlands running

---

* *Hiſt. Kamtſch.* 107 to 109.   † *Elliſ's Nar.* ii. 217.   ‡ *Deſcr. Kamtſch.* 602.

ſouth

south from the low promontory *Lopatka*, in lat. 51; between which and *Shoomſka*, the moſt northerly, is only the diſtance of one league. On the lofty *Paramouſer*, the ſecond in the chain, is a high-peaked mountain, probably vulcanic * : on the fourth, called *Araumakutan*, is another vulcano † ; on *Uruſs* is another ; on *Storgu* two ; and on *Kunatir*, or *Kaunachir*, one. Theſe three make part of the group which paſs under the name of the celebrated land of *Jeſo* ‡. *Japan* abounds with vulcanoes § ; ſo that there is a ſeries of ſpiracles from *Kamtſchatka* to *Japan*, the laſt great link of this extenſive chain. Time may have been, when the whole was a continuation of continent, rent aſunder before the laboring earth gave vent to its inward ſtruggles, through the mouths of the frequent vulcanoes. Even with theſe diſcharges, *Japan* has ſuffered conſiderably by earthquakes ‖. Vulcanoes are local evils, but extenſive benefits.

The *Ruſſians* ſoon annexed theſe iſlands to their conqueſts. The ſea abounded with Sea Otters, and the land with Bears and Foxes ; and ſome of them ſheltered the Sable. Temptations ſufficient for the *Ruſſians* to invade theſe iſlands ; but the rage after the furs of the Sea Otters has been ſo great, that they are become extremely ſcarce, both here and in *Kamtſchatka*.

The iſlands which lie to the eaſt of that peninſula, and form a chain between it and *America*, muſt now engage our attention. They lie in the form of a creſcent, and are divided into three groupes ; the *Aleutian*, the *Andreanoffskie*, and the *Fox* iſles : but mention muſt firſt be made of BERING's iſle, and that of *Mednoi*, and one or two ſmall and of little note. Theſe lie about two hundred and fifty verſts to the eaſt of the mouth of *Kamtſchatka* river. BERING's is in lat. 55, where that great ſeaman was ſhipwrecked in *November* 1741, on his return from his *American* diſcoveries ; and, after enduring great hardſhips, periſhed miſerably. Numbers of his people died of the ſcurvy, with all the dreadful ſymptoms attendant on thoſe who periſhed by the ſame diſeaſe in Lord *Anſon's* voyage ¶ ; the ſurvivors, among whom was the philoſopher STELLER, reached *Kamtſchatka* in *Auguſt* 1742, in a veſſel conſtructed out of the wreck of their ſhip. The iſle is about ſeventy or eighty verſts long ; conſiſts of high granitical mountains, craggy with rocks and peaks, changing into free-ſtone towards the promontories. All the vallies run from north to ſouth : hills of ſand, formed by inundations of the ſea, floated wood, and ſkeletons of marine animals, are found at great diſtances from the ſhore, at thirty fathoms perpendicular height above the high-water level ; which ſerve as a monument of the violent inundations that the vulcanoes before mentioned

---

* *Voyage*, iii. 388. † *Decouvertes des Ruſſes*, i. 113. ‡ Theſe iſles are marked in a *Ruſſian* map, communicated to me by Doctor PALLAS, with MS. notes. § *Kæmpfer Hiſt. Japan.* i. 305. ‖ Same, 304. ¶ Book i. ch. x. and *Decouvertes*, &c. ii. 293.

produce

produce in thefe feas. Farther, the effect of the meteoric waters, and of the frofts, caufes the rocks very fenfibly to fhiver and fall down, and precipitates every year fome great mafs into the fea, and changes the form of the ifland. The others are in the fame cafe; fo nothing is more probable than their gradual diminution, and, by confequence, the more eafy communication formerly from one continent to the other, before the injuries of time, the effects of vulcanoes, and other cataftrophes, had infenfibly diminifhed the fize, and perhaps the number of thefe ifles, which form the chain; and had eaten in the coafts of *Afia*, which every where exhibit traces of the ravages they have undergone *.

The ifland fwarmed with Sea Otters, which difappeared in *March*. The Urfine Seal fucceeded them in vaft numbers, and quitted the coaft the latter end of *May*. The *Leonine* Seal, the *Lachtach* or Great Seal, and the *Manati*, abounded, and proved the fupport of the wrecked during their ftay. *Arctic* Foxes were feen in great multitudes, and completed the lift of Quadrupeds. The fame fpecies of water-fowl haunt the rocks, and the fame fpecies of fifh afcend the rivers, as do in *Kamtfchatka*. The tides rife here feven or eight feet. The bottom of the fea is rocky, correfpondent with the ifland.

The few plants of this ifland, which have not been difcovered in *Kamtfchatka*, are as follow:

Campanula, *Gm. Sib.* iii. 160, 28.	Senecio, *Gm. Sib.* ii. 136, N° 118.
Leontodon taraxacum, *A. E. Virg.*	Arnica montana.
Hieracium murorum, *β. E.*	Chryfanthemum leucanthemum, *A.*
Tanacetum vulgare, *E.*	*Virg.*
Gnaphalium dioicum, *A.*	

Thefe, with a few creeping Willows, added to thofe in the *Kamtfchatkan Flora*, form the fum of thofe obferved in *Bering*'s ifland.

MEDNOI.   *Mednoi*, or the copper ifland, lies a little to the fouth-eaft. A great quantity of native copper is found at the foot of a ridge of calcareous mountains on the eaftern fide, and may be gathered on the fhores in vaft maffes, which feems originally to have been melted by fubterraneous fires. This ifland is full of hillocks, bearing all the appearance of vulcanic fpiracles; which makes it probable, that thefe iflands were rent from the continent by the violence of an earth-

---

* I am indebted to Doctor PALLAS for the whole account of this chain of iflands, except where I make other references.—My extracts are made from a *French* Memoir, drawn up by my learned friend, and communicated to me.

3

quake.

quake. Among the float-wood off this island is camphor, and another sweet wood, driven by the currents from the isle of *Japan*.

The *Aleutian* group lies in the bend of the crescent, nearly in mid-channel between *Asia* and *America*, lat. 52. 30, and about two hundred versts distant from *Mednoi*. It consists of *Attok*, *Schemija*, and *Semitchi*. The first seems to surpass in size *Bering's* isle; but resembles it in its component parts, as do the other two. *Attok* seems to be the island which *Bering* called *Mount St. John*. These are inhabited by a people who speak a language different from the northern *Asiatics*; they seem emigrants or colonists from *America*, using a dialect of the neighboring continent. They were discovered in 1745, by *Michael Nevodtsikoff*, a native of *Tobolski*, who made a voyage, at the expence of certain merchants, in search of furs, the great object of these navigations, and the leading cause of discoveries in this sea. This voyage was marked with horrid barbarities on the poor natives. The marine animals must have swarmed about this period, and for some time after. Mention is made of adventurers who brought from hence to *Kamtschatka* the skins of 1,872 Sea Otters, 940 females, and 715 cubs. Another, on a small adjacent isle, killed 700 old, and 120 cub Sea Otters, 1,900 blue Foxes, 5,700 black Ursine Seals, and 1,310 of their cubs *. The blue Foxes abound in these islands, brought here on floating ice, and multiply greatly. The blue variety is ten times more numerous here than the white; but the reverse is observed in *Sibiria*. They feed on fish, or any carrion left by the tide. The natives bore their under lips, and insert in them teeth cut out of the bones of the Walrus; and they use boats covered with the skins of sea animals.

At a great distance from the first group is the second, or farthest *Aleutian* isles: of those we know no more than that the natives resemble those of the first. By the vast space of sea which Doctor PALLAS allows between the two groups, Captain COOK is fully vindicated for omitting, in his chart, the multitude of islands which, in the *Russian* maps, form almost a complete chain from BERING's isle to *America*. Dr. PALLAS's information must have been of the best kind; and he and our illustrious navigator coincide in opinion, that they have been needlessly multiplied, by the mistake of the *Russian* adventurers in the reckoning, or, on seeing the same island in different points of view, putting it down as a new discovery, and imposing on it a new name. The *Andreanoffskie*, so called from their discoverer (in 1761) *Andrean Tolstyk*, succeed. On two of them are vulcanoes. Lastly, are the *Fox* islands, so called from the number of black, grey, and red Foxes found on them; the skins of which are so coarse, as to be of little

* Coxe's Russ. Disc.

s 2   value.

ALEUTIAN ISLES.
THE NEAREST.

ALEUTIAN ISLES.
THE FARTHEST.

ANDREAN ISLES.

FOX ISLES.

value. The natives bore their nofes and under lips, and infert bones in them by way of ornament. Among the laft in this group is *Oonolafcha*, which was vifited by Captain Cook. This lies fo near to the coaft of *America*, as to clame a right to be confidered as an appurtenance to it. I fhall therefore quit thefe detached paths for the prefent, and, in purfuance of my plan, trace the coafts of the northern divifion of the great continent, from the place at which it is divided from *South America*.

CALIFORNIA.

After traverfing obliquely the *Pacific Ocean*, appears *California*, the moft foutherly part of my plan on this fide of the new world. This greateft of peninfulas extends from Cape *Blanco*, lat. 32, to Cape *St. Lucas*, lat. 23; and is bounded on the eaft by a great gulph, called the *Vermillion* fea, receiving at its bottom the vaft and violent river *Colorado*. The weft fide is mountanous, fandy, and barren *, with feveral vulcanoes on the main land and the ifles † : the eaftern, varied with extenfive plains, fine vallies watered with numbers of ftreams, and the country abounds with trees and variety of fruits. The natives, the moft innocent of people, are in a ftate of paradifaical nature, or at left were fo before the arrival of the *European* colonifts among them. The men went nearly naked, without the confcioufnefs of being fo. The head is the only part they pay any attention to; and that is furrounded with a chaplet of net-work, ornamented with feathers, fruits, or mother of pearl. The women have a neat matted apron falling to their knees: they fling over their fhoulders the fkin of fome beaft, or of fome large bird, and wear a head-drefs like the other fex. The weapons of the country are bows, arrows, javelins, and bearded darts, calculated either for war or the chace. In the art of navigation, they have not got beyond the bark-log, made of a few bodies of trees bound parallel together; and in thefe they dare the turbulent element. They have no houfes. During fummer they fhelter themfelves from the fun under the fhade of trees; and during nights fleep under a roof of branches fpread over them. In winter they burrow under ground, and lodge as fimply as the beafts themfelves : fuch however was their condition in 1697; I have not been able to learn the effect of *European* refinement on their manners. Numbers of fettlements have, fince that time, been formed there, under the aufpices of the Jefuits. The Order was of late years fupported by the Marquis *de Valero*, a patriotic and munificent nobleman ‡, who favored their attempts, in order to extend the power and wealth of the *Spanifh* dominions;

---

* *Shelvoke*, in *Harris's Coll.* i. 233.   † *Hackluyt*, iii. 401.—*Hift. California*, i. 140.

‡ This is the nobleman whom the writer of Lord *Anfon's* Voyage ftigmatifes with the epithet of *munificent bigot*. It was not by a reverend author, as is generally fuppofed, but by a perfon whofe principles were unhappily in the extreme of another tincture.—Having from my youth been honored

with

dominions; and I believe with fuccefs. The land and climate, particularly *Monterey*, in lat. 36, is adapted for every vegetable production; and a good wine is made from the vines introduced by the colonifts.

The natives are a fine race of men, tall, brawny, and well made; with black hair hanging over their fhoulders, and with copper-colored fkins. We have a moft imperfect account of the animals of this peninfula. It certainly poffeffes two wool-bearing quadrupeds. As to birds, I doubt not but the Jefuits are right, when they fay, that it has all that are found in *New Mexico* and *New Spain*. The capes of *Florida* and cape *St. Lucas* lie nearly under the fame latitudes, and form the fouthern extremities of *North America*; but our ignorance of the productions of the vaft provinces of *New Mexico*, will leave ample fubject to a future naturalift to fupply my deficiencies.

This country was difcovered under the aufpices of the great *Cortez*, and Don *Antonio de Mendoça*, cotemporary viceroy of the new conquefts: each, actuated by a glorious fpirit of emulation, fent out commanders to advance the welfare of their country to the utmoft; and *Francifco Ulloa*, in 1539, and *Fernando Alarchon*, in 1540, foon difcovered this peninfula, and other adjacent regions, fources of immenfe wealth to their country *. The *Spanifh* adventurers of thefe early times failed as high as lat. 42; and named, in honor of the viceroy, the fartheft point of their difcovery *Cabo di Mendoça*.

Our celebrated navigator, Sir *Francis Drake*, on *June* 5th 1578, touched on this coaft, firft in lat. 43; but was induced, from the feverity of the cold, to fail to lat. 38, where he anchored in a fine bay. He found the natives to be a fine race of men, naked as the *Californians*, with the fame kind of head-dreffes; and the females habited like their fouthern neighbors. He was treated like a deity. The chief of the country, by the refignation of his crown or chaplet, his fceptre, *i. e.* calumet, and other infignia of royalty, vefted in Sir *Francis* the whole land; which he named *New Albion*, from its white cliffs, and took formal poffef-

---

with the friendfhip of the *Anfon* family, I can give a little hiftory of the compilation of the Voyage:—A Mr. *Paman* firft undertook the work. It was afterwards taken out of his hands, and placed in thofe of the reverend Mr. *Walters*, chaplain of the *Centurion*; but he had no fhare in it, farther than collecting the materials from the feveral journals: thofe were delivered to Mr. *Benjamin Robins*, a moft able mathematician, and the moft elegant writer of his time. He was fon of a quaker-taylor at *Bath*, whom I have often feen: a moft venerable and refpectable old man. Mr. *Robins* unfortunately forgot that he was writing in the character of a divine; and it was not thought proper to affront Mr. *Walters*, by omitting his name in the title-page, as he had taken in fubfcriptions: this, therefore, will account for the conftant omiffion of the word *Providence*, in a voyage which abounded with fuch fignal deliverances.

* A full account of thefe voyages may be feen in *Hackluyt*, iii. 397, &c.

fion of in the name of his royal miftrefs. We may be thankful that we never clamed the ceffion : it forms at prefent part of *New Mexico* ; and probably is referved for future contefts between the *Spaniards* and the offspring of our late colonifts. Sir *Francis* found this country a warren of what he calls, 'a ftrange ' kind of Conies, with heads as the heads of ours ; the feete of a Want, *i. e.* a ' Mole, and the tail of a Rat, being of a great length : under her chinne is on ' either fide a bag, into the which fhe gathereth her meat when fhe hath filled her ' bellie abroad.' The common people feed on them, and the king's coat was made of their fkins *. This fpecies is to be referred to the divifion of Rats with pouches in each jaw ; and has never been obferved from that period to this.

CAPTAIN COOK.   Exactly two hundred years from that time the coaft was again vifited by an *Englifhman*, who in point of abilities, fpirit, and perfeverance, may be compared with the greateft feaman our ifland ever produced. Captain JAMES COOK, on *March* 7th 1778, got fight of *New Albion*, in lat. 44. 33 north, and long. 235. 20 eaft, about eight leagues diftant. The fea is here (as is the cafe the whole way from *California*) from feventy-three to ninety fathoms deep. The land is moderately high, diverfified with hills and vallies, and every where covered with wood, even to the water's edge. To the moft fouthern cape he faw he gave the name of Cape *Gregory*, its latitude 43. 30 : the next, which was in 44. 6, he called Cape *Perpetua* ; and the firft land he faw, which was in 44. 55, Cape *Foul-weather*. The whole coaft, for a great extent, is nearly fimilar, almoft ftrait, and harborlefs, with a white beach forming the fhore. While he was plying off the coaft, he had a fight of land in about lat. 43. 10, nearly in the fituation of Cape *Blanco de St. Sebaftian*, difcovered by *Martin d'Aguilar* in 1603. A little to the north, the *Oregon*, or great river of the *Weft*, difcharges itfelf into the *Pacific Ocean*. Its banks were covered with trees ; but the violence of the currents prevented *D'Aguilar* from entering into it †. This, and the river of *Bourbon*, or *Port Nelfon*, which falls into *Hudfon's Bay* ; that of *St. Laurence*, which runs to the eaft ; and the *Miffifipi*, which falls into the bay of *Mexico*, are faid to rife within thirty miles of each other. The intervening fpace muft be the higheft ground in *North America*, forming an inclined plane to the difcharges of the feveral rivers. An ill-fated traveller, of great merit, places the fpot in lat. 47, weft long. from *London* 98, between a lake from which the *Oregon* flows, and another called *White Bear* lake, from which the *Miffifipi* ‡.

---

* *Hackluyt*, iii. 738.   † *Hift. California*, ii. 292.

‡ *Carver's Travels*, 76, 121.——Mr. *Carver*, captain of an independent company, penetrated far inland into *America* ; and publifhed an interefting account of his travels. This gentleman was fuffered to perifh for want, in *London*, the feat of literature and opulence.

This

This exalted situation is part of the *Shining Mountains*, which are branches of the vast chain which pervades the whole continent of *America*. It may be fairly taken from the southern extremity, where *Staten Land* and *Terra del Fuego* rise out of the sea, as insulated links, to an immense height, black, rocky, and marked with rugged spiry tops, frequently covered with snow. *New Georgia* may be added, as another, horribly congenial, rising detached farther to the east. The mountains about the streights of *Magellan* soar to an amazing height, and infinitely superior to those of the northern hemisphere, under the same degree of latitude. From the north side of the streights of *Magellan*, they form a continued chain through the kingdoms of *Chili* and *Peru*, preserving a course not remote from the *Pacific Ocean*. The summits, in many places, are the highest in the world. There are not less than twelve which are from two thousand four hundred toises high, to above three thousand. *Pichincha*, which impends over *Quito*, is about thirty-five leagues from the sea, and its summit is two thousand four hundred and thirty toises above the surface of the water; *Cayambé*, immediately under the equator, is above three thousand; and *Chimborazo* higher than the last by two hundred. Most of them have been vulcanic, and in different ages marked with eruptions far more horrible than have been known in other quarters of the globe. They extend from the equator, through *Chili*; in which kingdom is a range of vulcanoes, from lat. 26 south, to 45. 30 *, and possibly from thence into *Terra del Fuego* itself, which, forming the streights of *Magellan*, may have been rent from the continent by some great convulsion, occasioned by their laborings; and *New Georgia*, forced up from the same cause. An unparalleled extent of plain appears on their eastern side. The river of *Amazons* runs along a level cloathed with forests, after it bursts from its confinement at the *Pongo of Borjas*, till it reaches its sea-like discharge into the *Atlantic Ocean*.

In the northern hemisphere, the *Andes* pass through the narrow isthmus of *Darien*, into the kingdom of *Mexico*, and preserve a majestic height and their vulcanic disposition. The mountain *Popocatepec* made a violent eruption during the expedition of *Cortez*, which is most beautifully described by his historian, *Antonio de Solis* †. This, possibly, is the same with the vulcano observed by the Abbé *d'Auteroche*, in his way from *Vera Cruz* to *Mexico*, which, from the nakedness of the lavas, he conjectured to have been but lately extinguished ‡. From the kingdom of *Mexico*, this chain is continued northward, and to the east of *California*; then verges so greatly towards the west, as to leave a very

---

* *Ovalli, Hist. Chili,* in *Churchill's Coll.* iii. 13.   † *Conquest of Mexico,* book iii. ch. iv.

‡ *Voy. to California,* 33.

inconsiderable

inconfiderable fpace between it and the *Pacific Ocean*; and frequently detached branches jut into the fea, and form promontories; which, with parts of the chain itfelf, were often feen by our navigators in the courfe of their voyage. Some branches, as we have before obferved, extend towards the eaft, but not to any great diftance. A plain, rich in woods and favannas, fwarming with Bifons or Buffaloes, Stags, and *Virginian* Deer, with Bears, and great variety of game, occupies an amazing tract, from the great lakes of *Canada*, as low as the gulph of *Mexico*; and eaftward to the other great chain of mountains, the *Apalachian*, which are the *Alps* of that fide of northern *America*. I imagine its commencement to be about lake *Champlain* and lake *George*, with branches pointing obliquely to the river *St. Laurence* eaftward, and rifing on its oppofite coafts: others extending, with lowering progrefs, even into our poor remnant of the new world, *Nova Scotia*. The main chain paffes through the province of *New York*, where it is diftinguifh-ed by the name of the *Highlands*, and lies within forty miles of the *Atlantic*. From thence it recedes from the fea, in proportion as it advances fouthward; and near its extremity in *South Carolina* is three hundred miles diftant from the water. It confifts of feveral parallel ridges *, divided by moft enchanting vallies, and generally cloathed with variety of woods. Thefe ridges rife gradually from the eaft one above the other, to the central; from which they gradually fall to the weft, into the vaft plains of the *Miffifipi*. The middle ridge is of an enormous bulk and height. The whole extends in breadth about feventy miles; and in many places leaves great chafms for the difcharge of the vaft and numerous rivers which rife in the bofoms of the mountains, and empty themfelves into the *Atlantic* ocean, after yielding a matchlefs navigation to the provinces they water. In p. xcv, I have given a view of the immenfe elevated plain in the *Ruffian* em-pire. Beyond the branch of the *Apalachian* mountains, called *The Endlefs*, is another of amazing extent, nearly as high as the mountains themfelves †. This plain, (called the *Upper Plains*) is exceedingly rich land; begins at the *Mohock's* river; reaches to within a fmall diftance of lake *Ontario*; and to the weftward forms part of the extenfive plains of the *Ohio*, and reaches to an unknown diftance beyond the *Miffifipi*. Vaft rivers take their rife, and fall to every point of the compafs; into lake *Ontario*, into *Hudfon's* river, and into the *Delawar* and *Suf-quehanna*. The tide of the *Hudfon's* river flows through its deep-worn bed far up, even to within a fmall diftance of the head of the *Delawar*; which, after a

---

* Doctor *Garden*.   See alfo Mr. *Lewis Evans's* Effays and map. *Philadelphia*, 2d ed. p. 6, &c.
† Mr. *Lewis Evans*, p. 9, and map.

furious courfe down a long defcent, interrupted with rapids, meets the tide not very remote from its difcharge into the ocean *

LOW GROUNDS.

Much of the low grounds between the bafe of the *Apalachian* hills and the fea (efpecially in *Virginia* and *Carolina*) have in early times been occupied by the ocean. In many parts there are numbers of fmall rifings, compofed of fhells, and in all the plains incredible quantities beneath the furface. Near the *Miſſiſipi* again, in lat. 32. 28, from the depth of fifty to eighty feet, are always found, in digging, fea-fand and fea-fhells, exactly fimilar to what are met with on the fhores near *Penfacola* †. This is covered with a ftratum of deep clay or marle, and above that with a bed of rich vegetable earth. All this proves the propriety of applying the epithet of NEW to this quarter of the globe, in a fenfe different to that intended by the novelty of its difcovery. Great part of *North America* at left became but recently habitable: the vaft plains of the *Miſſiſipi*, and the tract between the *Apalachian Alps* and the *Atlantic*, were once poffeffed by the ocean. Either at this period *America* had not received its population from the old world, or its inhabitants muft have been confined to the mountains and their vallies, till the waters ceafed to cover the tracts now peopled by millions.

COMPONENT PARTS.

The compofition of the northern mountains agrees much with thofe of the north of *Afia*, and often confifts of a grey rock ftone or granite, mixed with glimmer and quartz; the firft ufually black, the laft purplifh. Near the river *St. Laurence*, a great part of the mountains refts on a kind of flaty limeftone. Large beds of limeftones, of different colors, are feen running from the granitical mountains; and are filled with *Cornua Ammonis*, and different forts of fhells, particularly with a fmall fpecies of fcallop, together with various forts of corals, branched as well as ftarry. The ftrata of limeftone alfo appear near the bafe of different parts of the *Apalachian* chain ‡. Without doubt, the fchiftous band, confifting of variety of ftone, fplit and divided by fiffures horizontal and perpendicular (in *Afia* the repofitory of metallic veins) is alfo found attendant on the granitical mountains of *North America*, and like them will be found rich in ores §: but that country has not yet been furveyed by a philofophical eye. The labor will be amply repayed to the proprietors, by the difcovery of mineral fources of wealth, perhaps equal to thofe already difcovered in the fimilar fecondary chains of mountains in the *Ruſſian* empire ‖.

Captain COOK continued his voyage to the northward; but, by reafon of fqually weather and fogs for a few degrees, or from lat. 50 to 55. 20, was deprived

---

* Mr. *Lewis Evans*, p. 9, and map.　　† *J. Lorimer*, efq.　　‡ *Kalm*, iii. 21, 198, 216.—*Bartram's Travels*, 10, 38.　　§ In fuch feem to be lodged the lead and filver ores found in *Canada*. See *Kalm*, iii. 212.　　‖ See Dr. PALLAS's *Obf. fur la formation de Montagnes*, &c.

of the opportunity of making the obfervations he wifhed. In lat. 48. 15, he in vain looked for the pretended ftreights of *Juan de Fuca*, who impofed on a *Michael Lock*, an *Englifhman* he met with at *Venice*, an account of having found, in 1592, an entrance in this latitude, and failed through it, till he arrived in the North fea, *i. e. Hudfon's Bay* *. Of equal credibility is the pretended paffage of Admiral *de Fontes*, in 1640, which is placed in lat. 50. 1 ; and, according to one map, falls into that of *De Fuca*: according to another, into a vaft inland fea, called *Mer de l'Oueft* †. Diligent fearch was alfo made after this in the *Spanifh* expedition of 1775 ; which ended in difproving thefe ftrange fictions ‡. It had likewife the farther importance of filling up the gap in the charts, by furnifhing us with a furvey of that tract of coaft which Captain COOK was obliged to quit.

In lat. 49, Captain COOK found a fecure fhelter in an harbor called by him *King George's Sound*; by the natives, *Nootka*. The fhores are rocky § ; but within the Sound appears a branch of the range I before mentioned. It is here divided into hills of unequal heights, very fteep, with ridged fides, and round blunted tops ; in general cloathed with woods to the very fummits. In the few exceptions, the nakednefs difcovers their compofition, which is rocky, or in parts covered with the adventitious foil of rotten trees or moffes.

The trees were the *Pinus Canadenfis*, or *Canada* Pine ; the *P. Sylveftris*, or *Scotch* Pine, and two or three other forts ; *Cupreffus Thyoides*, or the White Cedar. The Pines of this neighborhood are of a great fize : fome are a hundred and twenty feet high, and fit for mafts or fhip-building ‖ ; but the dimenfions of fome of the canoes in *Nootka Sound* beft fhew their vaft bulk—they are made of a fingle tree, hollowed fo as to contain twenty perfons ; and are feven feet broad, and three deep. They are the fame with the *monoxyla* of the antient *Germans* and *Gauls* ¶, but conftructed with much more elegance. The old *Europeans* were content if they could but float. They probably were formed on the fame rude model as thofe of the old *Virginians* **, or of the antient *Britons*, fimilar to one I have feen dug up in a morafs in *Scotland*, as artlefs as a hog-trough ††. Thofe of *Nootka Sound* are at the head tapered into a long prow, and at the ftern they decreafe in breadth, but end abrupt.

The day-tides rife here, two or three days after the full and new moon, eight

---

* North-weft *Fox*, 163.   † See *Jefferies's Obf. on the Letter of Adm. de Fontes*, and his map ; alfo *de L'Ifle*'s map.   ‡ *Maurelle's* Voy. *in* 1775, in Mr. *Barrington's Mifcellanies*, 508.
§ *Voyage*, ii. 290. tab. 86, 87.   ‖ *Barrington's Mifcell.* 290.   ¶ *Polyæn. Stratagem.* lib. v. c. 23.—*Vel. Paterc.* lib. ii. c. 107.   *Brevis et fida Narratio Virginiæ*, in which are engraven the canoes of the country, taken from the drawing of *John With* ; fent there with *Tho. Harriot* for that purpofe, by Sir *Walter Raleigh*, who communicated them to *De Bry*.—See tab. xii. and xlii. of the *Account of Florida*.   †† *Tour Scotl.* ii. p. 106.

feet

feet nine inches. The night-tides, at the same periods, rise two feet higher. Pieces of drift wood, which the navigators had placed during day out of the reach (as they thought) of the tides, were in the night floated higher up, so as to demonstrate the great increase of the nocturnal flux *.

I have described, to the best of my power, the quadrupeds and birds of the *American* part of this voyage. In p. 12 I have given my suspicions of certain animals of the Sheep kind being natives of this neighborhood and *California*; but am not sufficiently warranted to pronounce them to be the same with the *Argali* or wild Sheep. Woollen garments are very common among the people of this Sound, and are manufactured by the women. The materials of many of them seem taken from the Fox and the Lynx; others, I presume, from the exquisite down of the Musk Ox, N° 2. The only peculiar animal of these parts is the Sea Otter, N° 36: it extends southward along the coast, as far as lat. 49, and as high as 60 †. The other quadrupeds observed by the navigators are common to the eastern side of *North America.*

BIRDS.

I may mention, that small Perroquets, and Parrots with red bills, feet, and breasts, were seen by M. *Maurelle* about Port *Trinidada*, in lat. 41. 7; and great flocks of Pigeons in the same neighborhood ‡. This was in *June*: possibly they were on their migration when our navigators reached the coasts, which was on *March* 29th. As to the Parrots, it is possible that those birds may not extend so far north as *Nootka*; for on the eastern side of the continent they do not inhabit higher, even in summer, than the province of *Virginia*, in lat. 39; or, in the midland parts, than lat. 41. 15, where they haunt in multitudes the southern sides of the lakes *Erie* and *Michigam*, and the banks of the rivers *Illinois* and *Ohio*. Another delicate species of bird was seen here in plenty, a kind of *Honey-sucker* or *Humming-bird*, a new species; which I have described, N° 177, under the title of the *Ruffed*. Among the water-fowl were seen the Great Black Petrel, p. 536. A. or the *Quebrantahueffos*, or *Bone-breaker* of the *Spaniards*, which seems to be found from the *Kuril* isles to *Terra del Fuego*; the *Northern Diver*, N° 439; a great flock of Black Ducks with white heads; a large species of White Ducks with red bills; and Swans flying northward to their breeding-places: common Corvorants were also very frequent.

MEN.

The inhabitants of this Sound alter in their appearance from those who live more southern. They are in general below the middle stature; plump, but not muscular: their visage round, full, and with prominent cheeks; above which the face is compressed from temple to temple: the nostrils wide: nose flat, with a rounded point; through the *septum narium* of many is introduced a ring of iron,

---

* *Voyage*, ii. 339.   † In p. 89, for lat. 44, read 49.   ‡ See *Barrington's*
Miscell. 489, 502.

brass,

brafs, or copper: eyes fmall, black, languifhing: mouth round: lips large and thick: hair of the head thick, ftrong, black, long, and lank; that on the eyebrows very thin: neck fhort and thick: limbs fmall and ill-made: fkin a pallid white, where it can be viewed free from dirt or paint. The women are nearly of the fame form and fize as the men, but undiftinguifhable by any feminine foftnefs. Many of the old men have great beards, and even muftachios; but the younger people in general feem to have plucked out the hair, except a little on the end of the chin.

Their drefs confifts of mantles and cloaks, well manufactured among themfelves, and either woollen, matting, or fome material correfpondent to hemp. Over their other cloaths the men frequently throw the fkin of fome wild beaft, which ferves as a great cloak. The head is covered with a cap made of matting, in form of a truncated cone, or in that of a flower-vafe, with the top adorned with a pointed or round knob, or with a bunch of leathern taffels. Their whole bodies are incrufted with paint or dirt, and they are a moft fquallid offenfive race; filent, phlegmatic, and uncommonly lazy; eafily provoked to violent anger, and as foon appeafed. The men are totally deftitute of fhame: the women behave with the utmoft modefty, and even bafhfulnefs *. I fhall not repeat what has been faid of the infinite variety of hideous mafques this nation poffeffes, and feems particularly fond of, was not the ingenious Editor of the Voyage at a lofs for their intent, whether for religious or for mafquerading purpofes †. Mr. *Bartram* ‡ proves that thefe mafques extend to the eaftern fide of the continent, and that their ufe was fportive; for he was plagued part of a night with the buffoonery of a fellow, who came into his lodgings while he was on his travels, and, after playing a thoufand antic tricks, vanifhed in a manner as if he meant to be taken for a hobgoblin. The *Oftiaks* have exactly the fame cuftom §.

Thefe people have made fome progrefs in the imitative arts; for, befides their fkill in the fculpture of their mafques, which they cut into the fhape of the heads of various fpecies of beafts and birds, they are capable of painting with tolerable exactnefs: accordingly, they often reprefent on their caps the whole progrefs of the Whale-fifhery. I have feen a fmall bow made of bone, which was brought by the navigators from this fide of *North America*, on which was engraven, very intelligibly, every object of the chace. I have caufed this fingular bow to be engraven, and in the fame plate, that moft terrific *Tomahawk* of *Nootka Sound*, called the *Taaweefh*, or *Tfufkeeah*. The offenfive part is a ftone projecting out of the mouth of a fculpture in wood, refembling a human face, in which are ftuck human and other teeth: long locks of fcalped hair are placed on feveral parts of the head,

* *Voyage*, ii. 319.　　† Same, 307.　　‡ *Travels*, 43　　§ *Ruffian Nations*, i. 193.

waving

P. Mazell sculp.

waving when brandifhed in a moft dreadful manner.   I could diftinguifh the Elk, the Rein, the *Virginian* Deer, and the Dog; birds, probably of the Goofe kind; the Whale-fifhery, the Walrus, and the Seal.—With what facility might be reclamed and civilized a people fo ftrongly poffeffed with a difpofition towards the liberal arts !

From lat. 55. 20, towards. the north, the country increafes in height, efpecially inland, where a range of very lofty mountains, moftly covered with fnow, is feen nearly parallel with the coaft, a branch of thofe I have before mentioned.   Above lat. 56 the coaft is broken into bays and harbours.   In this neighborhood Captain *Tfchirikow*, confort to the great navigator BERING, who was feparated from his commander by a ftorm, was fo unfortunate as to. touch on an open part of the coaft, in about lat. 55, in which he anchored in a moft dangerous fituation, full of rocks.   Having loft his fhallop, and after that his fmall boat, with part of his crew, which he had fent on fhore to water, and which were deftroyed by the natives, he was obliged to return from his ineffectual voyage *.   A vaft conic mountain, called by Captain COOK Mount *Edgecumbe* †, rifes pre-eminent above all the others.   This is in lat. 57. 3, long. 224. 7.   Not remote from hence is the *Bay of Iflands*, the fame as the *Port los Remedios*, nearly the *ne plus* of the *Spanifh* expedition of 1775.   The adventurers comforted themfelves with having reached lat. 58, and having attained the higheft latitude ever arrived at in thefe feas ‡. This coaft, as well as the reft, continued covered with woods.

A high peaked mountain, Mount *Fair-weather*, and the inlet *Crofs Sound*, next appear.   The firft is the higheft of a chain of fnowy mountains, which lie inland about five leagues, in lat. 58. 52.   The land between them and the fea was very low, for the trees feemed to arife out of the water.   Several fea-birds, with a black ring round the head ; the tip of the tail, and upper part of the wings, marked with black ; the body bluifh above, white beneath, came in view; and on the water fat a brownifh Duck, with a deep blue or black head §.

In lat. 59. 18, is a bay, with a wooded ifle off its fouth point, named by Captain COOK, BERING's ; in honor of the illuftrious *Dane* who firft difcovered this part of *America*, and, as was conjectured, anchored there for a fmall fpace.   The appearance of the country was terrific ; it confifted of lofty mountains (in *July*) covered with fnow: but the chain is interrupted near this port by a plain of a few miles in extent; beyond which the view was unlimited, having behind it a continuance of level country, or fome great lake.   He had not leifure to make obfervations ; he only named a cape, which advanced into the fea, Cape *Elias* ‖ : this is not at prefent known ; but the name of Mount *Elias* was beftowed by Captain

* *Voy. & Decouvertes de Ruffes*, i. 250.   † COOK's *Voy*. ii. 344, tab. 86.   ‡ *Barrington s Mifcel.* 507.   § COOK's *Voy*. ii. 347.   ‖ *Voy. & Decouvertes*, i. 254.—COOK, ii. 347, 383.

COOK

Marginal notes:
RUSSIAN VOYAGE.

MOUNT EDGECUMBE.

Cook on a very conspicuous mountain *, which lay inland to the north-west of the bay, in lat. 60. 15.

Bering, during the short stay he made on the coast, sent his boat on shore to procure water. That great naturalist, *Steller*, companion of the voyage, took the opportunity of landing. The whole time allotted him was only six hours; during which he collected a few plants, and shot that beautiful species of Jay, N° 139, to which I have given his name. He returned on board with the regret a man of his zeal must feel at the necessity of so slight an examination in so ample a field. What he could have done, had circumstances permitted, is evident from the excellent collection he formed of natural history respecting *Kamtschatka*, and some of its islands †.

PLANTS.   Among the plants found by him on the *American* continent were, *Plantago major*, Sp. Pl. i. 163; Great Plantane, *Fl. Scot.* i. 117. *K. Virg.* : *Plantago Asiatica*, Sp. Pl. i. 163. *K.* : *Polemonium Cæruleam*, Sp. Pl. i. 230 : Greek Valerian, *Hudson*, i. 89. *K.* : *Lonicera Xylosteum*, Fl. Sib. iii. 129. *K.* : *Ribes Alpinum*, Sp. Pl. i. 291. Fl. Scot. i. 146. *K.* : *Ribes grossularia*, Sp. Pl. i. 291 ; Gooseberries, *K. Virg.*: *Claytonia Virginica?* Sp. Pl. i. 294. *K. Virg.* : *Heuchera Americana?* Sp. Pl. i. 328. *K.*: *Heracleum Panaces*, Sp. Pl. i. 358 ; or Cow Parsnep, *K.* which he found in one of the habitations of the natives, tied up in bundles ‡ ready for use. (I have mentioned, at p. cxvii. the application of it in *Kamtschatka*, for the purposes of distilling an intoxicating liquor; but the *Americans* are fortunate enough to be ignorant of that art, and only use it as a food.) *Vaccinium Myrtillus*, Sp. Pl. i. 498; Bilberries, *Fl. Scot.* i. 200. *K.*: *Vaccinium Vitis Idæa*, *Virg.* Sp. Pl. i. 500; Red Whortle-berries, *Fl. Scot.* i. 202. *K.*: *Erica*, Fl. Sib. 131, N° 22. *K.*: *Adoxa Moschatellina*, Sp. Pl. i. 527 ; tuberous Moschatel, *Fl. Scot.* i. 209. *K.*: *Rubus Idæus*, Sp. Pl. i. 706; Rasberry-bush, *Fl. Scot.* i. 263. *K.* : *Fragaria Vesca*, Sp. Pl. i. 708; Wood Strawberry, *Fl. Scot.* i. 267. *Virg. K.* : the *Leontodon Taraxicum*, *Virg. B.* Sp. Pl. ii. 1122; or common Dandelion, *Fl. Scot.* i. 433 : *Absinthium*, Sp. Pl. ii. 1188; or common Wormwood, *Fl. Scot.* i. 467 : *Artemisia Vulgaris*, Sp. Pl. ii. 1188 ; or Mugwort, *Fl. Scot.* i. 468 : *Gnaphalium Dioicum*, Sp. Pl. ii. 1199 ; Mountain Cudweed, or Cat's-foot, *Fl. Scot.* i. 470. *K.*: *Aster seu potiùs Helenium fruticosum*, Fl. Sib. ii. 175, *B. K.* with beautiful yellow flowers : *Erigeron acre*, Sp. Pl. ii. 1211; Blue Fleabane, *Fl. Scot.* i. 474. *K.*: *Chrysanthemum Leucanthemum*, ii. 1251 ; Great Daisy, or Ox-eye, *Fl. Scot.* i. 488. *B. K. Virg.*: *Pyrethrum*, Fl. Sib. ii. 203, N° 170. *B. K.*: *Achillea Millefolium*,

---

* Cook, ii. tab. 86.   ·† *Voy. & Decouvertes*, i. 257.   ‡ *Decouvertes faites par les Russes*, i. 256.—*Voyage*, ii. tab. 86.

Sp.

Sp. Pl. ii. 1267; Milfoil or Yarrow, *Fl. Scot.* i. 490. *K. Virg.*: *Empetrum nigrum*, Sp. Pl. ii. 1450; Black-berried Heath, Crow-berries, *Fl. Scot.* ii. 612. *K. Virg.*: *Menispermum Canadense?* Sp. Pl. ii. 1468. *K. Virg.*—I retain the mark of *British* vegetables, to shew the vast dilatation of plants; and that of *Virg.* to shew those which spread to the eastern side of *America.*

To these may be added a few trees and plants observed by our navigators; such as the *Pinus Strobus*, Sp. Pl. ii. 1490, the white or *Weymouth* Pine, which grows to an enormous size; *Pinus Canadensis*, Sp. Pl. ii. 1421, the *Canada* Pine; three or four other Pines, which we cannot determine; the *Cupressus Disticha?* Sp. Pl. ii. 1422, the deciduous Cypress; *Cupressus Thyoides*, Sp. Pl. ii. 1422, or white Cedar; some Birch, Alders, and Willows; wild Rose-bushes; and several plants, the species of which are unknown to us. Probably that useful Lily, the *Lilium Kamtschatchense*, or *Saranne*, extends to the continent, for it is found in abundance in the adjacent island *Oonalaschka*, where it serves as a food, as it does in *Kamtschatka*.

In this neighborhood, in lat. 59. 49, about *Kaye's* island †, off Cape *Suckling*, Captain Cook observed variety of birds; among them some Albatrosses, the snowy Gulls, and the common Corvorant: and in the poor woods which encircled the island like a girdle, were seen a Crow, the white-headed Eagle, and another species equally large, of a blacker color, with a white breast, which proves to be the kind described by Mr. *Latham*, i. p. 33. N° 72, under the name of the *white-bellied Eagle* ‡.

After doubling a cape, called by our great navigator, *Hinchinbroke* §, he anchored in a vast sound, named by him *Prince William's*, in lat. 61. 30, secured by a long island, called *Mountague's*, stretching obliquely across from north-east to south-west. The land round this harbour rose to a vast height, and was deeply covered with snow ‖. Vegetation in these parts seemed to lessen. The principal trees were the *Canadian* and Spruce Firs, and some of them moderately large.

Besides the quadrupeds found at *Nootka*, there is a variety of Bear of a white color; I will not call it the *Polar*, as that animal inhabits only the severest climates, where it can find dens of snow and isles of ice. An animal of the ermine kind, varied with brown, but the tail scarcely tipt with black. Wolverenes were here, of a very brilliant color; and the earless Marmot, N° 47, was very common. None of these were seen living, but their skins were brought

---

Voyage, ii. 501.   † Same, tab. 85.   ‡ Same, p. 352.   § Same, tab. 86.
‖ See the picturesque view of *Snug Corner Cove*, tab. 45.

in abundance as articles of commerce. The skin of the head of the male leonine Seal was also offered to sale : in the Voyage it is called the *Ursine*; but from the great shagginess of the hair I presume I am not wrong in my conjecture. This is the only place in the northern hemisphere in which it was found by the navigators *.

BIRDS.

Among the birds were the black Sea Pies with red bills, observed before in *Van Diemen's Land* and *New Zealand*. A Duck, equal in size to our Mallard, with a white bill tinged with red near the point, and marked with a black spot on each side near the base : on the forehead a large white triangular spot, and a larger on the hind part of the neck : the rest of the plumage dusky : the tail short and pointed : the legs red. The female was of duller colors, and the bill was far less gay. Another species resembled the small one found at *Kerguellen's Land*. A Diver (Grebe ?) of the size of a Partridge ; with a black compressed bill : head and neck black : upper part of the body deep brown, obscurely waved with black ; the lower part dusky, speckled minutely with white. Honey-suckers, probably migratory in this high latitude, frequently flew round the ships †.

To give all the additions I am able to my zoologic part, I shall here mention certain species of Petrels, observed on the western coast of *North America :* such as numberless brown Petrels near the entrance of COOK's river, flying round a remarkable sugar-loaf hill ‡. A species seen near *Nootka Sound*, about eleven inches long, with the nostrils scarcely tubular : bill and plumage above dusky, beneath white : legs back. This is common to *Turtle Isle*, lat. 19. 48, south, long. 178. 2, west ; and *Christmas Isle*, lat. i. 59, north, long. 202. 30, east. Another, about thirteen inches long, with the forehead, space between the eyes and bill, the chin, and throat, of a greyish white, varied with specks of dusky : crown and upper part of the body dusky : under parts hoary lead-color : legs pallid §. I may add a fourth, seen off the coast of *Kamtschatka*, which Mr. *Ellis* mentions as being small, and of a bluish color ‖.

MEN.

MANKIND here shew a variation from the last described. The natives are generally above the common stature, but many below it : square-built or strong-chested : their heads most disproportionably large ; their faces flat, and very broad : their necks short and thick : their eyes small, in comparison to the vast breadth of their faces : their noses had full round points, turned up at the end : their hair long, thick, black, and strong : their beards either very thin, or extirpated ; for several of the old men had large, thick, but strait beards : their countenances generally full of vivacity, good-nature, and frankness, not unlike the *Cristinaux*,

---

* *Voy.* ii. 377.   † Same, 378.   ‡ *Ellis's Narrative*, i. 251.   § This, and the preceding, in the LEVERIAN *Museum.*   ‖ *Narrative*, ii. 246.

a people

a people who live far inland, between the little and the great lakes *Ouinepique*. On the contrary, the inhabitants of *Nootka* in their dulnefs refemble the *Affinibouels*, who live on the weftern fide * : and thefe two nations may have been derived from a common ftock with the maritime tribes whom we have had occafion to mention. The fkins of the natives of this found were fwarthy, poffibly from going often naked; for the fkins of many of the women, and the children, were white, but pallid. Many of the women were diftinguifhable from the men by the delicacy of their features, which was far from the cafe with thofe of *Nootka*.

In thefe parts, within the diftance of ten degrees, is a change of both drefs and manners. The cloak and mantle are here changed for a clofe habit, made of the fkins of different beafts, ufually with the hair outwards; or of the fkins of birds, with only the down remaining; fome with a cape, others with a hood: over which, in rainy weather, is worn a garment like a carter's frock, with large fleeves, and tight round the neck, made of the inteftines probably of the whale, and as fine as gold-beater's leaf. On the hands are always worn mittens, made of the paws of a bear; and the legs are covered with hofe, reaching to midway the thigh. The head is generally bare; but thofe who wear any thing, ufe the high truncated conic bonnet, like the people of *Nootka* †. In this place only was obferved the *Calumet*; a ftick about three feet long, with large feathers, or the wings of birds, tied to it. This was held up as a fign of peace.

I leave the reader to amufe himfelf in the Voyage, by the account of the ftrange cuftom of the natives in cutting through their under lip, and giving themfelves the monftrous appearance of two mouths ‡: in the orifice they place a bit of bone or fhell by way of ornament. This cuftom extends to the diftant *Mofquitos*, and even to the *Brafilians* §, but feems unknown in other parts of *America*.—I endeavour to confine myfelf to paffages which may lead to trace the origin of the people. Thefe paint their faces, and puncture or tattow their chins. They are moft remarkably clean in their food, and in their manner of eating it, and even in the keeping of their bowls and veffels. In their perfons they are equally neat and decent, and free from greafe or dirt ‖ : in this they feem an exception to all other favages.

They have two kinds of boats; one large, open, and capable of containing above twenty people. It is made of the fkins of marine animals, diftended on ribs of wood, like the *vitilia navigia* of the *Britons*, at the time in which they were on a level with thefe poor *Americans*; or like the woman's boat of the *Greenlanders* and *Efkimaux*. The canoes are exactly of the fame conftruction with thofe of the latter; and the difference of both is very trivial. The canoes of thefe

BOATS.

---

* *Dobbs*, 24. † *Voyage*, ii. 368, 369. ‡ Same, 369, tab. 46, 47. § *Dampier*, i. 32, *de Bry. Brafil*, 165. ‖ *Voyage*, ii. 374.

Americans

Americans are broader than those of the eastern side of the continent; and some have two circular apertures, in order to admit two men *. Every weapon which these people have for the chace of quadrupeds or fish, is the same with those used by the *Greenlanders:* there is not one wanting.

From *Prince William's* found the land trends north-west, and terminates in two headlands, called Cape *Elizabeth* and Cape *Bede;* these, with Cape *Banks* on the opposite shore, form the entrance into the vast estuary of Cook's river; in the midst of which are the naked isles, distinguished by the name of the *Barren.* Within, to the west, is a lofty two-headed mountain, called Cape *Douglas;* which is part of a chain of a vast height, in which was a vulcano, at the time this place was visited, emitting white smoke: and in the bottom of a bay, opposite to it, is an island, formed of a lofty mountain, on which was bestowed the name of Mount St. *Augustine* †. The estuary is here of a great breadth, owing to a bay running opposite to Mount *Augustine* deeply to the east.

The estuary of Cook's river is of great length and extent. The river begins between *Anchor Point* and the opposite shore, where it is thirty miles wide: the depth very considerable, and the ebb very rapid. Far within, the channel contracts to four leagues, through which rushes a prodigious tide, agitated like breakers against rocks. The rise of the tide in this confined part was twenty-one feet. It was examined seventy leagues from the entrance, as far as lat. 61. 30, long. 210, and its boundaries were found to be flat, swampy, and poorly wooded, till they reached the foot of the great mountains. Towards the north, it divides into two great branches, or perhaps distinct rivers. That to the east is distinguished by the name of *Turn-again* river. The first is a league wide, and navigable, as far as was tried, for the largest ships, and continued very brackish; there is therefore the greatest probability of its having a very long course, and being, in after times, of considerable use in inland navigation: that it is of some even at present is very certain; for here, as well as in *Prince William's* found, the *Indians* were possessed of glass beads and great knives of *English* manufacture, which the *Hudson's* bay company annually send in great quantities, and exchange for furs with the natives, who travel to our settlements very far from the west. The company also send copper and brass vessels; but neither copper or iron in bars. There does not seem to be any direct dealings with the *Indians* of this coast: the traffic is carried on by intermediate tribes, who never think of bringing furs to a people so amply supplied as the *Indians* are who deal with our factories. Nations who use the most pretious furs merely as a defence from the cold, make no distinction of kinds: if they could get more beads or more knives for the skins of Sea Otters

---

* *Voyage,* ii. 371.　　　　　† See the chart, ii. tab. 44.

than

than any other, they would inftantly become articles of commerce, and find their way acrofs the continent to the *European* fettlements.

From *Turn-again* river to the neareft part of *Hudfon*'s bay, is fifty-five degrees, or about fixteen hundred miles; but from the moft weftern part of *Arapathefcow* lake (which is intermediate) is only twenty-fix degrees, or about feven hundred and fifty miles. There is no difcharge out of that vaft water but what runs into *Hudfon*'s bay. We have fome obfcure accounts of rivers * which take a weftern courfe from the countries eaft of this coaft: fome of which may be thofe which have been feen by our navigators, and which, by means of lakes or other rivers falling into them, may prove a channel of intercourfe between thefe *Indians* and the *Hudfon*'s bay company, as foon as our friendly *Indians* become acquainted with the value of thefe maritime furs.

The inhabitants of *Cook*'s river differed very little from thofe of *Prince William*'s found. They had Dogs, which were the firft feen on the coafts; Sea Otters, Martins, and white Hares: and they were plentifully fupplied with Salmon and Holibut.

After leaving the entrance into the river, appears Cape St. *Hermogenes*, difco-vered firft by BERING. It proved a naked lofty ifland, about fix leagues in cir-cuit, and divided from the coaft by a channel a league broad. This lies in lat. 58. 15, off the vaft peninfula *Alafchka*, which begins between the eftuary of *Cook*'s river and *Briftol* bay, which bound its ifthmus. It points fouth-weft, and continues the crefcent formed by the ifles which crofs the fea from *Kamtfchatka*. *Alafchka* is the only name given by the natives to the continent of *America*. The land to the weft of COOK'S river rifes into mountains, with conoid tops thickly fet together. The coaft is frequently bold, and the rocks break into pinnacles of picturefque forms: the whole is fronted by groups of ifles and clufters of fmall rocks. In a word, the country and fhores are the moft rugged and disjointed imaginable, and bear evident marks of having undergone fome extraordinary change.

Among the ifles, thofe of *Schoumagin* are the moft important, which received their name from having been the place of interment of one of *Bering*'s crew, the firft which he loft in thefe feas. The principal lies the fartheft to the weft, and is called *Kadjak*: it is about a hundred verfts long, and from twenty to thirty broad; and, from the account of *Demetrius Bragin*, who vifited it from *Oonalafhka* in 1776, is very populous. The inhabitants fpoke a language different from thofe

DOGS.

CAPE ST. HERMO-GENES.

ALASCHKA, CON-TINENT OF AME-RICA.

KADJAK.

---

* Particularly from one *Jofeph de la France*, who in 1739 made a very long journey to the weft, and was a very obfervant man. See *Dobbs, Hudfon's Bay*, 21, 34, 35.

of that iſland : it ſeemed a dialeƈt of the *Greenlanders.* They called their wooden ſhields *Kuyaky,* probably becauſe they reſemble a *kaiak,* or a little canoe, a *Greenland* word for that ſpecies of boat ; and themſelves *Kanagiſt,* as the others ſtyle themſelves *Karalit.* They have likewiſe the woman's boat, like the people of *Prince William's* found : in faƈt, they ſeem to be the ſame people, but more refined. They were armed with pikes, bows and arrows, and wooden ſhields. Their ſhirts were made of the ſkins of birds ; alſo of the earleſs Marmot (*Arƈt. Zool.* i. N° 47), Foxes, and Sea Bears, and ſome of fiſhes ſkins. Dogs, Bears, common Otters, and Ermines, were obſerved here. Their dwellings were made with timber, and were from fifteen to twenty fathoms long, covered with a thatch and dried graſs. Within they were divided into compartments for every family, and every compartment lined neatly with mats. The entrance was on the top, covered with frames, on which were ſtretched the membranes of dried inteſtines inſtead of glaſs*. Theſe people ſeemed to have made far greater progreſs in the arts than their neighbors. They worked their carpets in a very curious manner ; on one ſide cloſe ſet with beaver wool. The Sea Otters ſkins which they brought for ſale were in ſome parts ſhorn quite cloſe with ſharp ſtones, ſo that they gliſtened and appeared like velvet. They ſhewed ſtrong proofs of genius in their invention to preſerve themſelves from the effeƈts of the *Ruſſian* fire-arms. They had the ſpirit to make an attack, and formed ſkreens with three parallel perpendicular rows of ſtakes, bound with ſea-weeds and oſiers ; their length was twelve feet, and thickneſs three : under the ſhelter of theſe they marched ; but their ſucceſs was not correſpondent to their plan † : a ſally of the *Ruſſians* diſconcerted them, and put them to the rout.

The iſland conſiſts of hills mixed with lowlands. It abounds with bulbs, roots, and berries, for food ; with ſhrubs, and even trees ſufficiently large to be hollowed into canoes capable of carrying five perſons ‡. In this kind of boat they differ from thoſe of the *Greenlanders.*

HOLIBUT ISLE.   Off the extremity of the peninſula of *Alaſchka* is *Holibut* iſland, in lat. 54, riſing into a lofty pyramidal mountain, lying oppoſite to the narrow ſhallow ſtreight which lies between the iſle *Oonemaka* and *Alaſchka.* The chain on the continent is ſeen to riſe into ſtupendous heights, covered with ſnow : among them ſeveral of the hills appear to riſe inſulated, and of a conic form. One

---

* From a MS. communicated to me by Dr. PALLAS, *Bragin* was commander of a veſſel which was fitted out by the merchants on a voyage to the new-diſcovered iſlands, and ſailed from *Ochotſk* in 1772. About ten years prior to this, another voyage was made to *Kadjak* by *Stephen Glottoff.*— See *Coxe's Diſc.* 108.   † *Coxe's Ruſſ. Diſc.* 12.   ‡ MS.

was

was a *vulcano*, flinging up volumes of black fmoke to a great height *, then ftreaming before the wind with a tail of vaft length and picturefque appearance. It often took a direction contrary to the point the wind blew from at fea, notwithftanding there was a frefh gale. It lies in lat. 54. 48 north, long. 195. 45 W. and is evidently a link in the vulcanic chain, which extends, in the fouthern hemifphere, as low at left as that of St. *Clement* in *Chili*, in lat. 45. 30.

The extremity of *Alafchka* ends abrupt, and has oppofite to it an ifland called *Oonemak* or *Unmak*, of nearly a correfpondent breadth, feparated from it by a very narrow and fhallow channel, fituated in lat. 54. 30, and leading into *Briftol* bay, pervious only by boats or very fmall veffels. The ifle is a hundred verfts long, and from feven to fifteen broad; and has in the middle a vulcano. In the low parts feveral hot fprings burft forth, to which the iflanders carry the fifh or flefh they want to boil; and they are alfo fond of bathing in the temperate parts †.

To the weft are the fmall ifles of *Oonella* and *Acootan*: at a fmall diftance from them is *Oonalafhka* or *Aghôun-alaifka* ‡, a name evidently referring to the continent. My MS. calls its length a hundred and twenty verfts, its breadth from ten to eighteen. It is the moft remote of the *Ruffian* colonies, who have now made fettlements on moft of the ifles between *Afia* and *America*; all under the care of private adventurers. The voyage from *Ochotfk* or *Kamtfchatka* lafts three or four years; and is folely undertaken for the fake of the fkins of Sea Otters. Poffibly other reafons will, in a little time, induce them to attempt the colonization of the continent. Timber may be one; for their northern *Afiatic* dominions and their iflands yield none. I forefee docks and timber-yards in all convenient places. At prefent, the natives of thefe ifles have only the fkin-covered canoes §, and even for the ribs they are obliged to the chance of drift-wood. In thefe, in drefs, and in weapons, they refemble the *Efkimaux*. The language is a dialect of the *Efkimaux*. They are rather of low ftature. They have fhort necks, fwarthy chubby faces, black eyes, and ftraight long black hair. The fafhion of wearing feathers or bits of fticks in their nofes is ufed in *Oonalafhka*. Both fexes cut their hair even over their foreheads: the men wear theirs loofe behind; the females tie theirs in a bunch on the top of their head: the firft wear long loofe frocks, of the fkins of birds; the laft of the fkins of Seals. The men fling over their frocks another, of the guts of the cetaceous animals, dried and oiled, to keep out the water ‖; and, to

Marginal notes: OONEMAK.   OONALASHKA.   NATIVES.

---

* See the plate, N° 87, vol. ii. for the feveral views.   † *Bragin's Voy.* MS.   ‡ Doctor
PALLAS, MS.   § See their boats, tab. 50.   ‖ See their dreffes, tab. 48, 49, 56, 57.

defend

defend their faces from the weather, they wear a piece of wood, like the front of the bonnet of an *English* lady *. Some ufe the bonnet in the form of the truncated cone. The women flightly tattow their faces, and often wear a ftring of beads pendent from their nofes; both fexes perforate their under lip, but it is very uncommon to fee any except the females ftick in it the ornamental bone. The nofe-ornaments extend far inland on the continent; for the *Americans*, who trade with the *Hudfon*'s bay company, ufe them: but from the figures given by *De Brie*, they do not feem ever to have reached the people of *Virginia* and *Florida*. They inhabit jourts, or fubterraneous dwellings, each common to many families, in which they live in horrible filthinefs: but they are remarkably civilized in their behaviour; and have been taught by the *Ruffians* to pull off their caps, and to bow, in their falutations.

BARROWS.   They bury their dead on the fummits of hills, and raife over the fpot a barrow of ftones †, in the manner cuftomary in all the north of *Europe* in very early days.

    On the north fide of the promontory *Alafchka*, the water decreafes confiderably in depth, and the mountains recede towards the bottom far inland, and leave a large tract of low land between them and the fea. Here it forms a great bay, called BRISTOL BAY AND *Briftol*; with a vaft river at the end, with an entrance a mile broad, feated in RIVER. lat. 58. 27. Cape *Newenham*, lat. 58. 42, a rocky promontory, is the northern horn of the bay, eighty-two leagues from Cape *Oonemak*, its fouthern: an univerfal barrennefs, and want of vegetation, appeared in the neighborhood of the former. The *Walrufes* (N° 71) began, the 15th of *July*, to fhew themfelves in great numbers about this place: a proof that ice is not effential to their exiftence. The inhabitants of this coaft were dreffed much more fqualidly than thofe before feen; but, like the others, deformed their nofes and lips. They fhaved their head or cut the hair clofe, and only left a few locks behind or on one fide, fomewhat in the *Chinefe* fafhion. From Cape *Newenham*, the continent runs due north. GORE'S ISLE. To the weft is *Gore*'s ifland, diftinguifhed by a vaft cliff, in lat. 60. 17, long. 187. 30, called *Point Upright*; and near it a moft rugged, high, rocky iflet, named the *Pinnacles* ‡. Myriads of the Auk tribe haunted thefe precipices. This feems the SEA OTTERS. extreme northern refort of the *Sea Otter*.

    From *Shoal-nefs*, in lat. 60, long. 196, there is a gap in the *American* geography, as far as *Point Shallow Water*, lat. 62. 50; and not far from thence were the fymptoms of the difcharge of fome great river, from the uninveftigated part. Be-

---

* *Voyage*, ii. 510.   † Same, 521.   ‡ See tab. 87.

yond *Point Shallow*, in lat. 63. 33, is Cape *Stephens*; and before it, at a fmall diftance, *Stuart's* ifle. Thefe make the fouthern points of *Norton's Sound*, formed by a vaft recefs of the land to the eaft. All the land near the fea is low and barren, bounded inland by mountains. The trees, which were Birch, Alder, Willow, and Spruce, very fmall; none of the laft above fix or eight inches in diameter: but the drift-wood, which lay in plenty on the fhore, much larger; having been brought down the rivers from land more favorable to its growth. Towards the bottom of the found, Cape *Denbigh* juts far to the weft into the water, and forms a peninfula. It has been an ifland; for there are evident marks on the ifthmus, that the fea had once poffeffed its place: a proof of the lofs of the element of water in thefe parts, as well as in other remote parts of the globe.

The found, from Cape *Denbigh*, is fuddenly contracted, and is converted into a deep inlet, feemingly the reception of a large river. The continent, in thefe parts, confifts of vaft plains, divided by moderate hills; the former watered by feveral rivers meandering through them. Vegetation improves in proportion to the diftance from the fea, and the trees increafe in bulk. A promontory, called *Bald Head*, bounds the northern entrance into this inlet. Farther to the weft Cape *Darby*, in lat. 64. 21, makes the northern horn of this great found.

Numbers of people inhabit this coaft. The men were about five feet two inches high; and in form and features refembled all the natives feen by the navigators fince they left *Nootka Sound*. They had, in their under lip, two perforations. The color of their fkin was that of copper: their hair fhort and black: the beard of the men fmall: their language a dialect of the *Efkimaux*. Their clothing is chiefly of Deer fkins, with large hoods, made in the form of loofe jackets, fcarcely reaching lower than half the thigh; where it was almoft met by a great wide-topped boot. The *Efkimaux* occafionally ftick their children in the top: the women of this country place them more commodiously within the upper part of the jacket, over one fhoulder *. In language there feems confiderable conformity. They had, like them, the woman's boat, and the *Kaiack*: the firft they fometimes made ufe of as a protection from the weather, by turning it upfide down, and fheltering beneath. But their hovels were the moft wretched of any yet feen; confifting of only a floping roof (without any fide walls) compofed of logs; a floor of the fame; the entrance at one end, and a hole to permit the efcape of the fmoke. Thefe poor people feem very fufceptible of feelings for the misfortunes of each other, which would do honor to the moft polifhed ftate. A family appeared, one of which was a moft diftorted figure, with fcarcely the human form:

* See tab. 54.

another,

another, feemingly the chief, almoft blind : the third, a girl : the laft, the wife. She made ufe of Captain KING to act as a charm to reftore her blind hufband to his fight *. He was firft directed to hold his breath ; then to breathe on, and afterwards to fpit on his eyes. We are not without fimilar fuperftitions. The *Romans* † applied the fame remedy to difeafes of the fame part : but I doubt whether they, or our polifhed nation, ever expreffed the fame feelings as this poor woman did. She related her ftory in the moft pathetic manner ; fhe preffed the hands of the Captain to the breaft of her hufband, while fhe was relating the calamitous hiftory of her family ; pointed fometimes to the hufband, fometimes to the cripple, and fometimes to the poor child. Unable to contain any longer, fhe burft into tears and lamentation. She was followed by the reft of her kindred in an unifon, which, I truft, filled the eyes of the civilized beholders, as their relation has mine.

From Cape *Darby* the land trends to the weft, and ends in *Point Rodney* ; low land, with high land far beyond, taking a northerly direction inland. Off this point, in lat. 64. 30, is *Sledge ifland*, fo called from a fledge being found on it, refembling thofe which the *Ruffians* ufe in *Kamtfchatka* to carry goods over the fnow. It was ten feet long, twenty inches broad, with a rail on each fide, and fhod with bone ; all neatly put together, in fome parts with wooden pins, but moftly with thongs of whalebone : a proof of the ingenuity of the natives. Whether it was to be drawn with dogs or rein-deer, does not appear ; for the ifland was deferted, and only the remains of a few *jourts* to be feen. In lat. 64. 55, long. 192, is KING's ifland, named in honor of the able and worthy continuator of the voyage. The continent oppofite to it bends towards the eaft, and forms a fhallow bay ; then fuddenly runs far into the fea, and makes the moft weftern extremity yet known, and probably the moft weftern of all. On it were feveral huts ; and ftages of bone, fuch as had been obferved in the *Tfchutfki* country. This cape forms one fide of BERING's ftreights, and lies nearly oppofite to *Eaft Cape*, on the *Afiatic* fhore, at the fmall diftance of only thirty-nine miles. This lies in lat. 65. 46 ; is named *Cape Prince of Wales* ; is low land, and the heights, as ufual, appeared beyond ; among which is a remarkable peaked hill. It would be unjuft to the memory of paft navigators, not to fay, that there is the greateft probability that either this cape, or part of the continent adjacent to it, was difcovered, in 1730, by *Michael Gwofdew*, a land furveyor attendant on the *Coffack*,

---

* See *Voyage*, ii. 481.   † Mulieris falivam quoque jejunæ potentem dijudicant oculis cruentatis.—*Plin. Hift. Nat.* lib. xxviii. c. 7.

Colonel

Colonel *Scheſtakow*, in the unfortunate expedition undertaken by him to render the *Tſchutſki* tributary *.

Here begins the *Icy Sea* or *Frozen Ocean*. The country trends ſtrongly to the eaſt, and forms, in lat. 67. 45, long. 194. 51, *Point Mulgrave* ; the land low, backed inland with moderate hills, but all barren, and deſtitute of trees. From hence it makes a ſlight trend to the weſt. *Cape Liſburn* lies in lat. 69 ; and *Icy Cape*, the moſt extreme land ſeen by any navigators on this ſide, was obſerved in lat. 70. 29, long. 198. 20, by our illuſtrious ſeaman, on *Auguſt* 18th 1778. The preceding day he had made an advance as high as 70. 41 ; but, baffled by impenetrable ice, upon the juſteſt reaſoning was obliged to give up all thoughts of the north-eaſt paſſage : which reaſons were confirmed, in the following year, by his ſucceſſor in command, Captain CLERKE. All the trials made by that perſevering commander could not attain a higher latitude than 70. 11, long. 196. 15. He found himſelf laboring under a lingering diſeaſe, which he knew muſt be fatal, unleſs he could gain a more favorable climate ; but his high ſenſe of honor, and of his duty to his orders, determined him to perſiſt, till the impoſſibility of ſucceſs was determined by every officer. He gave way to their opinion, ſailed towards the ſouthward on *July* 21ſt, and on *Auguſt* 22d honorably ſunk, at the age of thirty-eight, under a diſorder contracted by a continued ſcene of hardſhips, endured from his earlieſt youth in the ſervices of his country †.

To ſuch characters as theſe we are indebted for the little we know, and probably all that can be known, of the ICY SEA. The antients had ſome obſcure notion of its coaſts, and have given it the name of *Scythicum Mare* ; a cape jutting into it was ſtyled *Scythicum Promontorium* ; and an iſland at the bottom of a deep bay to the weſt of it, *Scythica Inſula*. It is following the conjectures of the ingenious to ſay, that the firſt may be the Cape *Jalmal*, and the laſt, *Nova Zemlja*, which ſome will make the *Inſula Tazata* of *Pliny*, as it reſembles in name the river *Tas*, which flows almoſt oppoſite to it into the gulph of *Ob* ‡. The knowlege which the antients had of theſe parts muſt have been from traffic. The old *Ladoga* was, in very early times, a place of great commerce, by aſſiſtance of rivers and ſeas, even from the fartheſt parts of the *Mediterranean* ; the coins of *Syria*, *Arabia*, *Greece*, and *Rome*, having been found in the burial-places adjoining to that antient city §. Another channel of knowlege was formed from the great traffic carried on by the merchants, from even the remote *India*, up the *Volga* and the *Kama*, and from thence to *Tſcherdyn*, an emporium on the river *Kolva*,

* *Decouvertes*, &c. i. 166.   † See the particulars of his ſervices, *Voyage*, iii. 280.
‡ *Strahlenberg Hiſt. Ruſſia*, 113.   § Same, 110.

x

ſeated

feated in the antient *Permia* or *Biormia*, and not far fouth of the river *Peczora*.
From thence the *Biormas*, who feem to have been the factors, embarked with the
merchandize on that river, went down with it to the coafts of the *Frozen Sea*; and,
after obtaining furs in exchange, they returned and delivered them, at *Tfcherdyn*, to
the foreign merchants * : and from them the antients might pick up accounts.

The ICY SEA extends from *Nova Zemlja* to the coaft of *America*. We have
feen how unable even the *Ruffians* have been to furvey its coafts, except by interrupted detail, notwithftanding it formed part of their own vaft empire. To
our navigators was given the honor not only of fettling parts of its geography
with precifion, but of exploring the whole fpace between the moft northern promontory of *Afia* and the fartheft accefiible part of *America*. This was a tract
of one hundred leagues †. The traverfing it was a work of infinite difficulty
and danger. The fea fhallow ; and the change from the greateft depth, which did
not exceed thirty fathoms, to the leaft, which was only eight, was fudden : the
bottom muddy, caufed by the quantity of earth brought down from the vaft rivers which pour into it from the *Afiatic* fide. We fufpect that it receives but
few from the *American*, their general tendency being eaft and weft. The *Icy Sea*
is fhallow, not only becaufe its tides and currents are very inconfiderable ; but its
outlet through the ftreights of *Bering* very narrow, and even obftructed in the
middle by the iflands of *St. Diomedes*. both which circumftances impede the
carrying away of the mud. The current, fmall as it is, comes chiefly from the
fouth-weft, and is another impediment. The land of each continent is very low near
the fhores, and high at a fmall diftance from them : the former is one inftance of
a correfpondent fhallownefs of water. The foundings off each continent, at the
fame diftances from the fhore, were exactly the fame.

The ice of this fea differs greatly from that of *Spitzbergen*. It probably is entirely generated from the fea-water. The *Icy Sea* feems to be in no part bounded
by lofty land, in the valleys of which might have been formed the ftupendous
icebergs, which, tumbling down, form thofe lofty iflands we had before occafion
to mention. The ice here is moveable, except about the great headlands, which
are befet with a rugged mountanous ice. It is notorious, that a ftrong gale
from the north in twenty-four hours covers the whole coaft, for numbers of miles
in breadth ; will fill the ftreights of BERING, and even the *Kamtfchatkan* feas ;
and in fmaller pieces extend to its iflands. In the *Icy Sea* it confifts chiefly
of field ice. Some fields, very large, and furrounded with leffer, from forty

DEPTH.

ICE.

---

* *Nichols's Ruffian Nations*, i. 176.   † *Voyage*, iii. 277.

to fifty yards in extent, to four or five; the thicknefs of the larger pieces was about thirty feet under water; and the greateft height of others above, about fixteen or eighteen. It was tranfparent, except on the furface, which was a little porous, and often very rugged: the reft compact as a wall. At times it muft pack; for the mountanous ice which the *Coffack Morkoff* afcended (fee p. c.) muft have been of that nature. The deftruction of the ice is not effected by the fun, in a climate where fogs reign in far greater proportion than the folar beams; neither will the ftreights of BERING permit the efcape of quantity fufficient to clear the fea of its vaft load. It muft, in a little time, become wholly filled with it, was it not for the rage of the winds, which dafhes the pieces together, breaks and grinds them into minute parts, which foon melt, and refolve into their original element.

The animals of this fea are very few, and may be reduced to the *Polar Bear*, N° 18; the *Walrus*, N° 71; and Seals. The firft does not differ from thofe of other arctic countries: it is beautifully engraven in tab. LXXIII. of the *Voyage*. Amidft the extraordinary fcenery in tab. LII. is given the only accurate figure of the *Walrus* I have ever feen. I cannot but fufpect it to be a variety of the fpecies found in the *Spitzbergen* feas. The tufks are more flender, and have a flight diftinguifhing flexure: the whole animal is alfo much lefs. The length of one (not indeed the largeft) was only nine feet four inches; its greateft circumference feven feet ten; weight, exclufive of the entrails, about eleven hundred pounds. They lay on the ice by thoufands; and in the foggy weather cautioned our navigators, by their roaring, from running foul of it. They are ufually feen fleeping, but never without fome centinels to give notice of approaching danger: thefe awakened the next to them, they their neighbors, till the whole herd was roufed. Thefe animals are the objects of chace with the *Tfchutfki*, who eat the flefh, and cover their boats and hovels with the fkins. Whales abound in this fea. Fifh, the food of Seals, and partly of the polar Bears, muft be found here, notwithftanding they efcaped the notice of the navigators. Shells and fea-plants, the food of the *Walrus*, cannot be wanting.

Many fpecies of birds (which will occur in their place) were feen traverfing this fea. Geefe and Ducks were obferved migrating fouthward in *Auguft*; whether from their breeding-place in a circum-polar land, or whether from the probably far-extending land of *America*, is not to be determined. Drift-wood was very feldom feen here. Two trees, about three feet in girth, with their roots, were once obferved, but without bark or branches; a proof that they had been brought from afar, and left naked by their conteft with the ice and elements.

The fea, from the fouth of BERING's ftreights to the crefcent of ifles between

ANIMALS.

FISH.

BIRDS.

Afia and *America*, is very fhallow. It deepens from thefe ftreights (as the *Britifh* feas do from thofe of *Dover*) till foundings are loft in the Pacific Ocean.; but that does not take place but to the fouth of the ifles. Between them and the ftreights is an increafe from twelve to fifty-four fathom, except only off St. *Thaddeus Nofs*, where there is a channel of greater depth. From the vulcanic difpofition I am led to believe not only that there was a feparation of the continents at the ftreights of BERING, but that the whole fpace, from the ifles to that fmall opening, had once been occupied by land; and that the fury of the watery element, actuated by that of fire, had in moft remote times, fubverted and overwhelmed the tract, and left the iflands monumental fragments.

Whether that great event took place before or after the population of *America*, is as impoffible, as it is of little moment, for us to know. We are indebted to our navigators for fettling the long difpute about the point from which it was effected. They, by their difcoveries, prove, that in one place the diftance between continent and continent is only thirty-nine miles, not (as a celebrated cavilift * would have it) eight hundred leagues. This narrow ftreight has alfo in the middle two iflands, which would greatly facilitate the migration of the *Afiatics* into the New World, fuppofing that it took place in canoes, after the convulfion which rent the two continents afunder. Befides, it may be added, that thefe ftreights are, even in the fummer, often filled with ice; in winter, often frozen: in either cafe mankind might find an eafy paffage; in the laft, the way was extremely ready for quadrupeds to crofs, and ftock the continent of *America*. I may fairly call in the machinery of vulcanoes to tear away the other means of tranfit farther to the fouth, and bring in to my affiftance the former fuppofition of folid land between *Kamtfchatka* and *Oonalafcha*, inftead of the crefcent of iflands, and which, prior to the great cataftrophe, would have greatly enlarged the means of migration; but the cafe is not of that difficulty to require the folution. One means of paffage is indifputably eftablifhed.

But where, from the vaft expanfe of the north-eaftern world, to fix on the firft tribes who contributed to people the new continent, now inhabited almoft from end to end, is a matter that baffles human reafon. The learned may make bold and ingenious conjectures, but plain good fenfe cannot always accede to them. As mankind encreafed in numbers, they naturally protruded one another forward. Wars might be another caufe of migrations. I know no reafon why the *Afiatic* north might not be an *officina virorum*, as well as the *European*. The overteeming country, to the eaft of the *Riphaean* mountains, muft find it neceffary to difcharge its inhabitants: the firft great wave of people was forced forward by the next to

* The author of *Recherches Philofophiques fur les Americains*, i. 136.

it,

it, more tumid and more powerful than itself: fucceffive and new impulfes continually arriving, fhort reft was given to that which fpread over a more eaftern tract; difturbed again and again, it covered frefh regions; at length, reaching the fartheft limits of the Old World, found a new one, with ample fpace to occupy unmolefted for ages; till *Columbus* curfed them by a difcovery, which brought again new fins and new deaths to both worlds.

The inhabitants of the New do not confift of the offspring of a fingle nation: different people, at feveral periods, arrived there; and it is impoffible to fay, that any one is now to be found on the original fpot of its colonization. It is impoffible, with the lights which we have fo recently received, to admit that *America* could receive its inhabitants (at left the bulk of them) from any other place than eaftern *Afia*. A few proofs may be added, taken from cuftoms or dreffes common to the inhabitants of both worlds: fome have been long extinct in the old, others remain in both in full force.

The cuftom of fcalping was a barbarifm in ufe with the *Scythians*, who carried about them at all times this favage mark of triumph: they cut a circle round the neck, and ftripped off the fkin, as they would that of an ox*. A little image, found among the *Kalmucs*, of a *Tartarian* deity, mounted on a horfe, and fitting on a human fkin, with fcalps pendent from the breaft, fully illuftrates the cuftom of the *Scythian* progenitors, as defcribed by the *Greek* hiftorian. This ufage, as the *Europeans* know by horrid experience, is continued to this day in *America*. The ferocity of the *Scythians* to their prifoners extended to the remoteft part of *Afia*. The *Kamtfchatkans*, even at the time of their difcovery by the *Ruffians* †, put their prifoners to death by the moft lingering and excruciating inventions; a practice in full force to this very day among the aboriginal *Americans*. A race of the *Scythians* were ftyled *Anthropophagi* ‡, from their feeding on human flefh. The people of *Nootka Sound* ftill make a repaft on their fellow creatures §: but what is more wonderful, the favage allies of the *Britifh* army have been known to throw the mangled limbs of the *French* prifoners into the horrible cauldron, and devour them with the fame relifh as thofe of a quadruped ‖.

The *Scythians* were fayed, for a certain time, annually to transform themfelves into wolves, and again to refume the human fhape ¶. The new-difcovered *Americans* about *Nootka Sound*, at this time difguife themfelves in dreffes made of the fkins of wolves and other wild beafts, and wear even the heads fitted to their

CUSTOMS COMMON TO AMERICA AND THE NORTH OF ASIA.

---

* *Herodotus*, lib. iv.—Compare the account given by the hiftorian with the *Tartarian icunculus*, in Dr. PALLAS's *Travels*, i. tab. x. *a*.   † *Hift. Kamtfchatka*, 57.   ‡ *Mela*, lib. ii. c. 1.
§ *Voyage*, ii.   ‖ *Colden's Five Indian Nations*, i. 155.   ¶ *Heredotus*, lib. iv.

own.

own *. These habits they use in the chace, to circumvent the animals of the field. But would not ignorance or superstition ascribe to a supernatural metamorphosis these temporary expedients to deceive the brute creation?

In their marches the *Kamtschatkans* never went abreast, but followed one another in the same track †. The same custom is exactly observed by the *Americans.*

The *Tungusi*, the most numerous nation resident in *Sibiria*, prick their faces with small punctures, with a needle, in various shapes; then rub into them charcoal, so that the marks become indelible ‡. This custom is still observed in several parts of *America.* The *Indians* on the back of *Hudson*'s bay, to this day perform the operation exactly in the same manner, and puncture the skin into various figures; as the natives of *New Zealand* do at present, and as the antient *Britons* did with the herb *Glastum*, or Woad §; and the *Virginians*, on the first discovery of that country by the *English* ‖.

The *Tungusi* use canoes made of birch-bark, distended over ribs of wood, and nicely sewed together ¶. The *Canadian*, and many other *American* nations, use no other sort of boats. The paddles of the *Tungusi* are broad at each end; those of the people near *Cook*'s river, and of *Oonalascha*, are of the same form.

In burying of the dead, many of the *American* nations place the corpse at full length, after preparing it according to their customs; others place it in a sitting posture, and lay by it the most valuable cloathing, wampum, and other matters. The *Tartars* did the same: and both people agree in covering the whole with earth, so as to form a *tumulus*, barrow, or carnedd **

Some of the *American* nations hang their dead in trees. Certain of the *Tungusi* observe a similar custom.

I can draw some analogy from dress: conveniency in that article must have been consulted on both continents, and originally the materials must have been the same, the skins of birds and beasts. It is singular, that the conic bonnet of the *Chinese* should be found among the people of *Nootka.* I cannot give into the notion, that the *Chinese* contributed to the population of the New World; but I can readily admit, that a shipwreck might furnish those *Americans* with a pattern for that part of the dress.

SIMILAR FEATURES.

In respect to the features and form of the human body, almost every tribe found along the western coast has some similitude to the *Tartar* nations, and still retain the little eyes, small noses, high cheeks, and broad faces. They vary in size,

---

* *Voyage*, ii. 311, 329.—A very curious head of a Wolf, fitted for this use, is preserved in the LEVERIAN *Museum.*   † *Hist. Kamtsch.* 61.   ‡ *Bell's Travels*, oct. ed. i. 240. § *Herodian in Vita Severi*, lib. iii.   ‖ *De Bry, Virginia*, tab. iii. 111.   ¶ *Ysbrandt Ides*, in *Harris's Coll.* ii. 929.   ** Compare *Colden*, i. 17; *Lafitau*, i. 416; and *Archæologia*, ii. 222, tab. xiv.

from

from the lufty *Calmucs* to the little *Nogaians*. The internal *Americans*, fuch as the *Five Indian* nations, who are tall of body, robuft in make, and of oblong faces, are derived from a variety among the *Tartars* themfelves. The fine race of *Tfchutfki* feem to be the ftock from which thofe *Americans* are derived. The *Tfchutfki* again, from that fine race of *Tartars*, the *Kabardinfki*, or inhabitants of *Kabarda*.

But about *Prince William's Sound* begins a race, chiefly diftinguifhed by their drefs, their canoes, and their inftruments of the chace, from the tribes to the fouth of them. Here commences the *Efkimaux* people, or the race known by that name in the high latitudes of the eaftern fide of the continent. They may be divided into two varieties. At this place they are of the largeft fize. As they advance northward they decreafe in height, till they dwindle into the dwarfifh tribes which occupy fome of the coafts of the *Icy Sea* *, and the maritime parts of *Hudfon's* bay, of *Greenland*, and *Terra de Labrador*. The famous *Japanefe* map † places fome iflands feemingly within the ftreights of BERING, on which is beftowed the title of *Ya Zue*, or the kingdom of *the dwarfs*. Does not this in fome manner authenticate the chart, and give us reafon to fuppofe that *America* was not unknown to the *Japanefe*, and that they had (as is mentioned by *Kæmpfer* and *Charlevoix* ‡) made voyages of difcovery, and, according to the laft, actually wintered on the continent? That they might have met with the *Efkimaux* is very probable ; whom, in comparifon of themfelves, they might juftly diftinguifh by the name of dwarfs. The reafon of their low ftature is very obvious : thefe dwell in a moft fevere climate, amidft penury of food ; the former in one much more favorable, abundant in provifions ; circumftances that tend to prevent the degeneracy of the human frame. At the ifland of *Oonalafcha* a dialect of the *Efkimaux* is in ufe, which was continued along the whole coaft, from thence northward. I have before mentioned the fimilarity in the inftruments between the *Americans* of this fide of the coaft and the *Efkimaux*, which is continued even to *Greenland*.

I cannot think the accounts well fupported, that *America* received any part of its firft inhabitants from *Europe*, prior to the fifteenth century. The *Welfh* fondly imagine that our country contributed, in 1170, to people the New World, by the adventure of *Madoc*, fon of *Owen Gwynedd*, who, on the death of his father, failed there, and colonized part of the country. All that is advanced in proof is, a quotation from one of our poets, which proves no more than that he had diftinguifhed himfelf by fea and land. It is pretended that he made two voyages : that failing weft, he left *Ireland* fo far to the north, that he came to a land unknown, where

---

* See Mr. *Hearne's Difcoveries*.   † Given by *Kæmpfer* to Sir *Hans Sloane*, and now preferved in the *Britifh Mufeum*   ‡ *Hift. Japan.* i. 67.—*Charlevoix, fafies Chronologiques*, ann. 168.

he faw many ftrange things: that he returned home, and, making a report of the fruitfulnefs of the new-difcovered country, prevaled on numbers of the *Welfh* of each fex to accompany him on a fecond voyage, from which he never returned. The favorers of this opinion affert, that feveral *Welfh* words, fuch as *gwrando*, to hearken or liften; the ifle of *Croefo* or *welcome*; *Cape Breton*, from the name of our own ifland; *gwynndwr*, or the white water; and *pengwin*, or the bird with a white head; are to be found in the *American* language *. I can lay little ftrefs on this argument, becaufe likenefs of found in a few words will not be deemed fufficient to eftablifh the fact; efpecially if the meaning has been evidently perverted: for example, the whole *Pinguin* tribe have unfortunately not only black heads, but are not inhabitants of the northern hemifphere; the name was alfo beftowed on them by the *Dutch*, a *Pinguedine*, from their exceffive fatnefs †: but the inventor of this, thinking to do honor to our country, inconfiderately caught at a word of *European* origin, and unheard of in the New World. It may be added, that the *Welfh* were never a naval people; that the age in which *Madoc* lived was peculiarly ignorant in navigation; and the moft which they could have attempted muft have been a mere coafting voyage.

NORWEGIANS. The *Norwegians* put in for fhare of the glory, on grounds rather better than the *Welfh*. By their fettlements in *Iceland* and in *Greenland*, they had arrived within fo fmall a diftance of the New World, that there is at leaft a poffibility of its having been touched at by a people fo verfed in maritime affairs, and fo adventurous, as the antient *Nortmans* were. The proofs are much more numerous than thofe produced by the *Britifh* hiftorians; for the difcovery is mentioned in feveral of the *Icelandic* manufcripts. The period was about the year 1002, when it was vifited by one *Biorn*; and the difcovery purfued to greater effect by *Leif*, the fon of *Eric*, the difcoverer of *Greenland*. It does not appear that they reached farther than *Labrador*; on which coaft they met with *Efkimaux*, on whom they beftowed the name of *Skrælingues*, or dwarfifh people, from their fmall ftature. They were armed with bows and arrows, and had leathern canoes, fuch as they have at prefent. All this is probable; nor fhould the tale of the *German*, called *Turkil*, one of the crew, invalidate the account. He was one day miffing; but foon returned, leaping and finging with all the extravagant marks of joy a *bon vivant* could fhew, on difcovering the inebriating fruit of his country, the grape ‡: *Torfæus* even fays, that he returned in a ftate of intoxication §. To convince his commander, he brought feveral bunches, who from that circumftance named the country *Vinland*. I do

---

* *Powel's Hift. Wales*, 228, 229.   † *Cluf. Exot.* 101.   ‡ *Mallet's Northern Antiq.*
Fngl. ed. i. 284.   § *Hift. Vinlandiæ antiq. per Thorm. Torfæum*, p. 8.

not deny that *North America* produces the true vine *; but it is found in far lower latitudes than our adventurers could reach in the time employed in their voyage, which was comprehended in a very small space. I have no doubt of the discovery; but, as the land was never colonized, nor any advantages made of it, it may be fairly conjectured, that they reached no farther than the barren country of *Labrador*.

The continent which stocked *America* with the human race, poured in the brute creation through the same passage. Very few quadrupeds continued in the peninsula of *Kamtschatka*. I can enumerate only twenty-five which are inhabitants of land; for I must omit the marine animals, which had at all times power of changing their situation: all the rest persisted in their migration, and fixed their residence in the New World. Seventeen of the *Kamtschatkan* quadrupeds are found in *America*: others are common only to *Sibiria* or *Tartary*, having, for unknown causes, entirely evacuated *Kamtschatka*, and divided themselves between *America* and the parts of *Asia* above cited. Multitudes again have deserted the Old World, even to an individual, and fixed their seats at distances most remote from the spot from which they took their departure; from mount *Ararat*, the resting-place of the ark, in a central part of the Old World, and excellently adapted for the dispersion of the animal creation to all its parts. We need not be startled at the vast journies many of the quadrupeds took to arrive at their present seats: Might not numbers of species have found a convenient abode in the vast *Alps* of *Asia*, instead of wandering to the *Cordilleras* of *Chili?* or might not others have been contented with the boundless plains of *Tartary*, instead of travelling thousands of miles, to the extensive flats of *Pampas?*—To endeavour to elucidate common difficulties is certainly a trouble worthy of the philosopher and of the divine; not to attempt it would be a criminal indolence, a neglect to

Vindicate the ways of God to man.

But there are multitudes of points beyond the human ability to explain, and yet are truths undeniable: the facts are indisputable, notwithstanding the causes are concealed. In such cases, faith must be called in to our relief. It would certainly be the height of folly to deny to that Being who broke open the great fountains of the deep to effect the deluge—and afterwards, to compel the dispersion of mankind to people the globe, directed the confusion of languages—powers inferior in their nature to these. After these wondrous proofs of Omnipotency,

* *Glover's Account of Virginia, Phil. Transf. Abr.* iii. 570.

y

it

it will be abfurd to deny the poffibility of infufing inftinct into the brute creation. DEUS *eft anima brutorum*; GOD himfelf is the foul of brutes: His pleafure muft have determined their will, and directed feveral fpecies, and even whole genera, by-impulfe irrefiftible, to move by flow progreffion to their deftined regions. But for that, the *Llama* and the *Pacos* might ftill have inhabited the heights of *Armenia* and fome more neighboring *Alps*, inftead of laboring to gain the diftant *Peruvian Andes*; the whole genus of *Armadillos*, flow of foot, would never have abfolutely quitted the torrid zone of the Old World for that of the New; and the whole tribe of Monkies would have gambolled together in the forefts of *India*, inftead of dividing their refidence between the fhades of *Indoftan* and the deep forefts of the *Brafils*. Lions and Tigers might have infefted the hot parts of the New World, as the firft do the deferts of *Africa*, and the laft the provinces of *Afia*; or the Pantherine animals of *South America* might have remained additional fcourges with the favage beafts of thofe antient continents. The Old World would have been overftocked with animals; the New remained an unanimated wafte! or both have contained an equal portion of every beaft of the earth. Let it not be objected, that animals bred in a fouthern climate, after the defcent of their parents from the ark, would be unable to bear the froft and fnow of the rigorous north, before they reached *South America*, the place of their final deftination. It muft be confidered, that the migration muft have been the work of ages; that in the courfe of their progrefs each generation grew hardened to the climate it had reached; and that after their arrival in *America*, they would again be gradually accuftomed to warmer and warmer climates, in their removal from north to fouth, as they had in the reverfe, or from fouth to north. Part of the Tigers ftill inhabit the eternal fnows of *Ararat*, and multitudes of the very fame fpecies live, but with exalted rage, beneath the Line, in the burning foil of *Borneo* or *Sumatra*; but neither Lions or Tigers ever migrated into the New World. A few of the firft are found in *India* and *Perfia*, but they are found in numbers only in *Africa*. The Tiger extends as far north as weftern *Tartary*, in lat. 40. 50, but never has reached *Africa*. I fhall clofe this account with obferving, that it could be from no other part of the globe except *Afia*, from whence the New World could receive the animal creation.

The late voyage of the illuftrious COOK has reduced the probable conjectures of philofophers into certainty. He has proved that the limits of the Old and New World approach within thirteen leagues of each other. We know that the intervening ftreights are frequently frozen up; and we have great reafon to fuppofe, that the two continents might have been once united, even as low as the *Aleutian* iflands, or lat. 52. 30. Thus are difcovered two means of paffage from *Afia* to *America*; the laft

5   in

in a climate not more rigorous than that which feveral animals might very well endure, and yet afterwards proceed gradually to the extreme of heat.

In fact, every other fyftem of the population of the New World is now over-thrown. The conjectures of the learned, refpecting the vicinity of the Old and New, are now, by the difcoveries of our great navigator, loft in conviction. The ftrained fyftems of divines, laudably indeed exerted in elucidating SACRED WRIT, appear to have been ill-founded; but, in the place of imaginary hypothe-fes, the real place of migration is uncontrovertibly pointed out. Some (from a paffage in *Plato*) have extended over the *Atlantic*, from the ftreights of *Gibraltar* to the coaft of *North* and *South America*, an ifland equal in fize to the continents of *Afia* and *Africa*; over which had paffed, as over a bridge, from the latter, men and animals; wool-headed Negroes, and Lions and Tigers *, none of which ever exifted in the New World. A mighty fea arofe, and in one day and night engulphed this ftupendous tract, and with it every being which had not com-pleted its migration into *America*. The whole Negro race, and almoft every Quadruped, now inhabitants of *Africa*, perifhed in this critical day. Five only are to be found at prefent in *America*; and of thefe only one, the Bear †, in *South America*. Not a fingle cuftom, common to the natives of *Africa* and *America*, to evince a common origin. Of the Quadrupeds, the Bear, Stag, Wolf, Fox, and Weefel, are the only animals which we can pronounce with certainty to be found on each continent. The Stag ‡, Fox, and Weefel, have made alfo no farther progrefs in *Africa* than the north; but on the fame continent the Wolf is fpread over every part, yet is unknown in *South America*, as are the Fox and Weefel. I fufpect, befides, that the Stag hath not advanced farther fouth than *Mexico*. In *Africa* and *South America* the Bear is very local, being met with only in the north of the firft, and on the *Andes* in the laft. Some caufe unknown arrefted its progrefs in *Africa*, and impelled the migration of a few into the *Chilian Alps*, and induced them to leave unoccupied the vaft tract from *North America* to the lofty *Cordilleras*.——My promifed Table of Quadrupeds will at once give a view of thofe which inhabit *North America*, and are either peculiar to it, or are met

---

* *Catcott on the Deluge*, edit. 2d. p. 139, 15, &c.

† On the reafoning of Mr. *Zimmerman* (*Zool. Geogr.* 476), and the opinion of Mr. *Erxleben* (*Syft. Regn. An.* 508), I give up my notion of the Panther (*Hift. Quad.* Nᵒ 153), being a native of *South America*. It is moft probable, that the fkin which I faw at a furrier's fhop, which was faid to have been brought from the *Brafils*, had originally been carried there from the weftern coaft of *Africa*, where the *Portuguefe* have confiderable fettlements, and a great flave-trade for their *American* colonies, and where thofe animals abound.

‡ *Shaw's Travels*, 243. Quere? whether exactly the fame with the *European*.

with in other countries. It certainly will point out the courſe they have taken in their migration; and, in caſe miſnomers are avoided, will reduce to the ſingle continent of *Aſia* the original country from whence they ſprung. Men of the firſt abilities, and firſt in learning, who have neglected the ſtudy of natural hiſtory, will give Lions and Tigers to *America*, miſled by the ignorance of travellers, who miſtake the Puma, N° 14 of this Work, for the firſt; and the ſpotted wild beaſts, allied to the Pantherine race, for the ſecond.

TABLE OF QUADRUPEDS.

HOOFED.

GENUS.			OLD WORLD.	NEW WORLD.
I. Ox.	Biſon,	N° 1.	In parts of *Lithuania*, and about mount *Caucaſus*; except there, univerſally domeſticated.	To the weſt of *Canada*, and as low as *Louiſiana*. In *New Mexico*, on the weſtern ſide of *North America*.
	Muſk,	N° 2.	A variety in the interior parts of *Guinea*, and the ſouth of *Africa*. See *Hiſt. Quad.* i. N° 9.	To the north of *Hudſon's Bay*, from *Churchil* river to lat. 73, and among the *Chriſtinaux*, and in *New Mexico*.
II. Sheep.	Argali,	p. 12.	*Sardinia. Corſica. Crete.* North of *India. Perſian Alps.* About the *Onon* and *Argun*, in *Sibiria. Mongalia*, to lat. 60. Eaſt of the *Lena*, and quite to *Kamtſchatka.*	Suſpected to be found in *California*; but not on the beſt authorities.
III. Deer.	Mooſe,	N° 3.	*Norway. Sweden*, to lat. 64. *Ruſſia. Sibiria*, as low as lat. 53. As far eaſt as Lake *Baikal*; and in the north of *China* to the north of *Corea*. lat. 45*.	*Hudſon's Bay. Canada. Nova Scotia. New England*; and near the northern part of the river *Ohio.*
	Rein,	N° 4.	*Lapland. Norway. Samoiedea.* Along the *Arctic* coaſts,	*Hudſon's Bay.* Northern parts of *Canada. Labrador*,

* Or lat. 42, according to Mr. *Zimmerman*'s new Map.

GENUS.			OLD WORLD.	NEW WORLD.
			coafts, to *Kamtfchatka*. In the *Urallian* mountains to *Kungur*, in lat. 57. 10. About Lake *Baikal*. *Spitzbergen*. *Greenland*.	*brador*. Ifland of *New-foundland*.
Stag,	N° 5.		*Norway*, and moft part of *Europe* to the fouth. In the north of *Afia*. *China*. *Barbary*. E.	From *Canada*, over all parts of *North America*. *Mexico*.
Virginian,	N° 6.		— —	From the provinces fouth of *Canada* to *Florida*. Perhaps in *Guiana*.
Mexican Roe,	N° 7.		— —	Interior north-weftern parts of *America?* *Mexico*.
Roe,	N° 8.		*Norway*. *Sweden*. Moft part of *Europe*, except *Ruffia*. *Scotland*.	According to *Charlevoix*, in *Canada?*

DIGITATED.

DIV. I.

IV. Dog.	Wolf,	N° 9.	From the *Arctic* circle to the moft fouthern part of *Europe*. In *Afia*, from the circle to *Perfia*. *Kamtfchatka*. All parts of *Africa*.	From *Hudfon's Bay* to the moft fouthern parts of *North America*.
	Arctic Fox,	N° 10.	Within the whole *Arctic* circle. *Iceland*. *Spitzbergen*. *Greenland*. *Finmark*. North of *Sibiria*. *Kamtfchatka*, and its ifles.	*Hudfon's Bay*. The ifles in the high latitudes on the weftern fide of *America*.
	Common Fox, N° 11.		In all parts of *Europe*, and the cold and temperate parts of *Afia*. *Kamtf-chatka*,	From *Hudfon's Bay*, crofs the continent to the *Fox Ifles*. *Labrador*. *New-foundland*.

GENUS.			OLD WORLD *chatka*, and its furthest isles. *Iceland*. E.	NEW WORLD. *foundland*. *Canada*. Not further south : a variety only, the Brandt Fox, in *Pensylvania*.
	Grey,	N° 12.	— —	From *New England* to the southern end of *North America*.
V. CAT.	Silvery,	N° 13.	— —	In *Louisiana*.
	Puma,	N° 14.	— —	From *Canada* to *Florida*; thence through *Mexico*, quite to *Quito* in *Peru*.
	Lynx,	N° 15.	Forests of the north of *Europe*, and many of the south. *Spain*. North of *Asia*, and the mountains in the north of *India* *.	From *Canada*, over most parts of *North America*.
	Bay Lynx, N° 16.		— —	In the province of *New York*.
	Mountain, N° 17.		— —	*Carolina*, and perhaps other parts of *North America*.
VI. BEAR.	Polar,	N° 18.	Within the whole polar circle of *Europe* and *Asia*.	The same in *America*; also as low as *Hudson's Bay* and *Labrador*.
	Black,	N° 19.	*Jeso Masima*, north of *Japan*; perhaps in *Japan*.	In all parts of *North America*.
	Brown,	N° 20.	In most parts of *Europe*, north and south. The same in *Asia*, even as far as *Arabia*. *Barbary* †. *Ceylon*. *Kamtschatka*.	To the north-west of *Hudson's Bay*, and on the western side of *America*. About *Nootka Sound*. On the *Andes* of *Peru* ‡.
	Wolverene, N° 21.		North of *Norway*. *Lapland*. North of *Sibiria*. *Kamtschatka*.	As far north as the *Copper River*, and south as the country between lake *Huron* and *Superior*. On the

* As I have been assured by Doctor PALLAS, since the publication of my *History of Quadrupeds*.
† *Shaw's Travels*, 249.   ‡ *Condamine's Travels*, 82.—*Ulloa's Voyage*, i. 461.

GENUS.		OLD WORLD.	NEW WORLD.
			the western side of *North America.*
	Raccoon, N° 22.	— —	From *New England* to *Florida. Mexico.* Isles of *Maria,* near Cape *Corientes,* in the *South Sea.*
VII. BADGER.	N° 23.	In the south of *Norway,* and all the more southern parts of *Europe.* In the temperate parts of *Asia,* as far as *China* eastward. *E.*	In the neighborhood of *Hudson's Bay. Terra de Labrador,* and as low as *Pensylvania.*
VIII. OPOSSUM.	*Virginian,* N° 24.	— —	As far north as *Canada,* and from thence to the *Brasils* and *Peru.*
IX. WEESEL.	Common, N° 25.	Most parts of *Europe. Sibiria. Kamtschatka. Barbary. E.*	*Hudson's Bay. Newfoundland.* As far south as *Carolina.*
	Stoat, N° 26.	All the northern parts of *Europe* and *Asia*; and as far as *Kamtschatka* and the *Kuril* isles. *E.*	*Hudson's Bay,* and as low as *Newfoundland* and *Canada.*
	Pine Martin, N° 27.	North of *Europe.* Rare in *France.* Only in the west of *Sibiria.* In *China. E.*	Northern parts of *North America,* quite to the *South Sea.*
	Pekan, N° 28.	— —	*Hudson's Bay. Canada.*
	Vison, N° 29.	— —	*Canada.*
	Sable, N° 30.	*Sibiria. Kamtschatka. Kuril* isles.	*Canada.*
	Fisher, N° 31.	— —	*Hudson's Bay. New England. Pensylvania.*
	Striated, N° 32.	— —	*Pensylvania* to *Louisiana.*
	Skunk, N° 33.	— —	From *Hudson's Bay* to *Peru.*
X. OTTER.	Common, N° 34.	Northern *Europe* and *Asia. Kamtschatka. E.*	From *Hudson's Bay* to *Louisiana.*

Lesser,

GENUS.			OLD WORLD.	NEW WORLD.
	Leſſer,	N° 35.	About the banks of the Yaik. Poland. Lithuania. Finland.	From *New Jerſey* to *Carolina.*
	Sea,	N° 36.	*Kamtſchatka. Kuril* iſles.	Weſtern coaſts of *America.*

D I V.   II.

GENUS.			OLD WORLD.	NEW WORLD.
XI. HARE.	Varying,	N° 37.	*Scandinavia. Ruſſia. Sibiria. Kamtſchatka. Greenland. E.*	Hudſon's Bay. About COOK's river.
	American,	N° 38.	— —	From *Hudſon's Bay* to the extremity of *North America.*
	Alpine,	N° 39.	From the *Altaic* chain to lake *Baikal*; thence to *Kamtſchatka.*	*Aleutian* iſles. Poſſibly the weſt of *North America.*
XII. BEAVER.	Caſtor,	N° 40.	*Scandinavia.* About the *Jeneſei* and *Kondu.* In *Caſan,* and about the *Yaik.*	From *Hudſon's Bay* to *Louiſiana.*
	Muſk,	N° 41.	— —	From *Hudſon's Bay* to *Louiſiana.*
XIII. PORCU-PINE.	*Canada,*	N° 42.	— —	From *Hudſon's Bay* to *Virginia.*
XIV. MARMOT.	*Quebec,*	N° 43.	— —	*Canada.*
	Maryland,	N° 44.	— —	From *Penſylvania* to the *Bahama* iſles.
	Hoary,	N° 45.	— —	North of *North America.*
	Tail-leſs,	N° 46.	— —	Hudſon's Bay.
	Earleſs,	N° 47.	*Bohemia. Auſtria. Hungary.* From the *Occa* over the temperate parts of *Sibiria.* About *Jakutz. Kamtſchatka.*	Weſtern ſide of *North America.*
XV. SQUIRREL.	*Hudſon,*	N° 48.	— —	Hudſon's Bay. Labrador.

Grey,

GENUS.			OLD WORLD.	NEW WORLD.
	Grey,	N° 49.	— —	*New England* to *Peru* and *Chili.*
	Black,	N° 50.	— —	*New England* to *Mexico.*
	Flying,	N° 51.	— —	From the southern part of *Hudson's Bay* to *Mexico.*
	Hooded,	N° 52.	— —	*Virginia.*
	Severn River, N° 53.		— —	*Hudson's Bay.*
XVI. DORMOUSE.	Striped,	N° 54.	*Sibiria,* as high as lat. 65.	*Hudson's Bay* to *Louisiana.*
	English ?	N° 55.	*Sweden,* and all *Europe* south. E. *Carolina ?*	
XVII. RAT.	Black,	N° 56.	All *Europe.* Many of the *South Sea* islands. E.	The rocks among the *Blue Mountains.*
	American,	N° 58.	*Mongolia.*	*North America.*
	Water,	N° 59.	From *Lapland* to the south of *Europe.* From *Petersburgh* to *Kamtschatka,* and as low as the *Caspian* sea, and *Persia.* E.	From *Canada* to *Carolina.*
	Mouse,	N° 60.	Universal. E.	Among the rocks, with the Black Rat.
	Field,	N° 61.	All *Europe.* Not beyond the *Urallian* chain. E.	*Hudson's Bay.* *New York.*
	Virginian,	N° 62.	— —	*Virginia.*
	Labrador,	N° 63.	— —	*Hudson's Bay. Labrador.*
	Hudson's,	N° 64.	— —	Same places.
	Meadow,	N° 65.	*Sweden.* All temperate *Russia.* In *Sibiria* only to the *Irtisch.* E.	*Hudson's Bay.* *Newfoundland.*
	Hare-tailed ? N° 66.		*Sibiria.*	*Hudson's Bay.*
XVIII. SHREW.	Fœtid,	N° 67.	*Europe. Sibiria. Kamtschatka.* E.	*Hudson's Bay. Carolina.*
XIX. MOLE.	Long-tailed, N° 68.		— —	*New York.* Interior parts of *Hudson's Bay.*
	Radiated,	N° 69.	— —	*New York.*
	Brown,	N° 70.	— —	*New York.*

                 D I V.

DIV. III.

GENUS.			OLD WORLD.	NEW WORLD.
XX. WALRUS.	*Arctic,*	N° 71.	*Spitzbergen. Greenland. No-va Zemlja.* The coast of the *Frozen Sea.* And on the *Asiatic* side, to the south of *Bering's* streights, as low as lat. 62. 50.	*Hudson's Bay.* Gulph of St. *Laurence.* On the western side of *America,* as low as lat. 58. 42.
XXI. SEAL.	Common,	N° 72.	All the *European* and northern *Asiatic* seas, even to the farthest north. *Kamtschatka.* E.	Northern seas of *America.*
	Great,	N° 73.	*Greenland* and *Kamtschatka.* E.	West of *North America.*
	Leporine,	N° 75.	*White Sea. Iceland. Spitzbergen. Kamtschatka.*	There can be no doubt that every species of Seal is found on the *American* coast.
	Harp,	N° 77.	*Spitzbergen. Greenland. Iceland. White Sea. Kamtschatka.*	
	Ursine,	N° 79.	*Kamtschatka. New Zealand.*	West of *America,* and from the isle of *Gallipagos* to *New Georgia.*
	Leonine,	N° 80.	*Kamtschatka.*	West of *America.* Streights of *Magellan. Staten land. Falkland* isles.
XXII. MANATI.	Whale-tailed, } N° 81. }		*Bering's* isle, and near the isle of St. *Mauritius.*	West of *America.*
	Sea Ape, p. 181.		— —	West of *America.*

DIV. IV.

XXIII. BAT.	*New York,* N 82.		*New Zealand.*	*New York.*
	Long haired, } N° 83. }		— —	*Carolina.*
	Noctule, N° 84.		*France.* E.	*Hudson's Bay.*

Some

Some years ago a very important difcovery was made, not very remote from the place where Captain COOK was obliged to defift from his northern voyage. Mr. *Samuel Hearne*, in the fervice of the *Hudfon's Bay* Company, by direction of the governors, began a journey, on *December* 7th 1770, towards the northern limits of *America*. He went attended only by *Indians*, with whom he had been long acquainted. He fet out from *Prince of Wales* fort, 58. 50, north lat. He for a long fpace took a north-weftern courfe, croffed *Menifchtic* lake, in lat. 61, a water thirty-five miles in breadth, full of fine iflands, and joining with the river *Namaffy*. He paffed over *Wiethen* and *Caffed* lakes, and from the laft kept due weft. In *April* he reached *Thleweyaza Yeth*, a fmall lake in long. 19, weft from *Churchil* fort, lat. 61. 30, near which he made fome ftay to build canoes, now requifite againft the breaking up of the froft. From that lake he began a courfe due north, and croffed a chain of lakes, of which *Titumeg* is one. In lat. 64. he went over *Pefhew* lake; after that, the great lake *Cogeed*, out of which iffues a river pointing north-eaft, which is fuppofed to fall into *Baffin*'s bay. About the middle of *June* he croffed the great river *Conge-catha-wha-chaga*, in lat. 68. 46; and from *Churchil* river weft long. 24. 2. About thofe parts are the *Stoney Mountains*, extending in longitude from 116 to 122 from *London*: craggy, and of a tremendous afpect. On *July* 7th he arrived at *Buffalo* lake, in lat. 69. 30 : here he firft faw the *Mufk Buffalo*, N° 2. Near the north end is *Grizzle Bear-hill*, in about lat. 70, fo called from its being the haunt of numbers of thofe animals. On *July* 13th he reached the banks of *Copper River*, which runs due north into the *Icy Sea*. About the fouth end is much wood, and very high hills. Its current is very rapid, and its channel choaked with fhoals, and croffed with ftoney ridges, which form three great cataracts. Its banks are high, the breadth about a hundred and eighty yards ; but in fome places it expands into the form of a lake. In an ifland of the river unfortunately happened to be a fummer encampment of five tents of *Efkimaux*. The *Indians* attendant on Mr. *Hearne* grew furious at the fight of them. It is their firm opinion, that thefe favages are magicians, and that all the evils they experience refult from their incantations. Mr. *Hearne* in vain folicited his *Indians* to forbear injuring thefe poor people. They, with their ufual cowardice, deferred the attack till night, when they furprifed and murdered every one, to the number of between twenty and thirty. A young woman made her efcape, and embraced Mr. *Hearne*'s feet ; but fhe was purfued by a barbarian, and transfixed to the ground. He obferved in their tents (which were made of deer-fkins with the hair on) copper veffels, and whale-bone, and the fkins of Seals, wooden troughs, and kettles made of a foft ftone (by his defcription a *lapis ollaris*), and difhes and fpoons formed from

Copper River.

Eskimaux.

the thick horns of the Buffalo. Their arms are spears, darts, and bows and arrows ; the last pointed with stone or copper, but most rudely made, for want of proper tools. In their dress they much resemble the *Eskimaux* of *Hudson's Bay*, but the tails of their jackets are shorter ; neither do the women, like them, stiffen out the tops of their boots. Their canoes differ in not having long projecting prows, but in other respects are of the same construction. In most circumstances these people resemble those of the *Bay* ; and differ materially only in one, for the men in these pull out by the roots all the hair of their heads.—Mr. *Hearne* first saw the sea on *July* 16, at the distance of eight miles. He went to the mouth of the river (in lat. 72 ; west long. from *London* 121) which he found full of shoals and falls, and inaccessible to the tide, which seemed to flow twelve or fourteen feet. The sea was at this time full of ice, and on many pieces he saw Seals. The land trended both to the east and to the west, and the sea was full of islands. The land about *Copper* river, for the space of nine or ten miles to the sea, consisted of fine marshes, filled in many places with tall Willow, but no sort of berry-bearing shrubs. There are no woods within thirty miles of the mouth of *Copper* river ; and those which then appear, consist of ill-shaped and stunted Pines.

The people who live nearest to this river, are the *Copper-mine Indians*, and the *Plat-cotes de Chiens*, or *Dog-ribbed Indians* ; these have no direct commerce with *Hudson's Bay*, but sell their furs to the more southern *Indians*, who come for them, and bring them down to the settlements. The *Dog-ribbed Indians* still make their knives of stones and bones, and head their arrows with slate. The *Copper Indians* have abundance of native copper in their country ; they make with it ice-chissels and arrow-heads. The mine is not known ; but I find that an *Indian* chief, who had many years ago communication with a Mr. *Frost*, one of the Company's servants, says, that the copper was struck off a rock with sharp stones ; and that it lay in certain islands far to the northward, where was no night during summer *.

Mr. *Hearne* set out on his return the 22d of *July*. He took, in some places, a route different from what he did in going, and got to the settlements in *June* 1772. I have perused the journal, and had frequent conversation with Mr. *Hearne* the last year. I took the liberty to question him about the waters he had crossed during winter upon the ice ; and whether they might not have been at that time obstructed streights, a passage to the *Pacific Ocean*? He assured me, that he could make no mistake : that he passed over many of them in canoes during the sum-

* *Dobbs's Account of Hudson's Bay, &c.* 47.

mer,

mer, and that the others had large rivers running out of them, almoſt every one to the weſt: that the *Indians*, who croſſed them annually, in their way to the north to trade for furs, were exceedingly well acquainted with them, and knew them to be freſh-water lakes; and in particular uſed to fiſh in them for Pikes, fiſh notoriouſly known never to frequent ſalt-water.

I muſt now take a blind unguided courſe along the *Icy Sea*. The charts give the land a turn to the ſouth, in lat. 81. long. 22 from *London*. This is the moſt northern extremity of the country called *Greenland*, if it reaches ſo far; but, beyond the diſcovery by Mr. *Hearne*, in lat. 72, the northern limits given in our charts appear to be merely conjectural. To the ſouth, on the eaſtern coaſt, in 1670, was ſeen land in lat. 79. Another part, in lat. 77. 30, called in the maps *the land of Edam*, was ſeen in 1655. The inlet named *Gael-hamkes*, in lat. 75, was diſcovered in 1664. A headland was obſerved, in 1665, a degree further ſouth: and in 1607 our celebrated *Hudſon* diſcovered what he named *Hold with Hope*, in lat. 73 *. Excepting the laſt, the reſt of the attempts were made by the *Danes*, for the recovery of *Old Greenland*. *Gael-hamkes* alone continues known to navigators, and is annually frequented by *European* Whale-fiſhers, who extend their buſineſs even to this coaſt. It is repreſented as a great ſtreight, twenty-five leagues wide, communicating with *Baffin's Bay*. A ſpecies of Whale, frequent in *Davis's Streights*, and not found on this ſide of the coaſts, is often ſeen here harpooned with the ſtone weapons of the inhabitants of the oppoſite country; which fiſh muſt have eſcaped through this paſſage †. The land to the north of *Gael-hamkes* is level, and not very high; and within five or ſix leagues from it are ſoundings. That to the ſouth is very lofty, and riſes into peaks like that of *Spitzbergen*; and the ſea oppoſite to it is fathomleſs ‡.

In lat. 71. long. 8. weſt from *London*, is *John Mayen*'s iſland, formerly much frequented by Whale-fiſhers; but thoſe animals have now left the neighboring ſea. The north end riſes into a prodigious mountain called *Beerenberg*, or *the Bears*, from its being the haunt of numbers; but it is ſo ſteep as to be inacceſſible to all human creatures. The ſea, within muſket-ſhot from ſhore, was ſixty fathoms deep; a little farther the depth is paſt the reach of the line ‖.

Oppoſite to *Iceland* begins the once-inhabited part of *Old Greenland*. A very deep ſtreight opens a little oppoſite to *Snæfelnas*, and runs acroſs *Greenland*, near *Jacob's Haven*, into *Davis's Streights*, ſo as quite to inſulate the country: it is

GREENLAND.

JOHN MAYEN'S ISLE.

OLD GREENLAND.

---

* *Purchas*, iii. 568. † *Voyages par de Pagès*, ii. 222. ‡ Same. ‖ *Marten's Spitzb.* 186.

now

now almoſt entirely cloſed with ice, and annually fills the ſea with the greateſt *icebergs*, which are forced out of it. A little to the north of the eaſtern entrance are two mountains of a ſtupendous height, called *Blaaſerk* and *Huitſerk*, caſed in perpetual ice. The whole country, to the ſouthern end, conſiſts of ſimilar mountains : a few exhibit a ſtoney ſurface; but the greater part are genuine *glacieres*, ſhooting into lofty peaks, or rugged ſummits : yet ſuch a country as

PEOPLED BY NOR-
WEGIANS.

this became the ſettlement of numbers of *Norwegians* during ſeveral centuries. The valiant *Eric Raude*, or the *Red*, having committed a murder in his own country (a common cauſe for ſeeking adventures, with the heroes of *Greece* as well as *Scandinavia*) fled here in the tenth century. Numbers of his countrymen followed him. *Leif*, his ſon, became a convert to Chriſtianity. Religion flouriſhed here : a biſhoprick was eſtabliſhed, and monaſteries founded. The cathedral was at *Gardar*, a little to the ſouth of the polar circle.

VOYAGE OF THE
ZENI.

In *Hackluyt* * is a relation of the voyage of the two *Zeni* (noble *Venetians*) who in 1380 viſited this country, and give evidence to the exiſtence of the convent, and a church dedicated to St. *Thomas*, poſſeſſed by friers preachers. It appears to have been built near a vulcano, and the materials were lava, cemented with a ſort of *pulvis puteolanus*, which is known to be a vulcanic attendant. A ſpring of boiling water was near the houſe, and was conveyed into it for all their culinary uſes. I am not averſe to giving credit to this account ; there being no reaſon to deny the former exiſtence of burning mountains, when ſuch numbers are to be found in the neighboring *Iceland*; and at this very time there is a fountain of hot water in the iſle of *Onortok*, not remote from *Cape Farewell*†. A ſtrange phraſeology runs through the voyage of theſe two brethren, and perhaps ſome romance; but ſo much truth is every where evident, that I heſitate not to credit the authenticity.

Torfæus enumerates ſeventeen biſhops who preſided over the dioceſe. The laſt prelate was appointed in 1408. The *black death* had almoſt depopulated the country not long before that period. Probably the ſurviving inhabitants fell victims to want, or were extirpated by the natives : for, after that year, we hear no more of them. It certainly had been well inhabited: the ruins of houſes and churches evince its former ſtate. In the fifteenth century the kings of *Denmark* attempted to diſcover whether any of the antient race remained; but all in vain : the adventurers were driven off the coaſt by the ice with which it was blocked up, which remains an invincible obſtacle to re-ſettle the eaſtern coaſt, even were there the leſt temptation. All is a dreadful tract from lat. 81 to *Staten Hook* or *Cape Farewell*, its

* Vol. iii. 123 ; and *Purchas*, iii. 610.   † *Crantz*, i. 18.

ſouthern

southern extremity, on an isle off that point, in lat. 59; on both sides deeply indented with bays, bounded by icy promontories. Many of these bays had been parts of pervious streights, which had divided the country into several islands; but are now totally obstructed with ice. Besides that I before mentioned, was one in lat. 63, called *Bär-sund*; and that in 62. 50, immortalized by the name of our celebrated sailor *Frobisher*, who penetrated into it sixty leagues, in his first voyage in 1576, in his search for a passage to *Cathaya*; but imagined that *Asia* bounded the right side, and *America* the left *. He met with inhabitants, describes them and their œconomy, and is particular about their great dogs, and their use of them in drawing their sledges. In his second voyage he found a *Narwhal* dead on the shore, and has given a figure of it. ' This horne,' says he, ' is to be seene and reserved as a jewel by the Queens Majesties commandemet, in her wardrop of robes †.' — The original map of his voyages is a singular sketch of erroneous supposition. He makes his streights reach to the *Icy Sea*, opposite to what he calls *Cathaya*, just to the north of what is made to resemble the new-discovered streights of *Bering*; which, in the map, are called those of *Anian*; and accidentally gives them a tolerably just form ‡. Those of *Anian* are equally fabulous with those of *de Fuca*, but of prior invention; and, like them, were sayed to have been a passage from the South to the North sea ‖. Queen *Elizabeth* bestowed on his discoveries the name of *Meta Incognita*.

FROBISHER'S STREIGHTS.

Greenland was re-settled with *Norwegians* in 1721, by the zeal of the Reverend Mr. *Hans Egede*, the *Arctic* apostle §. He continued, till 1735, preaching the Gospel to the poor natives; and had not only the happiness of seeing his labors blessed with effect, but his example followed by a numerous set of missionaries, who have formed (on the western side only) many settlements, which flourish even to this day. Mr. *Egede* returned to *Denmark*, founded a seminary for students in the *Greenland* language, from which missionaries were to be drawn; and finished his pious life in 1754.

NEW GREEN-LAND.

At *Cape Farewell* begins the vast opening between *Greenland* and *Terra de Labrador*, which leads to *Hudson's Bay*. Between the west side of *Greenland* and certain vast islands, are *Davis's Streights*, which lead to *Baffin's Bay*. These islands

* 'A true Discourse of the late Voyages of Discoverie for finding a Passage to *Cathaya* by the ' north-west, under the Conduct of *Martin Frobisher*, General. Printed by *Henry Bynnyman*, 1578.' First Voyage, p. 48.

† The Same, Second Voyage, p. 19.

‡ In the same book.

‖ See an account of these imaginary streights in *Drage's Voy.* to *Hudson's Streights*, vol. ii. 68,

§ *Crantz*, i. 279. 285.

in different maps bear different names, and in one are even confolidated ; fo little are thefe parts known *.

To defcribe *Greenland*, would be to ring changes on ice, and fnow, and lofty mountains (fome, according to Mr. *Crantz*, a thoufand fathoms high) rifing into broken crags or fharp fpires, or vallies with no other garniture than mofs and fome moor grafs ; and in fome parts are long flat mountains, clad with perpetual ice and fnow. Where the birds, by their dung, have formed a little foil, fome plants are found. Mr. *Crantz* † enumerates about twenty-four fpecies, befides the cryptogamious kinds. *Egede* obferved, in lat. 60 or 61, fmall Junipers,

TREES.

Willows, and Birch ; the laft two or three yards high, and as thick as a man's leg ‡ ; an amazing tree for this country. *Davis* alfo faw fome low Birch and Willows as high as about lat. 65 §. Nature here fuffers the reverfe of melioration ; the *glacieres* conftantly gain on the vallies, and deftroy all hopes of im-

ICE-BLINCK.

provement. That amazing *glaciere*, the *Ice Blinck* or *Ice Glance*, on the weftern coaft, is admirably defcribed by Mr. *Crantz*. I muft refer to him for the account, after faying, that it is a ftupendous aggregate at the mouth of an inlet, and of an amazing height ; the brilliancy of which appears like a glory to the navigators at many leagues diftance. It forms, beneath, a feries of moft magnificent arches, extending eight leagues in length, and two in breadth ; through thefe are carried, at the ebb of tide, great fragments of ice, which have fallen from various *icebergs*, and prove one fupply to the ocean of its floating ice ‖. The ftreights, now obftructed to navigation, are fuppofed to be open at bottom, by arches fimilar to thofe fpoken of ; for an immenfe quantity of ice is annually difcharged from their mouths **.

I have mentioned the iflands of ice at p. LXXXV ; for thofe of *Spitzbergen* have every thing in common with thofe of *Greenland*. Perhaps the colors in the laft may be more brilliant ; the green being as high as that of the emerald, the blue equal to that of the fapphir ; the firft, Mr. *Egede* attributes to the congelation of frefh, the latter to that of falt-water.†† Here are frequent inftances of the freezing of the fea-water. The froft often forms a pavement of ice from ifland to ifland, and in the confined inlets ‡‡.

TIDES.

The tides rife at the fouth of this country three fathoms, in lat 65 ; on the weft fide two, or in fpring-tides three ; at *Difco*, about lat. 69, only one ; further north it finks even to one foot. In great fpring-tides, efpecially in winter, is this ftrange phænomenon : fprings of frefh-water are forced up on the fhores in places where they were before unknown §§.

* Collate Mr. *Middleton*'s map, and others.   † Vol. i. 60.   ‡ *Hift Greenl.*

§ *Hacklyt.* iii. 101.   ‖ *Crantz.* i. 21 to 24.   ** Same, 19.   †† *Egede*, 55.

‡‡ *Crantz*, i. 43.   §§ Same, 41.

During the long day of the short summer is considerable heat. The long winter is a little cheared by the *Aurora Borealis*, which appears and radiates with unusual brilliancy and velocity in the spring, about the time of the new moon. Fogs give a gloom to the summer, and frost-smoke often adds horror to the winter. It rises out of the opening of the ice in the sea, and peels off the very skin from those who venture to approach it. The effect of the frost is very violent on the human body; but less so than in the north-east of *Sibiria*, where at times it is fatal to stir abroad, even when protected with every guard of cloathing *.

The *Greenlanders* fastidiously style themselves *Innuit*, i. e. *men*, as if they were the standard of the human race; yet few of them attain the height of five-feet; but are well made. Their hair is long and black; their faces flat; their eyes small. They are a branch of the *Eskimaux*, the small race which borders all the *Arctic* coasts. They originated from the *Samoied Asiatics*, who, passing over into the New World, have lined the coast from *Prince William's Sound* on the western side, in lat. 61, quite to the southern part of *Labrador* on the eastern. They crept gradually in their little canoes northward, and diminished in size in their progress, till they attained their full degeneracy in the *Eskimaux* and *Greenlanders*. Similar people, or vestiges of them, have been seen in different places, from *Prince William's Sound* to the north of BERING's streights. They were again seen by Mr. *Hearne* in lat. 72. By report of the *Greenlanders* of *Disco* bay, there are a few inhabitants in *Baffin*'s bay, in lat. 78. *Egede* says, that the country is peopled to lat. 76 †; but the highest colonized spot is at *Noogsook*, in lat. 71. They are a race made for the climate, and could no more bear removal to a temperate clime, than an animal of the torrid zone could into our unequal sky: seasons, and defect of habitual food, would soon bring on their destruction. This race has been found to agree in manners, habits, and weapons, and in many instances in language, from *Prince William's Sound* to the end of *Labrador*, a tract extending near fifteen hundred leagues‡. They only line the coasts; for the *Indians* persecute them with merciless hatred, and almost push them into the sea. They imagine these poor creatures to be magicians, and that to them they owe every ill success in life §. The numbers of the *Greenlanders* are now amazingly diminished. In 1730 there were thirty thousand souls, at present only ten thousand; a decrease chiefly owing to the ravage of the small-pox.

Greenland has been most happy in its Zoologist. The Reverend Mr. *Otto Fabricius*, whom a laudable zeal for enlightening the minds of the gross inhabitants,

* *Voyage en Siberie*, i. 381.   † As quoted in *Green*'s map of *America*.   ‡ COOK's *Voy*. i. Pref. LXXIV.   § Same, ii. 43.

led

led to thefe parts, hath given a moft ample and claffical account of the animals. His *Fauna Groenlandica* is among the firft works of the kind. I eagerly expect the performance of the promifed remainder of the work.

QUADRUPEDS.   The Quadrupeds of this country are, the *Rein-deer*, N° 4, which are here merely confidered as objects of the chace. Their number is leffened greatly, and they are now only found in the moft remote parts. The *Ukalcrajek* * is, I fufpect, an animal of imagination. It is faid, by the *Greenlanders*, to be long-eared, hare-lipped, and to refemble that animal ; to have a fhort tail ; to be of a white color, with a dark lift down the back, and of the fize of a Rein-deer. The DOGS, p. 41, refemble Wolves in figure, fize, and nature. Left to themfelves, they hunt in packs the few animals of the country, for the fake of prey. They exactly refemble the Dogs of the *Efkimaux* of Labrador. It is probable, that they might have been originally brought here by their mafters, who firft fled that country, and populated *Greenland*. ARCTIC FOXES, N° 10, abound here ; and, with POLAR BEARS, N° 18, infeft the country. Had I not fuch excellent authority, I fhould have doubted whether the *Wolverene*, N° 21, ufually an inhabitant of wooded countries, was found in *Greenland*; but it is certainly met with, yet rarely, in the fouthern parts, where it preys on the Rein-deer and White Hares. It muft have been originally wafted hither on the ice from *Terra de Labrador*, the neareft place to this, of which it is an inhabitant. The VARYING HARE, N° 37, is very common. The WALRUS, and five fpecies of Seals, inhabit thefe feas : the Common, N° 72 ; the Great, N° 73 ; the Rough, N° 74 ; the Hooded, N° 76 ; the Harp, N° 77 ; and an obfcure fpecies, called by the *Laplanders*, *Fatne Vindac*, with a round head and long fnout, bending like the probofcis of an elephant †. Mr. *Fabricius* adds to the marine animals, the Whale-tailed *Manati*, N° 81, of which he once faw the head partly confumed.

The Polar Bears, Seals, and Manati, were originally natives of thefe countries. The other Quadrupeds found their way here from either *Hudfon's Bay* or *Labrador*, on the iflands of ice. The *Arctic* Fox found the fame kind of conveyance from *Greenland* to *Iceland* as it did with the Rein-deer to *Spitzbergen*. To the laft was wafted, probably from *Labrador*, the Common Weefel, the Red or Common Fox ; and the Moufe, mentioned p. XLIX, miffed *Greenland*, but arrived at and ftocked *Iceland* ; and the Common Bat was originally tempeft-driven to the latter from *Norway* : the Wolverene and Varying Hare never reached farther than *Greenland*.—This feems the progrefs of Quadrupeds in the frigid zone, as high as land is found.

* *Faun. Groenl.* p. 26.   † Same, p. 17.—*Leems Lapm.* 214, 215.

The

The note * gives the sum of the Birds, land and water.

The numbers of Fish which frequent these icy seas are very considerable. They are, indeed, the great rendezvous of Whales. There is a fishery for them by the *Dutch*, in *Disko Bay*, as early as *April* †. The natives take them at other times, cut off the blubber in an awkward manner, and preserve that and the whale-bone as articles of commerce. It is certain that they do not drink train-oil, like the true *Eskimaux*, and some other congenerous people ‡. The species which frequent *Greenland* are, the MONODON MONOCEROS, or NARWHAL, *Lin. Syst.* 105: the MONODON SPURIUS, *Faun. Groenl.* N° 19; a rare species, with two teeth, about an inch long, projecting from the extremity of the upper jaw: the BALAENA MYSTECETUS, or COMMON WHALE, *Br. Zool.* iii. N° 16: BALAENA PHYSALUS, or FINFISH, N° 18; BALAENA MUSCULUS, or ROUND-LIPPED, N° 19: the BALAENA ROSTRATA, *Faun. Groenl.* N° 84; a very small species with a long snout: PHYSETER MACROCEPHALUS, *Faun. Groenl.* N° 25: PHYSETER CATODON, or ROUND-HEADED CACHALOT, *Br. Zool.* iii. N° 22: PHYSETER MICROPS, or BLUNT-HEADED CA-

A.					
* Cinereous Eagle,	p. 214 B.	Hebridal Sandpiper,	N° 382	Glaucous,	p. 532 B.
Greenland Falcon,	220 E.	Dunlin Sandpiper,	N° 391	Ivory Gull,	N° 457
Gyrfalcon,	221 F.	Alwargrim Plover,	N° 398	Tarrock,	p. 533 D.
Collared Falcon,	222 G.	Ringed Plover	N° 401	Arctic,	N° 459
Long-eared Owl ?	N° 117			Fulmar Petrel,	N° 461
Snowy Owl,	N° 121	**C.**		Shearwater P.	N° 462
Raven,	N° 134	Grey Phalarope	N° 412	Goosander	N° 465
Ptarmigan,	p. 315 B.	Red Phalarope,	N° 413	Red-breasted G.	N° 466
Snow Bunting,	N° 222			Canada Goose ?	N° 471
Lulean Finch,	p. 380 B.	**D.**		Grey lag Goose,	N° 473
Less Red-poll	N° 262	Great Auk,	N° 424	Brant,	N° 478
Wheat-ear	p. 420 P.	Razor-bill,	N° 425	Bernacle,	N° 479
Crested Titmouse ?	p. 427 F.	Black-billed,	N° 426	Eider Duck,	N° 480
		Puffin,	N° 427	King Duck,	N° 481
B.		Little,	N° 429	Golden Eye	N° 486
		Black Guillemot,	N° 437	Pin-tail	N° 500
Common Heron,	N° 433	Northern Diver,	N° 439	Long-tailed,	N° 501
Snipe,	N° 366	Red-throated D.	N° 443	Harlequin,	N° 490
Jadreka,	N° 375	Great Tern,	N° 448	Mallard,	N° 494
Striated Sandpiper,	N° 383	Black-backed Gull,	N° 451	Morillon,	p. 573 F.
		Cinereous G. (*Lin. Syst.*) 224			

The fifth species is very doubtful. Except the *Canada* Goose ? there is not a species of Bird which is not found in *Europe*. This induces me to place all those of *Greenland* in the appendages to the genera, as they seem to have little clame to *America*.

† *Crantz,* i. 118.　　　　‡ *Egede,* 134.—*Crantz,* i. 144.

CHALOT, *Br. Zool.* iii. N° 21 : DELPHINUS ORCA, or SPEKHUGGER, *Faun. Groenl.* N° 28 ; the tormentor of the greater Whales, whom they will fix on, as Bull-dogs will on a Bull, and tear out large pieces from their bodies : DELPHINUS PHOCA, the PORPESSE, *Br. Zool.* iii. N 25 : DELPHINUS DELPHIS, or the Dolphin, N° 24 : the DELPHINUS TURSIO, or the GRAMPUS, N° 26 : and finally, the DELPHINUS ALBICANS, or BELUGA WHALE, p. 182 of this Work, which enlivens those waters with its resplendent whiteness.

Among the cartilaginous species are the RAIA FULLONICA, *Lin. Syst.* 396 : the WHITE SHARK, *Br. Zool.* iii. N° 42; equally voracious from the equator to the *Arctic* circle; and, with fierceness unsubdued by climate, often bites in two the *Greenlanders* sitting in their Seal-skin canoes : the PICKED SHARK, *Br. Zool.* N° 40 : the BASKING SHARK, N° 41 : the SQUALUS PRISTIS, or SAW SHARK, *Lin. Syst.* 401 : the LUMP SUCKER, *Br. Zool.* iii. N° 57 ; a great article of food with the natives : CYCLOPTERUS SPINOSUS, or SPINY SUCKER, *Faun. Groenl.* N° 93 : CYCLOPTERUS MINUTUS, or the MINUTE, N° 94 : the UNCTUOUS SUCKER, *Br. Zool.* N° 58.

Of the boney Fishes, the EEL, *Br. Zool.* N° 63, is rarely found in the southern rivers. The WOLF-FISH, N° 65, appears here in the spring with the Lump Fish, and disappears in autumn. The *Greenland* Faunist describes a lesser variety, in N° 97, b. The LAUNCE, *Br. Zool.* iii. N° 66 : the OPHIDIUM VIRIDE, *Faun. Groenl.* N° 99 : the HADDOCK, *Br. Zool.* iii. N° 74, is plentiful here in winter. GADUS CALLARIAS, or VARIED COD, *Lin. Syst.* 436 ; and COMMON COD, *Br. Zool.* iii. N° 73, frequent the coasts in spring and autumn. The POUT, N° 75 : GADUS VIRENS, or GREEN COD, *Lin. Syst.* 438 : the HAKE, *Br. Zool.* N° 81 : the LING, N° 85 : and the GADUS BROSME, *Faun. Groenl.* N° 107, are species of Cod-fish found in these seas. The SPOTTED BLENNY, *Br. Zool.* iii. N° 93. A new species, the BLENNIUS PUNCTATUS, *Faun. Groenl.* N° 110; and that curious fish the CORYPHAENA RUPESTRIS, N° 111, *Act. Nidr.* iii. tab. 111.; the first rare, the last frequent in the deep southern bays. The ARMED BULL-HEAD, *Br. Zool.* iii. N° 98. The FATHER LASHER, N° 99, is a most common fish, and singularly useful. COTTUS SCORPIOIDES, *Faun. Groenl.* N° 114, or QUADRICORNIS, *Lin. Syst.* 451 ; and the RIVER BULL-HEAD, *Br. Zool.* iii. N° 97, are found here in salt-water. The ZEUS GALLUS, *Lin. Syst.* 454, a fish of the hottest parts of *South America,* is suspected to be found here. The HOLIBUT, *Br. Zool.* iii. N° 102, is very common ; as is the PLEURONECTES CYNOGLOSSUS, *Faun. Groenl.* N° 118; and the new species, PL. PLATESSOIDES, N° 119, is seen here in small numbers near the mouths of rivers. LABRUS EXOLETUS, *Faun. Groenl.*

N° 120 :

N° 120: Striped Wrasse? *Br. Zool.* iii. 119: Porca Norvegica, *Faun. Groenl.* N° 121: Three-Spined Stickleback, *Br. Zool.* iii. N° 129, not only in rivers but places overflowed by the sea. The Salmon, N° 143, is extremely scarce at present; yet in *Davis*'s time, was among the presents made to him by the savages; and *Baffin* * saw most amazing shoals of these fish in *Cockin's Sound,* on this western coast, in lat. 65. 45. The Salmo Carpio, *Faun. Groenl.* N° 124, is one of the most common and useful fishes; is frequent in the lakes, rivers, and estuaries. The Char, *Br. Zool.* iii. N° 149, consorts with the other, and is as common. The Salmo Stagnalis, *Faun. Groenl.* N° 126, a new species, found remote in the mountain lakes, and caught only by the hunters of Rein-deer. The Salmo Rivalis, N° 127, is another, inhabiting small brooks. The Salmo Arcticus, N° 128, or Capelin of t e *Newfoundland* fishers †, is the last of this genus, but the most useful; th daily bread, and the fish in highest esteem with the *Greenlanders*, and provid ntially given to them in the greatest abundance. The Common Herring, B . *Zool.* iii. N° 160, is a rare fish in these seas; as is the Anchovy, N° 163.

The same indefatigable Zoologist hath discovered in this country (including crustaceous) not fewer than ninety-one Insects, a hundred and twenty-six Vermes, fifty-nine shells, and forty-two Zoophytes.

John Davis, a most able seaman, was the first who examined the west side of *Greenland*. Before his time the eastern coast was the only part known to *Europeans*. He made there three different voyages, in 1585, 1586, and 1587. After doubling *Cape Farewell*, he sounded, and could not find bottom with three hundred fathoms of line. North of what he properly called *the Land of Desolation*, he arrived in a filthy, black, and stagnating water, of the depth of a hundred and twenty fathoms. He found drift-wood in lat. 65, and one entire tree sixty feet long, with its root; the species were Fir, Spruce, and Juniper ‡, which came down from remote places on the banks of the rivers of *Hudson's Bay*; for Mr. *Hutchins* assures me, that to this day, in certain years, vast quantities of timber are brought down with the ice at the opening of the rivers. He also met with black Pumices ‖, whether from neighboring vulcanoes, burning or extinct, remains unknown; or whether, which is most probable, conveyed there from *Iceland*. The stone of the country is mostly granitical. Some sand-stone, and many sorts of coarse marble. The *Lapis Ollaris* is found here in abundance, and of great use to the natives for making of pots. Talc is frequent here,

---

* *Purchas,* iii. 848.   † See it well engraven in M. *Du Hamel, Hist. de Poissons,* part ii. tab. xxvi.   ‡ *Davis's Voy.* in *Hackluyt,* iii. 101.   ‖ Same, 111.

Asbestos,

Afbeftos, and Gypfum.   Granates are not uncommon.   Sulphureous Marcafites, which have more than once deceived the navigators with the opinion of their being gold *.   The mineral fymptoms of copper, fuch as ftains of blue and green, are feen oh thefe rocks ; but avarice itfelf will never tempt adventurers to make here a trial.

DAVIS got as high as lat. 72, and called the country *London Coaft*.   The ftreight he paffed, between the weft of *Greenland* and the great iflands, is honored by his name.   He feems to have been engaged among the great iflands ; for he fays he failed fixty leagues up a found, found the fea of the fame color with the main fea, and faw feveral Whales.   He failed through another found to the fouth-weft, found ninety fathom water at the entrance ; but within could not touch ground with three hundred and thirty.   He had hopes of having found the long-fought-for paffage.   The tides rofe fix or feven fathoms ; but, as is frequent among iflands, the flood came from fuch variety of places, that he could not trace its principal origin †.

BAFFIN'S BAY.   At lat. 72. 30, I muft take as my pilot that great feaman *William Baffin*, who gave name to the great bay I now enter on.   His firft voyage was in 1613 ; his fecond, in which he made the moft effectual trial for the north-weft paffage, was in 1616.   He paffed through *Davis's Streights*.   In lat. 70. 20, on the *London Coaft*, he found the tides rife only eight or nine feet.   In *Horn Sound*, lat. 73. 45, he met with feveral people ‡.   To the north of that, in 75. 40, was a large and open bay ; *Cape Dudley Digges* forms its northern point ; within is *Weftenholme Sound* ; beyond that, *Whale Sound* ; and in the extreme north, or bottom of this great bay, is that named by *Baffin* after Sir *Thomas Smith*, lying in 78 degrees.   In thofe three founds were abundance of Whales ; but in the laft the largeft in all this bay.   It is highly probable, that there are one or more communications from hence to the *Icy Sea*, through which the Whales pafs at certain feafons ; and this (if I may collect from their numbers) might be that of their migration fouthward.   The diftance into the *Icy Sea* can be but very fmall, but probably blocked up with ice ; or if not, from the fudden fhifting of the ice in that fea by the change of wind, the paffage muft be attended with too great hazard to be attempted.   The ice prevented our great feaman from making trial of the tides in this bay, which would have brought the matter to greater certainty.   He faw multitudes of *Walrufes* and Seals in thefe parts, but no figns of inhabitants.   From hence the land trended wefterly,

* *Purchas*, iii. 833.—*Egede*, 32.   † *Hackluyt*, iii. 102.   ‡ Same, 846.

to a found he called by the name of *Alderman Jones*, in lat. 76. 40. Here the land ran due fouth to a great found in lat. 74. 20, which he called Sir *James Lancafter's*. From this place the land took an eaftern curvature, to the ftreights between the continent and *Cumberland* ifland. *Baffin* took his courfe between that ifle and the ifle of Saint *James*, left his name to the ftreight he paffed, and arrived fafe in *Cockin's Sound*, on the coaft of *Weft Greenland*, where he found the tide rife eighteen feet: this, and fimilar exceffes, arifing from the confined fituation of places *.

This is the only voyage ever made into *Baffin's Bay*. *Chriftian* IV. of *Denmark*, in 1619, fent *John Munck*, a moft able feaman, to make difcoveries in thefe parts; but, notwithftanding any furmifes of his having reached this famous bay, he got no farther than *Hudfon's Bay*; to which, in honor of his mafter, he gave the name of *Chriftian Sea*. He paffed a miferable winter in *Churchill* river, and returned home the next year, after lofing, during his ftay on fhore, every man but two †.

Before I quit thefe frozen regions, I muft once more return to *Spitzbergen*, to relate, what has but very lately been communicated to me, that the *Ruffians* have of late attempted to colonize thefe dreadful ifands. They have, for a few years paft, fent parties to continue there the whole year; who have eftablifhed fettlements on the ifle of *Spitzbergen*, at *Croon Bay*, *King's Bay*, *Magdalena Bay*, *Smeerenburgh*, and *Green Harbour*; where they have built huts, each of which is occupied by about two boats crews, or twenty-fix men. They bring with them falted fifh, rye-flour, and the ferum or whey of four milk. The whey is their chief beverage, and is alfo ufed in baking their bread. Each hut has an oven, which ferves alfo as a ftove; and their fuel is wood, which they bring with them from *Archangel*. The huts are above ground, and moft furprizingly warm; placed alfo in fituations which may guard them as much as poffible from the keennefs of the northern wind.

Mr. *Erfkine Tonnach*, furgeon of *Dunbar* (who, by the friendfhip of the worthy Mr. *George Paton*, of *Edinburgh*, favored me with this account) gives me the following particulars from his own knowledge.—" During our ftay on the ifland, my curiofity prompted me to go on fhore, that I might fee the œconomy of thefe arctic fettlers; and had an opportunity of feeing them dine: and though their fare appeared coarfe, the difpatch they ufe, faid a great deal for their health and

---

* For the account of this curious voyage, fee *Purchas*, iii. from p. 836 to 848.
† Clerk of the *California's Voy.* i. 106.—For a further account of this unfortunate voyage, fee *Churchill's* Collection, ii. 472.

    appetite.

appetite. They boil their fish with water and rye-meal : and this constitutes their diet during winter. In the summer they live chiefly on fowls, or their eggs ; but in general they forbear flesh, as the fasts prescribed by their religion are so numerous. They are dressed in the skins of the animals they kill, which they use with the fur side next to their bodies : their bedding is likewise composed of skins, chiefly of those of the Bear or Rein Deer. The skin of the Fox is the most valuable ; but these are preserved as articles of commerce in their own country. They catch the Beluga, or white Whale, in nets, being conversant in this species of fishery ; but are ignorant of that of the great Whale. They were very solicitous to get information on that subject ; which I endeavoured to instruct them in, in return for the information they so readily gave me. They are most excellent marksmen ; but, what is peculiar, in presenting their piece, they do not raise it to their shoulder, but place the butt-end between their arm and their side, fixing their eye on the object toward which they direct the barrel. I saw a Bear receive a considerable shot : it astonished me greatly to see the animal apply great quantities of snow to the part (which was bleeding freely) as if conscious of its styptic powers. It retreated with much slowness ; but at short intervals looked behind, and, with much art, threw abundance of snow with its hind-paws into the wound. Few of the *Russians* die from the severity of the cold, but are often frost-bitten, so as to lose their toes or fingers ; for they are so hardy as to hunt in all weathers. I naturally asked them, Had they a surgeon ? They replied, ' No ! no ! CHRIST is our doctor !' They quit the island in *September*, and are privileged to leave the place by the 22d of that month, whether they are relieved by a fresh party from *Russia* or not."—Let me remark, that the great exercise used by these volunteer adventurers ; their quantity of vegetable food ; their freshening their salt provision, by boiling it in water, and mixing it with flour ; their beverage of whey ; and their total abstinence from spirituous liquors—are the happy preservatives from the scurvy, which brought all the preceding adventurers, who perished, to their miserable end *.

HUDSON'S BAY.   We now proceed through a nameless streight, between the main land and the two great islands on the east ; and, after doubling *Cape Southampton*, enter into *Hudson's Bay*, in the gulph called *the Welcome*. This bay was discovered in 1610, by that able seaman *Henry Hudson*, from whom it takes its name. His view, in the voyage he made, was the discovery of a passage to the *East Indies*. The

---

* See this subject amply treated by Doctor *Aikin*, in a Treatise on the success, with respect to the health, of some attempts to pass the winter in high northern latitudes.

trial

trial has been vigorously pursued since his days, but without success. In 1742 an attempt was made, as low as the bottom of *the Welcome*, by Captain *Middleton*; and from the check he met with, he called that part *Repulse Bay*. In subsequent trials *Wager's Water* was suspected to be the passage into the Western ocean; but in 1747 its end was discovered, and found to terminate in two navigable rivers. The romantic scenery which the adventurers met with in the way is most admirably described by the elegant pen of Mr. *Henry Ellis*.

Chesterfield, or *Bowden's Inlet*, was likewise suspected to have been the desired streight; but in 1762 Messrs. *Norton* and *Christopher*, in a sloop and cutter belonging to the Company, went to the remotest end. At the distance of a hundred and twenty-eight miles from the mouth was scarcely any tide; thirty miles further it quite died away. The land here grew contracted into a very narrow passage. Here the adventurers entered with the cutter, and discovered that the end was in a magnificent fresh-water lake, to which was given the name of *Baker's*. The land was quite level, rich in grass, and abounding with Deer. They found the end quite innavigable, and to terminate in a small stream, with many shoals at its mouth, and three falls across it. After finding the water decrease to the depth of two feet, they returned fully satisfied with their voyage.

CHESTERFIELD INLET.

Hudson's Bay has been so frequently described, that I shall only give a general view of it and its adjacent parts. Its entrance from the ocean, after leaving to the north *Cape Farewell* and *Davis's Streights*, is between *Resolution* isles on the north, and *Button's* isles, on the *Labrador* coast, to the south, forming the eastern extremity of the streights distinguished by the name of its great discoverer. The coasts very high, rocky, and rugged at top; in places precipitous; but sometimes exhibit large beaches. The isles of *Salisbury*, *Nottingham*, and *Digges*, are also very lofty, and naked. The depth of water in the middle of the bay is a hundred and forty fathoms. From *Cape Churchill* to the south end of the bay are regular soundings; near the shore shallow, with muddy or sandy bottom. To the north of *Churchill*, the soundings are irregular, the bottom rocky, and in some parts the rocks appear above the surface at low water. From *Moose* river, or the bottom of the bay, to *Cape Churchill*, the land is flat, marshy, and wooded with Pines, Birch, Larch, and Willows. From *Cape Churchill* to *Wager's Water* the coasts are all high and rocky to the very sea, and woodless, except the mouths of *Pockerekesko*, and *Seal* rivers. The hills on their back are naked, nor are there any trees for a great distance inland.

The mouths of all the rivers are filled with shoals, except that of *Churchill*, in which the largest ships may lie; but ten miles higher, the channel is obstructed with sand-banks; and all the rivers, as far as has been navigated, are full of rapids and cataracts, from ten to fixty feet perpendicular. Down thefe rivers the *Indian* traders find, a quick paffage ; but their return is a labor of many months.

As far inland as the Company have fettlements, which is fix hundred miles to the weft, at a place called *Hudfon Houfe*, lat. 53. long. 106. 27, from *London*, is flat country : nor is it known how far to the eaftward the great chain, feen by our navigators from the *Pacific Ocean*, branches off.

CLIMATE.

The climate, even about *Haye*'s river, in only lat. 57, is, during winter, exceffively cold. The fnows begin to fall in *October*, and continue falling by intervals the whole winter ; and, when the froft is moft rigorous, in form of the fineft fand. The ice on the rivers is eight feet thick. Port wine freezes into a folid mafs ; brandy coagulates. The very breath fell on the blankets of the beds in form of a hoar froft, and the bed-cloaths often were found frozen to the wall *. The fun rifes, in the fhorteft day, at five minutes paft nine, and fets five minutes before three. In the longeft day the fun rifes at three, and fets about nine. The ice begins to difappear in *May*, and hot weather commences about the middle of *June* ; which, at times, is fo violent, as to fcorch the face of the hunters. Thunder is not frequent, but very violent. But there muft be great difference of heat and cold in this vaft extent, which reaches from lat. 50. 40, to lat. 63, north.

During winter the firmament is not without its beauties. Mock funs and halôs are not infrequent ; are very bright, and richly tinged with all the colors of the rainbow. The fun rifes and fets with a large cone of yellowifh light. The night is enlivened with the *Aurora Borealis*, which fpreads a thoufand different lights and colors over the whole concave of the fky, not to be defaced even by the fplendor of the full moon ; and the ftars are of a fiery rednefs †.

FISH.

Hudfon's Bay is very ill fupplied with Fifh. The common Whale is frequent there. The Company have attempted to eftablifh a fifhery ; and for that purpofe procured experienced people from the *Spitzbergen* fhips, and made confiderable trials between lat. 61 and 69 ; but, after expending twenty thoufand pounds, and taking only three fifh, were, in 1771, obliged to defift. The ice prevented the veffels from getting to a proper ftation in due time ; and the hard gales, and quick return of winter, always deprived them of an opportunity of making a fair trial. The fifhery of the *Beluga*, or White Whale, is attended with more fuccefs. It haunts the mouths of rivers in *June*, as foon as they have difcharged the ice, and are taken in great numbers. There are two varieties ; one with a blue caft, the other of a pure white. Thefe animals, probably, fuperfete ; a

* *Voy. to Hudfon's Bay*, 1746, written by the Clerk of the *California*, i. 159. His name was *Drage*; his account is fenfible and entertaining.   † *Ellis*, 172.

foetus

fœtus of fix inches in length having been extracted, at the fame time that a young one has been feen (as is their custom) mounted on the back of another.

Sturgeons of a fmall fize are found in the rivers, not far from the fea. They appear to me to be of the fame fpecies with the *Englifh*. Sturgeons are found in great plenty in the lakes far inland, and from the weight of fix to forty pounds. I fufpect thefe to be the fame with the Sturgeons of the great lakes of *Canada*, which, I am told, are fmooth, or free from tubercles; and probably the *Acipenfer Hufo* of *Linnæus*, and *Hanfen* of the *Germans*, a fifh of the *Danube* and *Wolga*.

The *Lophius Pifcatorius*, or Common Angler, *Br. Zool.* iii. N° 51, appears towards the furface only in windy weather; for which reafon it is called by the natives *Thutina-meg*, or the *Wind-fifh*.

The *Gadus Lota*, or Burbot, *Br. Zool.* ii. N° 86, is common in the rivers, and is caught with hooks after nine o'clock at night. It is called here *Marthy*; grows to the weight of eight pounds; is fo voracious as to feed even on the tyrant Pike; will devour dead Deer, or any carrion, and even fwallow ftones to fill its ftomach: one of a pound weight has been taken out of a fifh of this fpecies. It fpawns about *February* 8th, and is unhappily moft prolific. Mr. *Hutchins* counted, in a fingle fifh, 671,248 ovaria.

Allied to this is the *Mathemeg* of the natives, the Land Cod of the *Englifh*, a fifh abundant in the northernly lakes; it grows to the length of three feet, and the weight of twelve pounds: has three beards on the lower jaw; the middlemoft the longeft: the back is brownifh: the belly grey.

The *Perca Fluviatilis*, or common Perch, *Br. Zool.* iii. N° 124, is found in the rivers, but not in plenty; and fometimes grows to the weight of eight pounds. The *Gafterofteus aculeatus*, or three-fpined Stickleback, *Br. Zool.* iii. N° 129, is found here in great numbers.

Salmo Salar, or the common Salmon, *Br. Zool.* iii. N° 143, is taken in plenty from *June* to *Auguft*, in nets placed along the fea-fhores, and falted for ufe. Very few are caught to the fouth of *Churchill* river.

The *Namaycufh*, is a fpecies of Trout, with the head, back, dorfal fin, and tail of a dark blue: the fides dufky, marked with white and reddifh fpots: the belly filvery: the flefh white, and very delicate. It is caught with the hook in lakes far inland; and fometimes of the weight of thirty pounds. A *Trutta lacuftris generis*, p. 1012. *Wil. Icth.* 198?

Salmo Alpinus, or Char, *Br. Zool.* iii. N° 149, is common in the frefh waters, and weighs from two to fix pounds.

The *Salmo Lavaretus*, or Gwiniad, *Br. Zool.* iii. N° 152, is found here in vaft abundance; and grows to a fize far fuperior to thofe of *Europe*. There is a leffer kind, called here the *Sea Gwiniad*: the head is not so dusky : eyes smaller ; and back less arched.   The nose of the male is blunt ; and the stomach muscular, like a gizzard : the female has an arched nose.   They are very numerous in autumn just when the rivers are frozen over, and are called here *Tickomeg*.   The *Salmon Arcticus*, or *Capelin*, is observed to precede the Salmon, and is sometimes thrown on shore in amazing quantities by hard gales.

The *Omisco Maycus* is a new species of Trout, taken in *May* in *Albany* river not exceeding four inches and a half long.   It has five branchiostegous rays : first dorsal fin has eleven rays, ventral eight, anal seven, pectoral thirteen : tail forked : in the jaws are minute teeth : back, as low as the lateral line, is of a pale color, marked with two longitudinal rows of black stelliform spots : below the lateral line the color silvery : the belly white.

The Pike, *Br. Zool.* iii. N° 153, abounds in all the lakes.   It by no means arrives at the size of the *English*.   Mr. *Hutchins* does not recollect any above the weight of twelve pounds.

The *Cyprinus Catastomus* of Dr. *Forster* *, or Sucker Carp, is a new species : of which there are two varieties ; the *Mithco-Mapeth* of the *Indians*, marked with a broad stripe of red along the lateral line, and found on the sea-coast ; and the White, or *Namapeth*, with larger scales, and wholly of a whitish color : very scarce in the salt-water, but in such plenty in the inland lakes and rivers, as to be even burdensome to the nets.   They grow to the weight of two pounds and a half.   The form is oblong : the head boney, rugged, and decreasing to the tip of the nose : the mouth small, and placed beneath : the body scaly : the tail lunated.

Shell-fish are very scarce in this sea.   *Mytilus Edulis*, the Edible Mussel, *Br. Zool.* iv. N° 73, alone are plentiful ; but of Cockles, only the dead shells are seen. From the number of shells which are dug up, for the space of ten miles inland of this flat muddy country, may be collected a proof of the great retreat of the water ; but for want of inhabitants, the period of its loss cannot be ascertained.

BIRDS.

Among the birds, which escaped my notice while I was writing the zoologic part of this Work, are two of the Eagle kind, found in this country : the first is the YELLOW-HEADED, with a dusky bill, cere, and irides : head and neck yellowish : back dark brown ; each feather tipped with dirty yellow.   This species appears in *Hudson's Bay* in *April*.   Builds its nest in trees, with sticks and grass ; and

* By whom it is well described and figured, in vol. lxiii. p. 155. tab. vi. of *Ph. Trans.*

lays one egg. It preys on young Deer, Rabbits, and Fowls. Retires southward in *October*. Is called by the *Indians, Ethenesue Mickesue* †.

A variety of the GOLDEN EAGLE is also a native of the same place. The forehead is brown: crown and hind part of the neck striped with brown, white, and rusty yellow: lower part of the neck, breast, and belly, deep brown: coverts of the wings, back, secondaries, and scapulars, of the same color; the two last white towards their bottoms, and mottled with brown: primaries black: middle feathers of the tail brown, barred with two or three cinereous bands; exterior feathers brown, blotched with cinereous: legs cloathed with pale brown feathers to the toes, which are yellow. Length three feet. A specimen of this was presented to the *British Museum*.

To these may be added a genuine Falcon, communicated to me by Mr. *Latham*. The bill very sharp, and furnished with a large and pointed process in the upper mandible: cere yellowish: head, front of the neck, breast, and belly, white: each feather marked along the shaft with a line of brown, narrowest on the head: the back and coverts of the wings of a dirty bluish ash-color; the edges of the feathers whitish, and many of them tipped with the same: primaries dusky; exterior webs blotched with white; interior barred with the same color: tail of the same color with the back, barred with white; but the bars do not reach the shaft, and, like those in the *Iceland* Falcon, oppose the dark bars in the adverse web: the legs bluish. The length of this fine species is two feet two inches.

Multitudes of birds retire to this remote country, to *Labrador*, and *Newfoundland*, from places most remotely south, perhaps from the *Antilles*; and some even of the most delicate little species. Most of them, with numbers of aquatic fowls, are seen returning southward, with their young broods, to more favorable climates. The savages, in some respects, regulate their months by the appearance of birds; and have their *Goose month* from the vernal appearance of Geese from the south. All the Grous kind, Ravens, cinereous Crows, Titmouse, and *Lapland* Finch, brave the severest winter; and several of the Falcons and Owls seek shelter in the woods. The Rein Deer pass in vast herds towards the north, in *October*, seeking the extreme cold. The male Polar Bears rove out at sea, on the floating ice, most of the winter, and till *June*: the females lie concealed in the woods, or beneath the banks of rivers, till *March*, when they come abroad with their twin cubs, and bend their course to the sea in search of their consorts. Several are killed in their passage; and those which are wounded shew vast fury, roar hideously, and bite and throw up into the air even their own progeny. The females and the young, when not interrupted, continue their way to sea. In *June*, the

† The description and history of this species was communicated to me by Mr. *Hutchins*.

males

males return to fhore, and, by *Auguft*, are joined by their conforts, with the cubs, by that time of a confiderable fize *.

The eaftern boundary of the bay is *Terra de Labrador* ; the northern part has a ftrait coaft facing the bay, guarded with a line of ifles innumerable. A vaft bay, called the *Archiwinnipy Sea*, lies within it, and opens into *Hudfon's Bay* by means of *Gulph Hazard*, through which the *Beluga* Whales dart in great numbers. Here the Company had a fettlement, for the fake of the fifhery, and for trading with the *Efkimaux* ; but deferted it as unprofitable about the year 1758 or 1759. The eaftern coaft, fo admirably defcribed by that honored name, Sir ROGER CURTIS † ! is barren paft the efforts of cultivation. The furface every where uneven, and covered with maffes of ftone of an amazing fize. It is a country of fruitlefs vallies and frightful mountains, fome of an aftonifhing height: the firft watered by a chain of lakes, formed not from fprings but rain and fnow, fo chilly as to be productive of only a few fmall Trout. The mountains have here and there a blighted fhrub, or a little mofs. The vallies are full of crooked ftunted trees, Pines, Fir, Birch, and Cedars, or rather a fpecies of Juniper. In lat. 60, on this coaft, vegetation ceafes. The whole fhore, like that on the weft, is faced with iflands at fome diftance from land. The inhabitants among the mountains are *Indians* ; along the coafts, *Efkimaux*. The Dogs of the former are very fmall ; of the latter, large, and headed like a Fox. Notwithftanding they have Rein-deer, they never train them for the fledge ; but apply the Dogs to that ufe ‡. *Walrufes* vifit a place called *Nuchvûnk*, in lat. 60, during winter ; from thence they purchafe the teeth, with which they head their darts. *Davis* fufpected that he had found a paffage on this coaft, in 1586, to the Weftern ocean ; but it proves no more than a deep bay.

The laudable zeal of the *Moravian* clergy hath induced them to fend, in the year 1752, miffionaries from *Greenland* to this country. They fixed on *Nifbet*'s harbour for their fettlement ; but the firft party was partly killed, partly driven away. In 1764, under the protection of our government, another attempt was made. The miffionaries were well received by the *Efkimaux*, and the miffion goes on with fuccefs ‖. Thefe pious people, like the Jefuits, have penetrated almoft into every part of the knownworld ; and, for the fake of the Gofpel, dared the extremities of heat and cold. They endeavour to humanize the favages of *Greenland*, and improve the morals of the foft inhabitants of the unwholefome coafts of *Bengal*. They are not actuated by ambition, political views, or ava-

* See an ingenious and laudable Calendar of *Hudfon's Bay*, publifhed by Doctor *Macfait*, in his new Syftem of General Geography, 348 to 354.   † *Pb. Tranf.* lxiv. 372.   ‡ Same, 386.

‖ *Crantz, Hift. Morav.* 404, 608.

rice.

rice. Here my comparifon with the once-potent order of the *Roman* church fails.

Terra de Labrador, at *Cape Charles*, in lat. 52, trends towards the fouth-weft. Between that cape and the ifle of *Newfoundland* begin the ftreights of *Belleifle*, a paffage with from twenty to thirty fathoms water; but often choaked up with the floating ice from the north, even fo late as the middle of *June* *. They open into the vaft triangular gulph of St. *Laurence*, bounded to the north by *Terra de Labrador*; to the weft by *Nova Scotia*; to the eaft by *Cape Breton* and *Newfoundland*. In the weftern corner, the vaft river of St. *Laurence* difcharges itfelf; arifing from a thoufand ftreams which feed the fea-like lakes of *Canada*, and, after falling down the amazing catara&t of *Niagara*, and darting down the flopes of numberlefs foaming rapids, tremendous to all but *Britifh* battalions †, forms a matchlefs navigation of many hundred miles. *Jacques Cartier*, a native of St. *Maloes*, had, in 1534, the honor of being the firft difcoverer of this noble river.

MAGDALENE ISLES.

In the gulph are fcattered feveral important iflands, occupied by the *Englifh* and *French* for the fake of the fifheries. The fmall rocky ifles of St. *Magdalene* are ftill frequented by numbers of *Walrufes*. There is an annual chace during the feafon, and numbers are killed for the fake of the oil and fkins ‡. The water round the *Magdalenes* is only from three to nine fathoms deep, and the fhores flope moft conveniently into it for the afcent or defcent of thefe animals. The water round the other ifles is of one depth, except on the north fide of St. *John's*.

Newfoundland (a name, in the infancy of difcovery, common to all *North America*) was difcovered in 1496, by the celebrated *Venetians*, *Sebaftian Cabot* and his three fons; who, at their own charges, under a grant of *Henry* VII. giving them poffeffion (as vaffals of his) of all lands they might difcover §, coafted from lat. 67. 30, to the capes of *Florida*, and thus indifputably gave to ill-fated *Britain* the right, by pre-difcovery, of the whole continent of *North America*. The fhort-fighted avaricious prince, under whofe banners it was difcovered, had not the heart to make the proper advantage. He had before negle&ted the offer of *Columbus*, which would have given him that fpecies of right to the whole New World. 'But,' fays the courtier-like *Bacon* ¶, 'it was not a refufal on the king's 'part, but a delay by accident, which put by fo great an acqueft.' The *French* foon found out the gold mine of the *Newfoundland* difcovery, which offered itfelf in the fifheries. Of all minerals (twice fays the fame noble philofopher) there

---

* *Barrington's Mifcel.* 25.   † Read the account of Lord *Amherft's* defcent down this river, in 1760. ‡ See p. 148.   § *Rymer's Fœd.*   ¶ *Hift. King Henry* VII. *Bacon's Works*, iii. 89.

is none like the fisheries. In 1534 they were actually engaged in them. A private man, Sir *Humphry Gilbert*, brother-in-law to RALEIGH, or, what was better, animated by a congenial soul, sailed in 1583 with every provision for settling this important colony. On his return he was swallowed up by the ocean. His love of improvement, and his piety, never forsook him. He was seen sitting unmoved in the stern of his ship, with a book in his hand; and often heard to say, ' Courage, my lads ! we are as near heaven at sea as at land *.'

The isle of *Newfoundland* is of a triangular form, and lies between lat. 46. 40, and 51. 30: visited occasionally, but not inhabited, by savages from the continent.

The boasted mine of this island lies on the southern and western sides, on the great bank, which stretches from north-east to south-west, about two hundred leagues. The water on the bank is from twenty-two to fifty fathoms; on the outside from sixty to eighty; on the lesser banks much the same. A great swell and thick fog generally mark the place of the greater. The subject of the fishery has been often treated of; but the following short though clear account of so interesting a subject cannot fail being acceptable to the *British* reader.

NEWFOUNDLAND FISHERY.

" The boats or shallops are forty feet in the keel, rigged with a mainmast and foremast, and lugsails; furnished with four oars; three of which row on one side, and the other (which is twice as large) *belays* the other three, by being rowed sideways over the stern, by a man who stands up for that purpose, with his face towards the rowers, counteracting them, and steering at the same time as he gives way to the boat.

" Each of the men in this boat is furnished with two lines, one at each side of the boat, each furnished with two hooks; so here are sixteen hooks constantly employed; which are thought to make a tolerable good day's work of it, if they bring in from five to ten quintals of fish, though they have stowage for, and sometimes bring in thirty. Two hundred quintals is called a saving voyage; but not under. The bait is small fish of all kinds; Herring, Capelin, Lance, Tom Cod, or young Cod; the first of which they salt, and keep for some time, in case of scarcity of the rest; but these are not near so eagerly taken by the fish when salted. In case small fish cannot be got, they use sea-fowl, which are easily taken in vast numbers, by laying nets over the holes in the rocks where they come to roost in the night. If neither small fish nor birds are to be got, they are forced to use the maws of fish they catch, which is the worst bait of any.

" When the fish are taken, they are carried to the stage, which is built with one end over the water for the conveniency of throwing the offals into the sea, and

* *Hackluyt*, iii. 159.

for their boats being able to come clofe to difcharge their fifh. As foon as they come on the ftage a boy hands them to the header, who ftands at the fide of a table next the water end; whofe bufinefs it is to gut the fifh and cut off the head, which he does by preffing the back of the head againft the fide of the table, which is made fharp for that purpofe; when both head and guts fall through a hole in the floor into the water. He then fhoves the fifh to the fplitter, who ftands oppofite to him; his bufinefs is to fplit the fifh, beginning at the head, and opening it down to the tail; at the next cut he takes out the larger part of the backbone, which falls through the floor into the water. He then fhoves the fifh off the table, which drops into a kind of hand-barrow, which, as foon as filled, is carried off to the falt-pile. The header alfo flings the liver into a feparate bafket, for the making of train-oil, ufed by the curriers, which bears a higher price than Whale-oil.

"In the falt pile, the fifh are fpread upon one another, with a layer of falt between. Thus they remain till they have taken falt; and then are carried, and the falt is wafhed from them by throwing them off from fhore in a kind of float called a *Pound*. As foon as this is completed, they are carried to the laft operation, of drying them; which is done on ftanding flakes made by a flight wattle, juft ftrong enough to fupport the men who lay on the fifh, fupported by poles, in fome places as high as twenty feet from the ground: here they are expofed, with the open fide to the fun; and every night, when it is bad weather, piled up five or fix on a heap, with a large one, his back or fkinny part uppermoft, to be a fhelter to the reft from rain, which hardly damages him through his fkin, as he refts flanting each way to fhoot it off. When they are tolerably dry, which in good weather is in a week's time, they are put in round piles of eight or ten quintals each, covering them on the top with bark. In thefe piles they remain three or four days to fweat; after which they are again fpread, and when dry put into larger heaps, covered with canvas, and left till they are put on board.

"Thus prepared, they are fent to the *Mediterranean*, where they fetch a good price; but are not efteemed in *England*: for which place another kind of fifh is prepared, called by them Mud Fifh; which, inftead of being fplit quite open, like their dry fifh, are only opened down to the navel. They are falted, and lie in falt, which is wafhed out of them in the fame manner with the others; but inftead of being laid out to dry, are barrelled up in a pickle of falt boiled in water.

"The train-oil is made from the livers: it is called fo to diftinguifh it from Whale or Seal oil, which they call fat oil, and is fold at a lower price (being only

ufed

used for lighting of lamps) than the train-oil, which is used by the curriers. It is thus made:—They take a half tub, and, boring a hole through the bottom, press hard down into it a layer of spruce boughs; upon which they place the livers; and expose the whole apparatus to as sunny a place as possible. As the livers corrupt the oil runs from them, and, straining itself clear through the spruce boughs, is caught in a vessel set under the hole in the tub's bottom."

CAPE BRETON.

The barren island of *Cape Breton* forms one side of the great entrance into the gulph of St. *Laurence.* It is high, rocky, and dreary: rich in thick beds of coal, and may prove the *Newcastle* of *America.* This isle was first discovered by Sir *Humphry Gilbert,* in his fatal voyage. It was soon after frequented, on account of the Walruses, and the fishery of Whales. Among the earliest adventurers were the industrious *Biscayeners,* who seem to have been our masters in the art. Till of late years, it had been important by being the seat of the *French* fishery; but the strong fortress of *Louisbourg* is now demolished, and the place deserted.

NOVA SCOTIA.

The great peninsula of *Nova Scotia* is separated from *Cape Breton* by a narrow streight. It was, in 1616, possessed by the *French,* who attempted to colonize it from their new settlement in *Canada;* but they were soon expelled by the *English,* who deemed it part of *North Virginia;* the whole continent, at that time, going under the name of *Virginia,* so called, originally, in honor of our virgin queen. The *French* had given it the name of *Acadie.* *James* I. made a grant of the country to Sir *William Alexander* in 1621, on condition that he would form there a settlement. It then received the title of *Nova Scotia.* In order to encourage Sir *William,* he planned the order of baronets, which is called after the country. To every knight who would engage to colonize any part, a grant was to be made of certain portions of land. The order was not instituted till 1625, when a number were created, and they held their lands from the crown of *Scotland* as a free barony, with great privileges to all who would settle in the country *. The design almost instantly failed, and the *French* were permitted to repossess themselves of the province. Its value became known, and since that period it has frequently changed masters. It never was effectually settled till the year 1749, when a large colony was sent there under the auspices of the Earl of *Halifax.*

CLIMATE.

The climate of this province is, during the long winter, extremely severe, and the country covered with snow many months: the summer misty and damp. The face of it is in general hilly; but can scarcely be called moun-

* *Collins's Baronets,* iv. 330.

tanous,

tanous, being the lowered continuation of the great chain which pervades the whole continent. The ground is not favorable to agriculture, but may prove excellent for pasturage. Due attention to the breeding of cattle will not only repay the industry of the farmer, by the home consumption, but be an extensive benefit to our islands. The country cannot boast, amidst its vast forests, timber fit for large masts, nor yet for the building of large ships; yet it will prove an inexhaustible magazine for that species of timber called lumber, so essential to our sugar plantations.

Its situation, in respect to the fisheries, is scarcely inferior to that of *Newfoundland.* The vast banks, called *Sable Island's*, *Brown's*, and *St. George's*, with many others, are frequented by myriads of Cod-fish. It is the duty of the Parent State to encourage, with all diligence, this branch of commerce; and in a manner so expeditious and so frugal, as may anticipate and undersell foreign adventurers. Without that, our remnants of the New World will be but of little use. The fisheries, the staples of *Nova Scotia* and *Newfoundland,* are open to other nations; and if they are permitted to excel us in the articles expedition and frugality, our labors are truly vain. It is to the antient hardy colonists we must look up for the support of the toils of the sea, and the advantages we may expect to gain from them: they should have their encouragement. But there is another set of men who of late (a public calamity) have made hither an involuntary migration, who with sad hearts recollect their exiled land:

> Nos Patriæ fines, nos dulcia linquimus arva:
> Nos Patriam fugimus.

These sufferers are in general unused to the fatigues of a maritime life, and ought to be fostered, for their filial piety, at first, with a parental care; to be encouraged in the pastoral life, or in such arts as may supply the sailor and the fisherman with food, and with materials for their professions. If the climate is fit for corn, for flax and hemp, let due rewards be given for the successful efforts of their industry. The succeeding generation, hardened to the climate, and early habituated to another kind of life, may join the maritime adventurers, and give importance to themselves, and strength to the island from which they sprung.

The harbours of this province are frequent and excellent. The tides are in many places most uncommonly high. Those of the bay of *Fundy* are the most remarkable; for they force themselves into the great creeks with a bore or head

from

from fifty to feventy-two feet high, and with moſt amazing rapidity. Hogs, which feed along the ſhores, are much more ſenſible of its approach than mankind: they are obſerved to liſten, to prick up their ears for ſome time, and then ſuddenly to run off at full ſpeed.

MAGNIFICENT VIEWS.

The coaſts are, in general, rude and rocky, with ſome variations; but in many places exhibit moſt picturefque ſcenery. All the northern ſide is high, red, and rocky. The iſles of *Canſo* are varied with many low white rocks. From them to *Torbay* is a ſeries of lofty coaſt, broken and white. *Beaver Harbour* is guarded by moſt picturefque rounded iſles. South ſhore of *Chebucto* ſteep: the plaiſter cliffs in *George Bay* are remarkable for their precipitous face and white-nefs. *Sable* or *Sand Iſland* is diſtinguiſhed (as the name imports) by amazing ſand-hills of a ſugar-loaf form. The iſle of *Great Manan*, on the weſtern ſide of the entrance of the bay of *Fundy*, is very lofty, the ſtrata divided, and the top wooded. *St. Mary's Bay* is nobly bounded by high rocks, cloathed on their ſummits with woods: the entrance into it are the *Grand* and *Petit Paſſage*; the ſides of the laſt are either covered with hanging woods, ſloping to the water-edge, or broke into ſhort precipices. The entrance into the fine harbour of *Annapolis* is moſt auguſt: a narrow gut, bounded by enormous precipices, with lofty hills ſoaring above, the tops of which are even and cloathed with woods. The approach to the baſon of *Minas* is not lefs magnificent. The columnar rocks of *Cape Split* are very ſingular. The iſle of *Haute* is lofty and ſteep on every ſide. The whole neighborhood abounds with views of the moſt ſublime and romantic caſt. This peninſula joins the great continent by a very narrow iſthmus, beyond which we retain a wretched barren remnant of near half of the New World; the ſad reverſe of the ſhort ſpace of twenty years!—My eyes withdraw themſelves from the mortifying ſight. BRITAIN, which ſate (by the wiſdom of one man) as the Queen of Nations, now deplores her folly; and ought to confefs, that ‘ thoſe things which were for her wealth, proved to her ‘ an occaſion of falling.’ She ſunk under the deluſion of proſperity, by falſe ſecurity, and the pride of victories. If ſhe makes a proper uſe of adverſity, ſhe ſtill may riſe into glory and wealth, by honeſt induſtry, and by the repreſſion of rapacity and ſordid ambition.—Once more, gracious Heaven, endeavour to ſave an ungrateful people! once more raiſe up ſome great inſtrument to execute thy mercies!—Pour with full meaſure into our youthful Miniſter the virtues of his father!—Emulate, young Man, his virtues, and then—

Si qua fata aſpera rumpas;
Tu MARCELLUS eris.

INDEX

INDEX

TO THE

INTRODUCTION.

INTRODUCTION.

INTRODUCTION.

ARCTIC ZOOLOGY.

CLASS I. QUADRUPEDS.

DIV. I. HOOFED.

HIST. QUAD. Genus II.

American Ox, *Hist. Quad.* p. 19. H.—*Smellie*, vi. 198.

OX. With fhort, black, rounded horns; with a great fpace between their bafes : on the fhoulders a vaft bunch, compofed of a flefhy fubftance, much elevated : the fore part of the body thick and ftrong: the hind part flender and weak : tail a foot long, naked to the end, which is tufted : the legs fhort and thick.

The head and fhoulders of the BULL are covered with very long flocks of reddifh woolly hair, falling over the eyes and horns, leaving only the points of the latter to be feen : on the chin, and along the dewlaps, is a great length of fhaggy hairs : the reft of the body during fummer is naked, in winter is cloathed equally in all parts. The Cow is leffer, and wants the fhaggy coat, which gives the Bull fo tremendous an afpect.

It grows to a great fize, even to the weight of fixteen hundred or two thoufand four hundred pounds *. The ftrongeft man cannot lift the hide of one of thefe animals from the ground †.

* *Lawfon*, 116. † *Catefby* ii. App.

B

The

The *Bifon* and *Aurochs* of *Europe* is certainly the fame fpecies with this; the difference confifts in the former being lefs fhaggy, and the hair neither fo foft nor woolly, nor the hind parts fo weak. Both *European* and *American* kinds fcent of mufk.

WHERE
ANTIENTLY
FOUND.

In antient times they were found in different parts of the old world, but went under different names; the *Bonafus* of *Ariftotle*, the *Urus* of *Cæfar*, the *Bos ferus* of *Strabo*, the *Bifon* of *Pliny*, and the *Bifton* of *Oppian*, fo called from its being found among the *Biftones*, a people of *Thrace*. According to thefe authorities, it was found in their days in *Media* and in *Pæonia*, a province of *Macedonia*; among the *Alps*, and in the great *Hercynian* foreft, which extended from *Germany* even into *Sarmatia* *. In later days a white fpecies was a native of the *Scottifh* mountains; it is now extinct in its favage ftate, but the offspring, fufficiently wild, is ftill to be feen in the parks of *Drumlanrig*, in the South of *Scotland*, and of *Chillingham* Caftle in *Northumberland* †.

WHERE AT
PRESENT.

EUROPE.

ASIA.

In thefe times it is found in very few places in a ftate of nature; it is, as far as we know, an inhabitant at prefent only of the forefts of *Lithuania*, and among the *Carpathian* mountains, within the extent of the great *Hercynian* wood ‡, its antient haunts; and in *Afia*, among the vaft mountains of *Caucafus*.

It is difficult to fay in what manner thefe animals migrated originally from the old to the new world; it is moft likely it was from the north of *Afia*, which in very antient times might have been ftocked with them to its moft extreme parts, notwithftanding they are now extinct. At that period there is a probability that the old and the new continents might have been united in the narrow channel between *Tchutki nofs* and the oppofite headlands of *America*;

---

* *Ariftot. Hift. An.* lib. ii. c. 1.—*Cæfar Bel. Gall.* lib. vi.—*Plinii Hift. Nat.* lib. xv. c. 15.—*Oppian Cyneg.* ii. Lin. 160.

† *Br. Zool.* 1. N° 3.—*Voy. Hebrides*, 124.—*Tour. Scotl.* 1772, Part ii. p. 285.

‡ There is a very fine figure of the *European Bifon* in Mr. *Ridinger's Jagbere Thiere*.

and

and the many iſlands off of that promontory, with the *Aleutian* or *New Fox* iſlands, ſomewhat more diſtant, ſtretching very near to *America*, may with great reaſon be ſuppoſed to be fragments of land which joined the two continents, and formed into their inſular ſtate by the mighty convulſion which divided *Aſia* from *America. Spain* was probably thus disjoined from *Africa*; *Britain* from *France*; *Iceland* from *Greenland*; *Spitzbergen* from *Lapland*.

But that they paſſed from *Aſia* to *America* is far the more probable, than that they ſtocked the new world from the ſide of *Europe*, not only on account of the preſent narrowneſs of the ſtreight between the two continents, which gives a greater cauſe to ſuppoſe them to have been once joined; but that we are now arrived at a certainty, that theſe animals in antient days were natives of *Sibiria:* the ſculls, with the horns affixed, of a ſize far ſuperior to any known at this time, have been found foſſil not only on the banks of the *Ilga*, which falls into the *Lena*, but even in thoſe of the *Anadyr*, the moſt eaſtern of the *Sibirian* rivers, and which diſembogues north of *Kamtſchatka* into thoſe ſtreights: ſimilar ſkulls and horns have been diſcovered near *Dirſchau*, in *Poland*, alſo of a gigantic magnitude; and in my opinion of the ſame ſpecies with the modern *Biſons* *.

In *America* theſe animals are found in the countries ſix hundred miles weſt of *Hudſon's Bay*; this is their moſt northern reſidence: From thence they are met with in great droves as low as *Cibole* †, in lat. 33, a little north of *California*, and alſo in the province of *Mivera*, in *New Mexico* ‡; the ſpecies inſtantly ceaſes ſouth of thoſe countries. They inhabit *Canada*, to the weſt of the lakes; and in greater abundance in the rich *ſavannas* which border the river *Miſſiſipi*, and the great rivers which fall into it from the weſt, in the upper *Louiſiana* §.

AMERICA.

---

* *Nov. Com. Petrop.* xvii. 460. tab. xi. xii.—I am ſorry to diſſent from my eſteemed friend Doctor *Pallas*, who thinks them to be the horns of Buffaloes; which are longer, ſtraiter, and angular.

† *Purchas*, iv. 1560, 1566.   ‡ *Fernandez, Nov. Hiſp.* x. c. 30.—*Hernandez*, 58.
§ *Du Pratz*, ii. 50. i. 116. 286.

There

There they are seen feeding in herds innumerable, promiscuously with multitudes of stags and deer, during morning and evening; retiring in the sultry heats into the shade of tall reeds, which border the rivers of *America*.

TIMID.

They are exceedingly shy; and very fearful of man, unless they are wounded, when they pursue their enemy, and become very dangerous.

CHASE.

The chase of these animals is a favorite diversion of the *Indians*: it is effected in two ways; first, by shooting; when the marksman must take great care to go against the wind, for their smell is so exquisite that the moment they get scent of him they instantly retire with the utmost precipitation * He aims at their shoulders, that they may drop at once, and not be irritated by an ineffectual wound. Provided the wind does not favor the beasts, they may be approached very near, being blinded by the hair which covers their eyes. The other method is performed by a great number of men, who divide and form a vast square: each band sets fire to the dry grass of the savanna where the herds are feeding; these animals have a great dread of fire, which they see approach on all sides; they retire from it to the center of the square †; the bands close, and kill them (pressed together in heaps) without the left hazard. It is pretended, that on every expedition of this nature, they kill fifteen hundred or two thousand beeves.

ANOTHER METHOD.

The hunting-grounds are prescribed with great form, least the different bands should meet, and interfere in the diversion. Penalties are enacted on such who infringe the regulations, as well as on those who quit their posts, and suffer the beasts to escape from the hollow squares; the punishments are, the stripping the delinquents, the taking away their arms (which is the greatest disgrace a savage can undergo), or lastly, the demolition of their cabins ‡.

* *Du Pratz*, i. 49. ii. 227.　　　† *Charlevoix, N. France*, v. 192.
‡ *Charlevoix*, v. 192.

The

The ufes of thefe animals are various. Powder-flafks are made of their horns. The fkins are very valuable; in old times the *Indians* made of them the beft targets*. When dreffed, they form an excellent buff; the *Indians* drefs them with the hair on, and cloath themfelves with them; the *Europeans* of *Louifiana* ufe them for blankets, and find them light, warm, and foft. The flefh is a confiderable article of food, and the bunch on the back is efteemed a very great delicacy. The Bulls become exceffively fat, and yield great quantity of tallow, a hundred and fifty pounds weight has been got from a fingle beaft†, which forms a confiderable matter of commerce. Thefe over-fed animals ufually become the prey of Wolves; for, by reafon of their great unwieldinefs, they cannot keep up with the herd.

The *Indians*, by a very bad policy, prefer the flefh of the Cows; which in time will deftroy the fpecies: they complain of the ranknefs of that of the Bulls; but *Du Pratz* thinks the laft much more tender, and that the ranknefs might be prevented, by cutting off the tefticles as foon as the beaft is killed.

The hair or wool is fpun into cloth, gloves, ftockings, and garters, which are very ftrong, and look as well as thofe made of the beft fheeps wool; Governor *Pownall* affures us, that the moft luxurious fabrick might be made of it‡. The fleece of one of thefe animals has been found to weigh eight pounds.

Their fagacity in defending themfelves againft the attacks of Wolves is admirable: when they fcent the approach of a drove of thofe ravenous creatures, the herd flings itfelf into the form of a circle: the weakeft keep in the middle, the ftrongeft are ranged on the outfide, prefenting to the enemy an impenetrable front of horns: fhould they be taken by furprize, and have recourfe to flight, numbers of the fatteft or the weakeft are fure to perifh ‖.

Ufes.

Skin.

Tallow.

Hair.

Defence against Wolves.

* *Purchas*, iv. 1550. † *Du Pratz*. ‡ *Topog. Defcr. N. Am.* 8. ‖ *Du Pratz*, i. 288.

Attempts have been made to tame and domefticate the wild, by catching the calves and bringing them up with the common kind, in hopes of improving the breed. It has not yet been found to anfwer: notwithftanding they had the appearance for a time of having loft their favage nature, yet they always grew impatient of reftraint, and, by reafon of their great ftrength, would break down the ftrongeft inclofure, and entice the tame cattle into the corn-fields. They have been known to engender together, and to breed; but I cannot learn whether the fpecies was meliorated * by the intercourfe: probably perfeverance in continuing the croffes is only wanted to effect their thorough domeftication; as it is notorious that the *Bifons* of the old world were the original ftock of all our tame cattle.

Thefe were the only animals which had any affinity to the *European* cattle on the firft difcovery of the new world: before that period, it was in poffeffion of neither Horfe nor Afs, Cow nor Sheep, Hog, Goat, nor yet that faithful animal the Dog. Mankind were here in a ftate of nature; their own paffions unfubdued, they never thought of conquering thofe of the brute creation, and rendering them fubfervient to their will. The few animals which they had congenerous to thofe mentioned, might poffibly by induftry have been reclamed. This animal might have been brought to all the ufes of the *European* Cow; the *Pecari* might have been fubftituted for the Hog; the Fox or Wolf for the Dog: but the natives, living wholly by chafe, were at war with the animal creation, and neglected the cultivation of any part, except the laft, which was imperfectly tamed.

Such is the cafe even to the prefent hour; for neither the example of the *Europeans*, nor the vifible advantages which refult from an attention to that ufeful animal the Cow, can induce the *Indian* to pay any refpect to it. He contemns every fpecies of domeftic labour, except what is neceffary for forming a provifion of bread. Every

* *Kalm,* i. 207.

wigwam

wigwam or village has its plantation of *Mayz*, or *Indian* corn, and on that is his great dependence, fhould the chafe prove unfuccefsful.

Domefticated cattle are capable of enduring very rigorous climates ; Cows are kept at *Quickjock* in *Lecha Lapmark*, not far from the arctic circle ; but they do not breed there, the fucceffion being preferved by importation : yet in *Iceland*, a fmall portion of which is within the circle, cattle abound, and breed as in more fouthern latitudes : they are generally fed with hay, as in other places ; but where there is fcarcity of fodder, they are fed with the fifh called the Sea-Wolf, and the heads and bones of Cod beaten fmall, and mixed with one quarter of chopped hay : the cattle are fond of it, and, what is wonderful, yield a confiderable quantity of milk. It need not be faid that the milk is bad.

LAPMARK.

ICELAND.

Kamtfchatka, like *America*, was in equal want of every domeftic animal, except a wolf-like Dog, till the *Ruffians* of late years introduced the Cow and Horfe. The colts and calves brought from the north into the rich paftures of *Kamtfchatka*, where the grafs is high, grow to fuch a fize, that no one would ever fufpect them to be defcended from the Ponies and Runts of the *Lena* *. The *Argali*, the ftock of the tame Sheep, abounds in the mountains, but even to this time are only objects of chafe. The natives are to this hour as uncultivated as the good *Evander* defcribes the primary natives of *Latium* to have been, before the introduction of arts and fciences.

KAMTSCHATKA.

> Queis neque mos, neque cultus erat, nec jungere tauros,
> Aut componere opes norant, aut parcere parto :
> Sed rami atque afper victu venatus alebat.
>
> No laws they know, no manners, nor the care
> Of lah'ring Oxen, or the fhining Share ;
> No arts of gain, nor what they gain'd to fpare :
> Their exercife the chafe : the running flood
> Supplied their thirft ; the trees fupplied their food.
>
> *Dryden.*

* *Pallas, Sp. Zool.* fafc. xi. 76.

Mufk.

2. MUSK.

Muſk Ox, *Hiſt. Quad.* N° 9.
Le Bœuf muſqùe, *de M. Jeremie, Voy. au Nord*, iii. 314.—Charlevoix, *N. France*,
  v. 194.—LEV. MUS.

Bull. With horns cloſely united at the baſe; bending inwards
and downwards; turning outwards towards their ends, which
taper to a point, and are very ſharp: near the baſe are two feet in
girth; are only two feet long meaſured along the curvature: weight
of a pair, ſeparated from the head, ſometimes is ſixty pounds *

The hair is of a duſky red, extremely fine, and ſo long as to trail on
the ground, and render the beaſt a ſeeming ſhapeleſs maſs, without
diſtinction of head or tail †: the legs and tail very ſhort: the
ſhoulders riſe into a lump.

SIZE.

In ſize lower than a Deer, but larger as to belly and quarters ‡.
I have only ſeen the head of this animal; the reſt of the deſcription
is taken from the authorities referred to: but by the friendſhip of
Samuel Wegg, Eſq; I received laſt year a very complete ſkin of the
cow of this ſpecies, of the age of three years, which enables me to
give the following deſcription:

Cow. The noſtrils long and open: the two middle cutting teeth
broad, and ſharp-edged; the three on each ſide ſmall, and truncated:
under and upper lips covered with ſhort white hairs on their fore
part, and with pale brown on their ſides: hair down the middle of
the forehead long and erect; on the cheeks ſmooth and extremely
long and pendulous, forming with that on the throat a long beard:
the hair along the neck, ſides, and rump hangs in the ſame manner,
and almoſt touches the ground: from the hind part of the head
to the ſhoulders is a bed of very long ſoft hair, forming an upright
mane: in the old beaſts the ſpace between the ſhoulders riſes into a

---

* *M. Jeremie*, in *Voyages au Nord*, iii. 315.
† The ſame.   ‡ *Drage's Voy.* ii. 260.

hunch:

Miss Griffith pinx.

R. Havell sculp.

Musk Bull & Cow. Nº 2.

hunch: the legs are very short, covered with smooth whitish hairs; those which encircle the hoofs very long, and of a pure white: hoofs short, broad, and black: the false hoofs large in proportion: tail only three inches long, a mere stump, covered with very long hairs, so as to be undistinguishable to the sight. Of the tail, the *Eskimaux* of the north-west side of the bay make a cap of a most horrible appearance; for the hairs fall all round their head, and cover their faces; yet it is of singular service in keeping off the Musquetoes, which would otherwise be intolerable*.

Space between the horns nine inches: the horns are placed exactly on the sides of the head; are whitish; thirteen inches and a half long; eight inches and a half round at the base; of the same sort of curvature with those of the Bull: the ears are three inches long, quite erect; sharp-pointed, but dilate much in the middle; are thickly lined with hair of a dusky color, marked with a stripe of white.

The color of the hair black, except on these parts:—from the base of one horn to that of the other, is a bed of white and light rust-colored hair: the mane is dusky, tinged with red, which is continued in a narrow form to the middle of the back; on which is a large roundish bed of pure white, and the hairs in that space shorter than any of the rest, not exceeding three inches in length, and of a pale brown towards their roots.

The hairs are of two kinds, the longest measure seventeen inches; are very fine and glossy, and when examined appear quite flat: this is the black part, which cloaths most part of the animal.

The bed of hair between the horns, and that which runs along the top of the neck, is far finer and softer than any human hair, and appears quite round. The white bed is still finer, and approaches to the nature of wool.

Beneath every part of the hair grows in great plenty, and often in flocks, an ash-colored wool, most exquisitely fine, superior, I think,

HORNS.

EARS.

COLOR.

HAIR

WOOL.

* *Ellis's voy.* 232.

C

to

to any I have feen, and which might be very ufeful in manufactures if fufficient could be procured. I give full credit to *M. Jeremie*, who fays, that he brought fome of the wool to *France*, and got ftockings made with it, more beautiful than thofe of filk *. The fkin is thin.

The length of the whole hide, from nofe to tail, is about fix feet four inches: of the head alone fourteen inches. The legs could not be well meafured, but were little more than a foot long.

The fituation of thefe animals is very local. They appear firft in the tract between *Churchill* river and that of *Seals*, on the weft fide of *Hudfon's Bay*. They are very numerous between the latitudes 66 and 73 north, which is as far as any tribes of *Indians* go. They live in herds of twenty or thirty. Mr. *Hearn* † has feen in the high latitudes feveral herds in one day's walk. They delight moft in the rocky and barren mountains, and feldom frequent the woody parts of the country. They run nimbly, and are very active in climbing the rocks. The flefh taftes very ftrong of Mufk, and the heart is fo ftrongly infected as hardly to be eatable; but the former is very wholefome, having been found to reftore fpeedily to health the fickly crew who made it their food ‡.

They are fhot by the *Indians* for the fake of the meat and fkins, the laft from its warmth making excellent blankets. They are brought down on fledges to the forts annually during winter, with about three or four thoufand weight of the flefh. Thefe are called *Churchill* Buffaloes, to diftinguifh them from the laft fpecies, which are in *Hudfon's Bay* called *Inland* Buffaloes, of which only the tongues are brought as prefents ‖.

They are found alfo in the land of the *Cris* or *Criftinaux*, and the *Affinibouels*: again among the *Attimofpiquay*, a nation fuppofed to in-

---

* *Voy. au Nord*, iii. 314.

† The gentleman who undertook, in 1770, 1771, 1772, the arduous journey to the *Icy Sea*, from *Prince of Wales's Fort*, *Hudfon's Bay*. To him, through Mr. *Hegg's* intereft, I am indebted for the fkin and this information.

‡ *Drage's voy.* ii. 260.   ‖ *Mr. Graham's MS.*

habit

habit about the head of the river of *Seals* *, probably not very remote from the South Sea. They are continued from thefe countries fouthward, as low as the provinces of *Quivera* and *Cibola*; for Father *Marco di Niça*, and *Gomara*, plainly defcribe both kinds †.

Some of the fkulls of this fpecies have been difcovered on the moffy plains near the mouth of the *Oby* in *Sibiria*. It is not faid how remote from the fea; if far, they probably in fome period might have been common to the north of *Afia* and of *America*; if near the fhore, it is poffible that the carcafes might have floated on the ice from *America* to the places where the remains might have been found ‡. Of this fpecies was the head, and fuch were the means of conveyance, from the coaft of *Hudfon's* or *Baffin's*, mentioned by Mr. *Fabricius*, and which he faw fo brought to *Greenland* ‖; for it could not have been, as he conjectures, the head of the *grunting* Ox, an animal found only in the very interior parts of northern *Afia*.

* *Dobbs's Hudfon's Bay*, 19, 25. † *Purchas*, iv. 1561. v. 854. ‡ *Pallas*, in nov. com. *Petrop.* xvii. 601. tab. xvii. ‖ *Faun. Groenl.* 28.

SHEEP.

S H E E P. *Hiſt. Quad.* Genus III.

A R G A L I : Wild Sheep, *Hiſt. Quad.* Nº 11. H. p. 38.—*Smellie*, vi. 205.—
Lev. Mus.

THE Sheep, in its wild ſtate, inhabits the north-eaſt of *Aſia*,
beyond lake *Baikal*, between the *Onon* and *Argun*, to the
height of latitude 60, on the eaſt of the *Lena*, and from thence to
Kamtſchatka, and perhaps the *Kurili* iſlands. I dare not pronounce
that they extend to the continent of *America*; yet I have received
from Doctor *Pallas* a fringe of very fine twiſted wool, which had ornamented a dreſs from the iſle of *Kadjak*; and I have myſelf another
piece from the habit of the *Americans* in latitude 50. The firſt was of
a ſnowy whiteneſs, and of unparalleled fineneſs; the other as fine, but
of a pale brown color: the firſt appeared to be the wool which
grows intermixed with the hairs of the *Argali*; the laſt, that which
is found beneath thoſe of the Muſk Ox. Each of theſe animals may
exiſt on that ſide of the continent, notwithſtanding they might have
not fallen within the reach of the navigators in their ſhort ſtay off the
coaſt.

Certain quadrupeds of this genus were obſerved in *California* by
the miſſionaries in 1697; one as large as a Calf of one or two years
old, with a head like a Stag, and horns like a Ram : the tail and
hair ſpeckled, and ſhorter than a Stag's. A ſecond kind was larger,
and varied in color; ſome being white, others black, and furniſhed
with

with very good wool. The Fathers called both Sheep, from their great resemblance to them * Either the *Americans* of latitude 50 are possessed of these animals, or may obtain the fleeces by commerce from the southern *Indians*.

The *Argali* abound in *Kamtschatka*; they are the most useful of their animals, for they contribute to food and cloathing. The *Kamtschatkans* cloath themselves with the skins, and esteem the flesh, especially the fat, diet fit for the Gods. There is no labor which they will not undergo in the chase. They abandon their habitations, with all their family, in the spring, and continue the whole summer in the employ, amidst the rude mountains, fearless of the dreadful precipices, or of the *avelenches*, which often overwhelm the eager sportsmen.

CHASE IN KAMTSCHATKA

These animals are shot with guns or with arrows; sometimes with cross-bows, which are placed in the paths, and discharged by means of a string whenever the *Argali* happens to tread on it. They are often chased with dogs, not that they are overtaken by them; but when they are driven to the lofty summits, they will often stand and look as if it were with contempt on the dogs below, which gives the hunter an opportunity of creeping within reach while they are so engaged; for they are the shyest of animals.

The *Mongols* and *Tungusi* use a nobler species of chase: they collect together a vast multitude of horses and dogs, attempting to surround them on a sudden; for such is their swiftness and cunning, that if they perceive, either by sight or smell, the approach of the *chasseurs*, they instantly take to flight, and secure themselves on the lofty and inaccessible summits.

IN MONGOLIA.

Domesticated Sheep will live even in the dreadful climate of *Greenland*. Mr. *Fabricius* † says, they are kept in many places. They are very numerous in *Iceland*. Before the epidemical disease which raged among them from 1740 to 1750, it was not uncommon for a

SHEEP IN ICELAND.

---

* *Ph. Trans. abr.* v. part ii. 195.   † *Faun. Groenl.* p. 29.

single

single person to be possessed of a thousand or twelve hundred. They have upright ears, short tails, and often four or five horns *. They are sometimes kept in stables during winter, but usually left to take their chance abroad, when they commonly hide themselves in the caves of exhausted vulcanoes †. They are particularly fond of scurvy-grass, with which they grow so fat as to yield more than twenty pounds. The ewes give from two to six quarts of milk a day, of which butter and cheese is made. The wool is never shorn, but left on till the end of *May*, when it grows loose, and is stripped entirely off in one fleece; and a fine, short, and new wool appears to have grown beneath; this continues growing all summer, becomes smooth and glossy like the hair of Camels, but more shaggy ‡. With the wool the natives manufacture their cloth; and the flesh dried is an article of commerce.

In all parts of *European Russia* are found the common Sheep. Those of the very north, and of the adjacent *Finmark*, have short tails and upright ears, and wool almost as rude as the hair of Goats; but are seldom polycerous. They sometimes breed twice in a year, and bring twins each time ‖.

In the *Asiatic* dominions of *Russia*, from the borders of *Russia* to those of *China*, is a most singular variety of Sheep, destitute of tails, with rumps swelling into two great, naked, and smooth hemispheres of fat, which sometimes weigh forty pounds: their noses are arched: their ears pendulous: their throats wattled: their heads horned, and sometimes furnished with four horns. These are so abundant throughout *Tartary*, that a hundred and fifty thousand have been annually sold at the *Orenburg* fairs; and a far greater number at the fort *Troizkaja*, from whence they are driven for slaughter into diffe-

---

* *Smellie*, vi. 207, 219.   † *Horrebow*, 46.   ‡ *Troil's voy.* 138.
‖ *Leems*, 228.

rent

rent parts of *Ruffia* *. Sheep do not thrive in *Kamtfchatka*, by reafon of the wetnefs of the country.

Sheep abound in *New England* and its iflands: the wool is fhort,
and much coarfer than that of *Great Britain*; poffibly proper attention to the houfing of the Sheep may in time improve the fleece;
but the feverity of the climate will ever remain an obftacle to its
perfection. Manufactures of cloth have been eftablifhed, and a tolerable cloth has been produced, but in quantities in no degree
equal to the confumption of the country. *America* likewife wants
downs; but by clearing the hills of trees, in a long feries of years
that defect may be alleviated. As we advance further fouth, the
Sheep grow fcarcer, worfe, and the wool more hairy.

* *Pallas, Sp. Zool.* fafc. xi. 63. tab. iv.

GOAT.

G O A T. *Hift. Quad.* G E N U S IV.

I B E X, *Hift. Quad.* N° 13, is fuppofed to extend to the mountains of the eaftern part of *Sibiria,* beyond the *Lena,* and to be found within the government of *Kamtfchatka.*—Lev. Mus.

THE tame Goat inhabits northern *Europe* as high as *Wardbuys,* in latitude 71, where it breeds, and runs out the whole year, only during winter has the protection of a hovel : it lives during that feafon on mofs and bark of Fir-trees, and even of the logs cut for fuel. They are fo prolific as to bring two, and even three, at a time. In *Norway* they thrive prodigioufly, infomuch that 70 or 80,000 of raw fkins are annually exported from *Bergen,* befides thoufands that are fent abroad dreffed.

Goats are alfo kept in *Iceland,* but not in numbers, by reafon of the want of fhrubs and trees for them to brouze. They have been introduced into *Greenland,* even to fome advantage. Befides vegetable food, they will eat the *Arctic* trouts dried; and grow very fat †.

The climate of *South America* agrees fo well with Goats, that they multiply amazingly : but they fucceed fo ill in *Canada,* that it is neceffary to have new fupplies to keep up the race ‡.

---

* *Smellie,* vi. 363.   † *Faun. Groenl.* p. 29.   ‡ *De Buffon,* ix. 71.

DEER.

D E E R. *Hiſt. Quad.* G E N U S VII.

Elk, Hiſt. Quad: N° 42:—*Smellie* vi. 315.—Lev. Mus.

DEER. With horns with ſhort beams, ſpreading into a broad palm, furniſhed on the outward ſide with ſharp ſnags; the inner ſide plain: no brow antlers: ſmall eyes: long ſlouching aſinine ears: noſtrils large: upper lip ſquare, great, and hanging far over the lower; has a deep furrow in the middle, ſo as to appear almoſt bifid: under the throat a ſmall excreſcence, with a long tuft of coarſe black hair pendent from it: neck ſhorter than the head; along the top an upright, ſhort, thick, mane: withers elevated: tail ſhort: legs long; the hind legs the ſhorteſt: hoofs much cloven.

Color of the mane a light brown; of the body in general a hoary brown: tail duſky above; white beneath. The vaſt ſize of the head, the ſhortneſs of the neck, and the length of the ears, give the beaſt a deformed and ſtupid look.

The greateſt height of this animal, which I have heard of, is ſeventeen hands, the greateſt weight 1229 pounds.

The largeſt horns I have ſeen are in the houſe of the *Hudſon's Bay* Company; they weigh fifty-ſix pounds: their length is thirty-two inches; breadth of one of the palms thirteen inches and a half; ſpace between point and point thirty-four.

The female is leſſer than the male, and wants horns.

Inhabits the iſle of *Cape Breton, Nova Scotia,* and the weſtern ſide of the Bay of *Fundy; Canada,* and the country round the great lakes, almoſt as far ſouth as the river *Ohio* *. Theſe are its preſent northern and ſouthern limits. In all ages it affected the cold and wooded regions in *Europe, Aſia,* and *America.* They are found in all the woody tracts of the temperate parts of *Ruſſia,* but not on the Arctic flats, nor yet in *Kamtſchatka.* In *Sibiria* they are of a monſtrous ſize, particularly among the mountains.

* *Du Pratz,* i. 301.

D

The

M O O S E.

The Elk and the Moose are the fame fpecies; the laft derived from *Mufu*, which in the *Algonkin* language fignifies that animal *. The *Englifh* ufed to call it the Black Moofe, to diftinguifh it from the Stag, which they named the Grey Moofe †. The *French* call it *L'Orignal.*

RESIDENCE AND FOOD.

Thefe animals refide amidft forefts, for the conveniency of broufing the boughs of trees, becaufe they are prevented from grazing with any kind of eafe, by reafon of the fhortnefs of their necks and length of their legs. They often have recourfe to water-plants, which they can readily get at by wading. *M. Sarrafin* fays, that they are very fond of the *anagyris fœtida,* or ftinking bean trefoil, and will uncover the fnow with their feet in order to get at it.

In paffing through the woods, they raife their heads to a horizontal pofition, to prevent their horns from being entangled in the branches.

GAIT.

They have a fingular gait : their pace is a fhambling trot, but they go with great fwiftnefs. In their common walk they lift their feet very high, and will without any difficulty ftep over a gate five feet high.

They feed principally in the night. If they graze, it is always againft an afcent; an advantage they ufe for the reafon above affigned.

RUMINATE.

They ruminate like the Ox.

YOUNG.

They go to rut in autumn; are at that time very furious, feeking the female by fwimming from ifle to ifle. They bring two young at a birth, in the month of *April*, which follow the dam a whole year. During the fummer they keep in families. In deep fnows they collect in numbers in the forefts of pines, for protection from the inclemency of the weather under the fhelter of thofe ever-greens.

They are very inoffenfive, except in the rutting-feafon; or except they are wounded, when they will turn on the affailant, and attack

---

* *Kalm,* i. 298. iii. 204.   † *Mr. Dudley's Phil. Tranf. Abridg.* vii. 447.

him

him with their horns, or trample him to death beneath their great hoofs.

Their flesh is extremely sweet and nourishing. The *Indians* say, that they can travel three times as far after a meal of Moose, as after any other animal food. The tongues are excellent, but the nose is perfect marrow, and esteemed the greatest delicacy in all *Canada*.

FLESH.

The skin makes excellent buff; is strong, soft, and light. The *Indians* dress the hide, and, after soaking it for some time, stretch and render it supple by a lather of the brains in hot water. They not only make their snow-shoes of the skin, but after a chase form the canoes with it : they sew it neatly together, cover the seams with an unctuous earth, and embark in them with their spoils to return home *.

SKIN.

The hair on the neck, withers, and hams of a full-grown Elk is of much use in making mattrasses and saddles; being by its great length well adapted for those purposes.

HAIR.

The palmated parts of the horns are farther excavated by the savages, and converted into ladles, which will hold a pint.

HORNS.

It is not strange that so useful an animal should be a principal object of chase. The savages perform it in different ways. The first, and the more simple, is before the lakes or rivers are frozen. Multitudes assemble in their canoes, and form with them a vast crescent, each horn touching the shore. Another party perform their share of the chase among the woods; they surround an extensive tract, let loose their dogs, and press towards the water with loud cries. The animals, alarmed with the noise, fly before the hunters, and plunge into the lake, where they are killed by the persons in the canoes, prepared for their reception, with lances or clubs †.

CHASE.

The other method is more artful. The savages inclose a large space with stakes hedged with branches of trees, forming two sides

* *La Hontan*, i. 59.   † *Charlevoix*, v. 188.

of

of a triangle·: the bottom opens into a fecond enclofure, completely triangular. At the opening are hung numbers of fnares,
made of flips of raw hides. The *Indians*, as before, affemble in great
troops, and with all kinds of noifes drive into the firft enclofure not
only the Moofes, but the other fpecies of Deer which abound in that
country: fome, in forcing their way into the fartheft triangle, are
caught in the fnares by the neck or horns; and thofe which efcape
the fnares, and pafs the little opening, find their fate from the arrows
of the hunters, directed at them from all quarters *.

They are often killed with the gun. When they are firft unhar-
boured, they fquat with their hind parts and make water, at which
inftant the fportfman fires; if he miffes, the Moofe fets off in a moft
rapid trot, making, like the Rein-deer, a prodigious rattling with its
hoofs, and will run for twenty or thirty miles before it comes to bay
or takes the water. But the ufual time for this diverfion is the winter. The hunters avoid entering on the chafe till the fun is ftrong
enough to melt the frozen cruft with which the fnow is covered,
otherwife the animal can run over the firm furface: they wait till
it becomes foft enough to impede the flight of the Moofe; which
finks up to the fhoulders, flounders, and gets on with great difficulty.
The fportfman purfues at his eafe on his broad rackets, or fnow-fhoes,
and makes a ready prey of the diftreffed animals,

> As weak againft the mountain heaps they pufh
> Their beating breaft in vain, and piteous bray,
> He lays them quivering on th' enfanguin d fnows,
> And with loud fhouts rejoicing bears them home.
> THOMPSON.

SUPERSTITIONS
RELATING TO
THE MOOSE.

The opinion of this animal's being fubject to the epilepfy feems *to*
have been univerfal, as well as the cure it finds by fcratching its ear
with the hind hoof till it draws blood. That hoof has been ufed in
Indian medicine for the falling-ficknefs; they apply it to the heart of

* *Charlevoix*, and *La Hontan*, i. 65.

the

the afflicted, make him hold it in his left hand, and rub his ear with it. They use it also in the colick, pleurisy, vertigo, and purple fever; pulverising the hoof, and drinking it in water. The *Algonkins* pretend that the flesh imparts the disease; but it is notorious that the hunters in a manner live on it with impunity.

The savages esteem the Moose a beast of good omen; and are persuaded that those who dream often of it may flatter themselves with long life *

Their wild superstition hath figured to them a Moose of enormous size, which can wade with ease through eight feet depth of snow; which is invulnerable, and has an arm growing out of its shoulder, subservient to the purposes of the human : that it has a court of other Mooses, who at all times perform suit and service, according to his royal will †.

FOSSIL HORNS NOT BELONGING TO THE MOOSE.

I lament that I am not able to discover the animal which owned the vast horns so often found in the bogs of *Ireland*, so long and so confidently attributed to the *Moose*. These have been found to be sometimes eight feet long, fourteen between tip and tip ‡, furnished with brow antlers, and weighing three hundred pounds : the whole skeleton is frequently found with them.

The fables delivered by *Joffelyn*, of the Moose being thirty-three hands, or twelve feet, high ; and by *Le Hontan*, of its horns weighing between three and four hundred pounds; occasioned the naturalists of past times to call the fossil horns those of the Moose ; and to flatter themselves that they had discovered the animal they belonged to : but recent discoveries evince the error. I once entertained hopes that the *Waskeffe* § of the *Hudson's Bay Indians* was the species ; but by some late information I received from Mr. *Andrew Graham*, factor in the *Bay*, I find it to be no other than the common Moose.

* *Charlevoix*, v. 186.   † The same.   ‡ *Wright's Louthiana*,
book iii. 20. tab. xxii.   § *Hift. Quad.* 45.

Hift.

Hiſt. Quad. N° 43.—*Smellie,* vi. 316.—*Hackluyt,* iii. 114.—LEV. MUS.

DEER. With large but ſlender horns, bending forward; with brow antlers broad and palmated, ſometimes three feet nine inches long; two feet ſix from tip to tip; weight, nine pounds twelve ounces avoirdupoiſe. The body is thick and ſquare: the legs ſhorter than thoſe of a Stag: the height of a full-grown Rein four feet ſix.

Color of the hair, at firſt ſhedding of the coat, of a browniſh aſh; afterwards changes to a hoary whiteneſs. The animal is admirably guarded againſt the rigor of the climate by the great thickneſs of the hairs, which are ſo cloſely placed as totally to hide the ſkin, even if they are put aſide with ever ſo much care.

Space round the eyes always black: noſe, tail, and belly white: above the hoofs a white circle: hair along the lower ſide of the neck very long: tail ſhort.

Hoofs, and falſe hoofs, long and black; the laſt looſely hung, making a prodigious clatter when the animal runs.

FEMALE. The female is furniſhed with horns; but leſſer, broader, and flatter, and with fewer branches than thoſe of the male. She has ſix teats, but two are ſpurious and uſeleſs. They bring two young at a time.

PLACE. The habitation of this Deer is ſtill more limited than that of the former, confined to thoſe parts where cold reigns with the utmoſt ſeverity. Its moſt ſouthern reſidence is the northern parts of *Canada,*
HUDSON'S-BAY. bordering on the territories of *Hudſon's Bay. Charlevoix* mentions a ſingle inſtance of one wandering as far as the neighborhood of *Quebec* *. Their true place is the vaſt tract which ſurrounds the

* V. 191.

Bay.

Bay. They are met with in *Labrador*, and again in *Newfoundland*, originally wafted thither acrofs the narrow ftraits of *Belleifle*, on iflands of ice.

They fpread northerly into *Greenland*, particularly on the weftern coaft, about *Difko* *. I can find no traces (even traditional) of them in *Iceland*; which is the more furprizing, as that ifland lies nearer to *Greenland* than *Newfoundland* does to the *Labrador* coaft. It is probable that they were deftroyed in very early times, when that ifland was fo infinitely more populous than it is at prefent; and the farther migration of thefe animals prevented by the amazing aggregate of ice, which in later ages blocked up and even depopulated the eaftern fide of *Greenland*. No vegetable, not even mofs, is to be found on that extenfive coaft to fupport thefe hardy animals. Their laft migration was from the weftern parts of *Greenland*, over unknown regions and fields of ice, to the inhofpitable *Alps* of *Spitzbergen*. Thefe, with the Polar Bear and Arctic Fox, form the fhort catalogue of its quadrupeds. They refide there throughout the year; and by wondrous inftinct do difcover their food, the *lichen rangiferinus*, beneath the fnow, which they remove to great depths by means of their broad and fpade-like antlers; and thus find fubfiftence thirteen degrees beyond the *Arctic* circle †.

To the weftern fide of *Hudfon's Bay* I trace the Rein as far as the nation called *Les Plat-coté des Chiens* ‡, the remoteft we are acquainted with in the parallel of that latitude. Beyond, are lands unknown, till we arrive at that new-difcovered chain of iflands, which extends to within a fmall diftance of *Afia*, or the northern cape of *Kamtfchatka*, where I again recover thefe animals. There is reafon to imagine that they are continued acrofs the continent of *America*, but not on the iflands which intervene between it and *Afia* ‖. But in the

---

* *Egede,* 59. *Crantz,* i. 70.—The *Canadians* call it *Le Caribou.*   † *Marten's Spitzbergen,* 99. *Phipps's voy.* 185.   ‡ *Dobbs's Hudfon's Bay,* 19. ‖ *Muller's voyages from Afia to America,* Preface xxv.

isle

ifle of *Kadjak*, and others of the eafternmoft *Fox iflands*, the inhabitants have fkins of them from the *American* continent, and border
their bonnets with the white hairs of the domeftic Rein-deers,
ftained red.   They are found again in the countries which border on
the Icy fea *; from which they retire, at approach of winter, towards
the woods, to feed on the mofs, not only that which grows on the
ground, but the fpecies pendulous from the trees.   The whole north-
eaft of *Sibiria* abounds with them.   They alfo are yet found wild in
the *Urallian* mountains; along the river *Kama*, as far as *Kungus*; and
about fome fnowy fummits more fouth: and again on the high chain
bordering on *Sibiria* on the fouth, and about lake *Baikal*.   Towards

SAMOIEDEA.
the weft they are continued in the land of the *Samoieds*; and finally
among the well-known *Laplanders*.   I here tranfgrefs the limits of
my plan, to give a flight comparative view of the progrefs of civilization among the inhabitants of thefe frozen climes.

LAPLANDERS,
THEIR USES OF IT.
With the *Laplanders* this animal is the fubftitute to the Horfe,
the Cow, the Sheep, and the Goat.   Thofe moft innocent of people
have, even under their rigorous fky fome of the charms of a paftoral
life.   They have fubdued thefe animals to various ufes, and reclamed them from their wild ftate.   They attend their herds of
Rein-deer, during fummer, to the fummits of their alps; to the
fides of their clear lakes and ftreams, often bordered with native rofes.
They know the arts of the dairy, milk thefe their cattle, and make
from it a rich cheefe.   They train them to the fledge, confider them
as their chief treafure, and cherifh them with the utmoft tendernefs.

SAMOIEDS.
The brutifh *Samoied* confiders them in no other view than as animals of draught, to convey them to the chafe of the wild Reins; which
they kill for the fake of the fkins, either to cloath themfelves, or to
cover their tents.   They know not the cleanly dc icacy of the milk
or cheefe; but prefer for their repaft the inteftines of beafts, or the
half-putrid flefh of a horfe, ox, or fheep, which they find dead on
the high road †.

* *Barentz voy.*   † *Le Bruyn*, i. 7, 8.

The

The *Koreki,* a nation of *Kamtfchatka,* may be placed on a level with the *Samoieds :* they keep immenfe herds of Reins ; fome of the richeft, to the amount of ten or twenty thoufand ; yet fo fordid are they as to eat none except fuch which they kill for the fake of the fkins ; an article of commerce with their neighbors the *Kamtfchatkans :* otherwife they content themfelves with the flefh of thofe which die by difeafe or chance. They train them in the fledge, but neglect them for every domeftic purpofe * Their hiftorian fays, they couple two to each carriage ; and that the Deer will travel a hundred and fifty verfts in a day, that is, a hundred and twelve *Englifh* miles. They caftrate the males by piercing the fpermatic arteries, and tying the fcrotum tight with a thong.

The inhabitants about the river *Kolyma* make ufe of the foft fkins of the Rein-deer, dreffed, for fails for a kind of boat called *Schit'ki,* caulked with mofs ; and the boards as if fewed together with thongs ; and the cordage made of flices of the fkin of the Elk †.

The favage and uninformed *Eskimaux* and *Greenlanders,* who poffefs, amidft their fnows, thefe beautiful animals, neglect not only the domeftic ufes, but even are ignorant of their advantage in the fledge. Their element is properly the water; their game the Seals. They feem to want powers to domefticate any animals unlefs Dogs. They are at enmity with all; confider them as an object of chafe, and of no utility till deprived of life. The flefh of the Rein is the moft coveted part of their food ; they eat it raw, dreffed, and dried and fmoked with the fnow lichen. The wearied hunters will drink the raw blood but it is ufually dreffed with the berries of the heath : they eagerly devour the contents of the ftomach, but ufe the inteftines boiled. They are very fond of the fat, and will not lofe the left bit ‡. The fkin, fometimes a part of their cloathing, dreffed with the hair on, is foft and pliant ; it forms alfo the inner lining of their tents, and moft

---

* *Hift. Kamtfchatka,* 226, 227.—The *Koreki* exchange their Deer with the neighboring nations for rich furs.   † *Muller's Summary, &c.* xviii.   ‡ *Faun. Groenl.* p. 28.

E

excellent

excellent blankets. The tendons are their bow-strings, and when split are the threads with which they sew their jackets *.

The *Greenlanders*, before they acquired the knowlege of the gun, caught them by what was called the *clapper-hunt* †. The women and children surrounded a large space, and, where people were wanting, set up poles capped with a turf in certain intervals, to terrify the animals; they then with great noise drove the Reins into the narrow defiles, where the men lay in wait and killed them with harpoons or darts. But they are now become very scarce.

MULTITUDES
IN
HUDSON'S BAY.

On the contrary, they are found in the neighborhood of *Hudson's Bay* in most amazing numbers, columns of eight or ten thousand are seen annually passing from north to south in the months of *March* and *April* ‡, driven out of the woods by the musketoes, seeking refresh-

MIGRATION.

ment on the shore, and a quiet place to drop their young. They go to rut in *September*, and the males soon after shed their horns; they are at that season very fat, but so rank and musky as not to be eatable. The females drop their young in *June*, in the most sequestered spots they can find; and then they likewise lose their horns. Beasts of prey follow the herds: first, the Wolves, who single out the stragglers (for they fear to attack the drove) detach and hunt them down: the Foxes attend at a distance, to pick up the offals left by the former. In autumn the Deer with the Fawns re-migrate northward.

USES.

The *Indians* are very attentive to their motions; for the Rein forms the chief part not only of their dress but food. They often kill multitudes for the sake of their tongues only; but generally they separate the flesh from the bones, and preserve it by drying it in the smoke: they also save the fat, and sell it to the *English* in bladders, who use it in frying instead of butter. The skins are also an article of commerce, and used in *London* by the Breeches-makers.

CHASE.

The *Indians* shoot them in the winter. The *English* make hedges, with stakes and boughs of trees, along the woods, for five miles in

* *Drage's voy.* i. 25.   † *Crantz*, i. 71.   ‡ *Dobbs*, 19, 22.

length.

length, leaving openings at proper intervals befet with fnares, in which multitudes are taken.

The *Indians* alfo kill great numbers during the feafons of migration, watching in their canoes, and fpearing them while paffing over the rivers of the country, or from ifland to ifland; for they fwim moft admirably well.

DEER. With long upright horns much branched : flender and fharp brow antlers : color a reddifh brown : belly and lower fide of the tail white : the horns often fuperior in fize to thofe of the *European* Stags, fome being above four feet high, and thirty pounds in weight.

Inhabits *Canada,* particularly the vaſt forefts about the lakes; are feen in great numbers grazing with the Buffaloes on the rich favannas bordering on the *Miſſiſipi,* the *Miſſouri,* and other *American* rivers; they are alfo found within our Colonies, but their numbers decreafe as population gains ground. An *Indian* living in 1748 had killed many Stags on the fpot where *Philadelphia* now ftands *.

They feed eagerly on the broad-leaved *Kalmia*; yet that plant is a poifon to all other horned animals; their inteſtines are found filled with it during winter. If their entrails are given to Dogs, they become ftupified, and as if drunk, and often are fo ill as hardly to efcape with life †.

Stags are alfo found in *Mexico,* where they are called *Aculliame* : they differ not from thofe of *Spain* in fhape, fize, or nature ‡. *South America* is deftitute of thefe animals : they can bear the extremes of heat but not of cold. They are found neither in *Hudſon's Bay,*

* *Kalm* i. 336.   † *Kalm* i. 338.   ‡ *Hernandez, Nov. Hiſp.* 325.

Kamtſchatka,

Kamtfchatka; nor in any country inhabited by the Rein—a line in a manner feparates them.

Their fkins are an article of commerce imported * by the *Hudfon's Bay* company; but brought from the diftant parts far inland by the *Indians,* who bring them from the neighborhood of the lakes. In moft parts of *North America* they are called the Grey Moofe, and the Elk; this has given occafion to the miftaken notion of that great animal being found in *Virginia,* and other fouthern provinces.

The Stags of *America* grow very fat: their tallow is much efteemed for making of candles. The *Indians* fhoot them. As they are very fhy animals, the natives cover themfelves with a hide, leaving the horns erect; under fhelter of which they walk within reach of the herd. *De Brie,* in the xxvth plate of the Hiftory of *Florida,* gives a very curious reprefentation of this artful method of chafe, when it was vifited by the *French* in 1564.

Stags are totally extirpated in *Ruffia,* but abound in the mountanous fouthern tract of *Sibiria,* where they grow to a fize far fuperior to what is known in *Europe.* The height of a grown Hind is four feet nine inches and a half, its length eight feet; that of its head one foot eight inches and a half.

The fpecies ceafes in the north-eaftern parts of *Sibiria,* nor are any found in *Kamtfchatka.*

6. VIRGINIAN.   *Hift. Quad.* N° 46.—LEV. MUS.

DEER. With round and flender horns, bending greatly forward; numerous branches on the interior fides: deftitute of brow antlers: color of the body a cinereous brown: head of a deep brown: belly, fides, fhoulders, and thighs, white, mottled with brown: tail

* In the fale of 1764, 1307 were entered.

ten

ten inches long, of a dufky color : feet of a yellowifh brown. Are not fo well haunched as the *Englifh* Buck, and are lefs active *.

Inhabits all the provinces fouth of *Canada*, but in greateft abundance in the fouthern; but efpecially·the vaft favannas contiguous to the *Miffifipi*, and the great rivers which flow into it. They graze in herds innumerable, along with the Stags and Buffaloes. This fpecies probably extends to *Guiana*, and is the *Baieu* of that country, which is faid to be about the fize of a *European* Buck, with fhort horns, bending at their ends †.

They are capable of being made tame; and when properly trained, are ufed by the *Indians* to decoy the wild Deer (efpecially in the rutting feafon) within fhot. Both Bucks and Does herd from *September* to *March*; after that they feparate, and the Does fecrete themfelves to bring forth, and are found with difficulty. The Bucks from this time keep feparate, till the amorous feafon of *September* revolves. The Deer begin to feed as foon as night begins; and fometimes, in the rainy feafon, in the day: otherwife they feldom or never quit their haunts. An old *Americam* fportfman has remarked, that the Bucks will keep in the thickets for a year, or even two ‡.

Thefe animals are very reftlefs, and always in motion, coming and going continually §. Thofe which live near the fhores are lean and bad, fubject to worms in their heads and·throats, generated from the eggs depofited in thofe parts ‖. Thofe that frequent the hills and favannas are in better cafe, but the venifon is dry. In hard winters they will feed on the long mofs which hangs from the trees in the northern parts.

Thefe and other cloven-footed quadrupeds of *America* are very fond of falt, and refort eagerly to the places impregnated with it. They are always feen in great numbers in the fpots where the ground

---

* The late ingenious Mr. *Ellis* fhewed me a Bezoar found in one of thefe Deer, killed in *Georgia*. It was of a fpheroid form, an inch and three quarters broad, half an inch thick in the middle; of a pale brown color; hard, fmooth, and gloffy.

† *Bancroft*.　　‡ Doctor *Garden*.　　§ *Du Pratz*, ii. 51.　　‖ *Lawfon*, 124.

has

has been torn by torrents or other accidents, where they are feen licking the earth. Such fpots are called *licking-places*. The huntf-men are fure of finding the game there; for, notwithftanding they are often difturbed, the Buffaloes and Deer are fo paffionately fond of the favory regale, as to bid defiance to all danger, and return in droves to thefe favorite haunts.

The fkins are a great article of commerce, 25,027 being imported from *New-York* and *Penfylvania* in the fale of 1764.

The Deer are of the firft importance to the Savages. The fkins form the greateft branch of their traffick, by which they procure from the colonifts, by way of exchange, many of the articles of life. To all of them it is the principal food throughout the year; for by drying it over a gentle but clear fire, after cutting it into fmall pieces, it is not only capable of long prefervation, but is very portable in their fudden excurfions, efpecially when reduced to powder, which is frequently done.

Hunting is more than an amufement to thefe people. They give themfelves up to it not only for the fake of fubfiftence, but to fit themfelves for war, by habituating themfelves to fatigue. A good huntfman is an able warrior. Thofe who fail in the fports of the field are never fuppofed to be capable of fupporting the hardfhips of a campaign; they are degraded to ignoble offices, fuch as dreff-ing the fkins of Deer, and other employs allotted only to flaves and women †.

When a large party meditates a hunting-match, which is ufually at the beginning of winter, they agree on a place of rendezvous, often five hundred miles diftant from their homes, and a place, perhaps, that many of them had never been at. They have no other me-thod of fixing on the fpot than by pointing with their finger. The preference is given to the eldeft, as the moft experienced ‡ .

† *Lawfon*, 208.   ‡ *Catefby*, App. xii.

When

When this matter is settled, they separate into small parties, travel and hunt for subsistence all the day, and rest at night; but the women have no certain resting-places. The Savages have their particular hunting countries; but if they invade the limits of those belonging to other nations, feuds ensue, fatal as those between *Percy* and *Douglas* in the famed *Chevy Chace*.

As soon as they arrive on the borders of the hunting country, (which they never fail doing to a man, be their respective routes ever so distant or so various) the captain of the band delineates on the bark of a tree his own figure, with a Rattlesnake twined round him with distended mouth; and in his hand a bloody tomahawk. By this he implies a destructive menace to any who are bold enough to invade their territories, or to interrupt their diversion *.

The chase is carried on in different ways. Some surprise the Deer by using the stale of the head, horns, and hide, in the manner before mentioned: but the general method is performed by the whole body. Several hundreds disperse in a line, encompassing a vast space of country, fire the woods, and drive the animals into some strait or peninsula, where they become an easy prey. The Deer alone are not the object; Foxes, Raccoons, Bears, and all beasts of fur, are thought worthy of attention, and articles of commerce with the *Europeans*.

The number of Deer destroyed in some parts of *America* is incredible; as is pretended, from an absurd idea which the Savages have, that the more they destroy, the more they shall find in succeeding years. Certain it is that multitudes are destroyed the tongues only preserved, and the carcases left a prey to wild beasts. But the motive is much more political. The Savages well discern, that should they overstock the market, they would certainly be over-reached by the *European* dealers, who take care never to produce more goods than are barely sufficient for the demand of the season, establishing their prices according to the quantity of furs brought by the natives. The hunters live in their quarters with the utmost festivity, and indulgence

* *Catesby,* App. ix.

in

in all the luxuries of the country.   The chafe rouzes their appetites ;
they are perpetually eating, and will even rife to obey, at midnight,
the calls of hunger.   Their viands are exquifite.   Venifon boiled
with red peafe ; turkies barbecued and eaten with bears fat ; fawns
cut out of the does belly, and boiled in the native bag ; fifh, and
crayfifh, taken in the next ftream ; dried peaches, and other fruits,
form the chief of their good living *.   Much of this food is carmina-
tive : they give loofe to the effects, and (reverfe to the cuftom of
the delicate *Arabs* †) laugh moft heartily on the occafion ‡.

They bring along with them their wives and miftreffes : not that
they pay any great refpect to the fair.   They make (like the *Cath-
nefians*) errant pack-horfes of them, loading them with provifions,
or the fkins of the chafe ; or making them provide fire-wood.   Love
is not the paffion of a Savage, at left it is as brief with them as with
the animals they purfue.

7. MEXICAN.   Mexican Roe ? *Hift. Quad.* N° 52.—*Smellie,* iv. 136.

DEER.   With horns near nine inches long, meafuring by the
curvature ; and near nine inches between tip and tip, and two
inches diftant between the bafes.   About an inch and a half from the
bottom is one fharp erect fnag.   This, and the lower parts of the
horns, are very rough, ftrong, and fcabrous.   The upper parts bend
forwards over the bafes ; are fmooth, flatted, and broad, dividing
into three fharp fnags.   Color of the hair like the *European* Roe; but
while young are rayed with white.   In fize fomewhat fuperior to the
European Roe.

Inhabits *Mexico* ‖ ; probably extends to the interior north-weftern
parts of *America,* and may prove the *Scenoontung* or *Squinaton,* defcribed
as being lefs than a Buck and larger than a Roe, but very like it,
and of an elegant form §.

* *Lawfon,* 207.   † *D'Arvieux's travels,* 147.   ‡ *Lawfon,* 207.
‖ *Hernandez.*   § *Dobbs's Hudfon's Bay,* 24.

Hift.

Hiſt. Quad. N° 51.—*Smellie,* iv. 120.—Lev. Mus.

DEER. With upright, round, rugged horns, trifurcated: hairs tawny at their ends, grey below: rump and under-ſide of the tail white. Length near four feet: tail only an inch.

According to *Charlevoix,* they are found in great numbers in *Canada.* He ſays they differ not from the *European* kind: are eaſily domeſticated. The Does will retreat into the woods to bring forth, and return to their maſter with their young*. They extend far weſt†. If *Piſo*'s figure may be depended on, they are found in *Brazil* ‡; are frequent in *Europe*; and inhabit as high as *Sweden* and *Norway* §: is unknown in *Ruſſia.*

Tail-less Roe, *Hiſt. Quad.* p. 109.

In its ſtead is a larger variety: with horns like the laſt, and color the ſame; only a great bed of white covers the rump, and extends ſome way up the back: no tail, only a broad cutaneous excreſcence around the anus.

Inhabits all the temperate parts of *Ruſſia* and *Sibiria,* and extends as far to the north as the *Elk.* Deſcends to the open plains in the winter. The *Tartars* call it *Saiga:* the *Ruſſians Dikaja Roza.*

Fallow Deer, *Hiſt. Quad.* N° 44.

Are animals impatient of cold: are unknown in the *Ruſſian* empire, except by importation: and are preſerved in parks in *Sweden* ‖. The *Engliſh* tranſlator of *Pontoppidan* mentions them (perhaps erroneouſly) among the deer of *Norway.*

* *Hiſt. Nouv. France,* v. 195.　　† *Dobbs's Hudſon's Bay,* 24.　　‡ 97.
§ *Faun. Suec.* N° 43, and *Pontop. Norway,* ii. 9.　　‖ *Du Pratz,* ii. 54.

F

MUSK.

M U S K. *Hift. Quad.* G E N U S X.

A.   TIBET M. *Hift. Quad.* N° 54.—Mofchus, *Pallas Sp. Zool.* fafc. xiii. LEV. MUS.

MUSK. With very fharp flender white tufks on each fide of the upper jaw, hanging out far below the under jaw: ears rather large: neck thick: hair on the whole body long, upright, and thick fet; each hair undulated; tips ferruginous; beneath them black; the bottoms cinereous: on each fide of the front of the neck is a white line edged with black, meeting at the cheft; another croffes that beneath the throat: limbs very flender, and of a full black: tail very fhort, and fcarcely vifible. The female wants the tufks and the mufk-bag.

The mufk-bag is placed on the belly, almoft between the thighs. A full-grown male will yield a drachm and a half of mufk; an old one two drachms.

SIZE.   The length of the male is two feet eleven; of the female, two feet three. The weight of a male from twenty-five to thirty pounds, Troy weight: of an old female, from thirty to thirty-five; but fome young ones do not exceed eighteen.

PLACE   Inhabits *Afia*, from lat. 20 to 60, or from the kingdoms of *Laos* and *Tong-King*, between *India* and *China*, and through the kingdom of *Tibet* * as high as *Mangafea*. The river *Jenefei* is its weftern boundary, and it extends eaftward as far as lake *Baikal*, and about the rivers *Lena* and *Witim*; but gradually narrows the extent of its refidence as it approaches the tropic. Lives on the higheft and rudeft mountains, amidft the fnows, or in the fir woods which lie between them: goes ufually folitary, except in autumn, when they collect in flocks to change their place: are exceffively active, and take amazing

* Correct in p. 113, *Hift. Quad.* 9. 44 or 45, read 20.

leaps

leaps over the tremendous chafms of their *alps*, or from rock to rock : tread fo light on the fnow, with their true and falfe hoofs extended, as fcarcely to leave a mark ; while the dogs which purfue them fink in, and are forced to defift from the chafe : are fo fond of liberty as never to be kept alive in captivity. They feed on *lichens, arbutus, rhododendron*, and *whortleberry*-plants. Their chafe is moft laborious : they are taken in fnares ; or fhot by crofs-bows placed in their tracks, with a ftring from the trigger for them to tread on and difcharge. The *Tungufi* fhoot them with bows and arrows. The fkins are ufed for bonnets and winter dreffes. The *Ruffians* often fcrape off the hair, and have a way of preparing them for fummer cloathing, fo as to become as foft and fhining as filk.

The two other hoofed animals of the north of *Afia*, the Two-bunched Camel, and the Wild Boar, do not reach as high as lat. 60 : the firft is found in great troops about lake *Baikal*, as far as lat. 56 or 57 ; but if brought as high as *Jakutfk*, beyond lat. 60, perifh with cold *. The Wild Boar is common in all the reedy marfhes of *Tartary* and *Sibiria*, and the mountanous forefts about lake *Baikal*, almoft to lat. 55 ; but none in the north-eaftern extremity of *Sibiria*.

CAMEL.

WILD BOAR.

* *Zimmerman*, 357.

DIV. II.
DIGITATED QUADRUPEDS.

SECT. I. With CANINE TEETH.

DIV. II.   Digitated Quadrupeds.

SECT I.   With CANINE TEETH.
Rapacious, Carnivorous.

DOG.   *HIST. QUAD.*   GENUS XVII.

9. WOLF.   *Hiſt. Quad.* Nº 137.—*Smellie*, iv. 196.—LEV. MUS.

DOG. With a long head : pointed noſe : ears ſharp and erect : legs long : tail buſhy, bending down : hair pretty long. Color uſually of a pale brown, mixed with dull yellow and black.

Inhabits the interior countries ſouth of *Hudſon's Bay* ; and from thence all *America*, as low as *Florida*. There are two varieties, a greater and a leſſer. The firſt uſually confines itſelf to the colder parts. The latter is not above fifteen inches high* In the more uninhabited parts of the country, they go in great droves, and hunt the deer like a pack of hounds, and make a hideous noiſe. They will attack the Buffalo ; but only venture on the ſtragglers. In the unfrequented parts of *America* are very tame, and will come near the few habitations in hopes of finding ſomething to eat. They are often ſo very poor and hungry, for want of prey, as to go into a ſwamp and fill themſelves with mud, which they will diſgorge as ſoon as they can get any food.

COLOR.   The Wolves towards *Hudſon's Bay* are of different colors ; grey and white ; and ſome black and white, the black hairs being mixed with the white chiefly along the back. In *Canada* they have been found entirely black†. They are taken in the northern parts in log-traps, or by ſpring guns ; their ſkins being an article of commerce.

In the LEVERIAN muſeum is the head and ſcull of a wolf: duſky and brown, formed by the natives into a helmet. The pro-

* *Du Pratz*, ii. 54.   † *Smellie*, iv. 212.
tection

tection of the head was the natural and firſt thought of mankind;
and the ſpoils of beaſts were the firſt things that offered. *Hercules*
ſeized on the ſkin of the Lion : the *Americans*, and ancient *Latians* that
of the Wolf.

> Fulvoſque Lupi de pelle galeros
> Tegmen habet capiti.

Wolves are now ſo rare in the populated parts of *America*, that the
inhabitants leave their ſheep the whole night unguarded : yet the
governments of *Penſylvania* and *New Jerſey* did ſome years ago allow
a reward of twenty ſhillings, and the laſt even thirty ſhillings, for the
killing of every Wolf. Tradition informed them what a ſcourge
thoſe animals had been to the colonies; ſo they wiſely determined
to prevent the like evil. In their infant ſtate, wolves came down in
multitudes from the mountains, often attracted by the ſmell of the
corpſes of hundreds of *Indians* who died of the ſmall-pox, brought
among them by the *Europeans :* but the animals did not confine their
inſults to the dead, but even devoured in their huts the ſick and dying
Savages *.

The Wolf is capable of being in ſome degree tamed and domeſ-
ticated †. It was, at the firſt arrival of the *Europeans*, and is ſtill in   Doe.
many places, the Dog of the *Americans* ‡. It ſtill betrays its ſavage
deſcent, by uttering only a howl inſtead of the ſignificant bark of the
genuine Dog. This half-reclamed breed wants the ſagacity of our
faithful attendant; and is of little farther uſe in the chaſe, than in
frightening the wild beaſts into the ſnares or traps,

The *Kamtſchatkans*, *Eſkimaux*, and *Greenlanders*, ſtrangers to the
ſofter virtues, treat theſe poor animals with great neglect. The former, during ſummer, the ſeaſon in which they are uſeleſs, turn
them looſe to provide for themſelves; and recall them in *October* into their uſual confinement and labor : from that time till ſpring they

---

* *Kalm*, i. 285.   † The ſame, 286. *Lawſon*, 119.   ‡ *Smith's Hiſt.*
Virginia, 27. *Crantz Greenland*, i. 74.

are fed with fifh-bones and *opana*, i. e. putrid fifh preferved in pits, and ferved up to them mixed with hot water. Thofe ufed for draught are caftrated; and four, yoked to the carriage, will draw five poods, or a hundred and ninety *Englifh* pounds, befides the driver; and thus loaden, will travel thirty verfts, or twenty miles, a day; or if unloaden, on hardened fnow, on fliders of bone, a hundred and fifty verfts, or a hundred *Englifh* miles *.

It is pretty certain that the *Kamtfchatkan* Dogs are of wolfifh defcent; for Wolves abound in that country, in all parts of *Sibiria*, and even under the *Arctic* circle. If their mafter is flung out of his fledge, they want the affectionate fidelity of the *European* kind, and leave him to follow, never ftopping till the fledge is overturned, or elfe ftopped by fome impediment †. I am alfo ftrengthened in my opinion by the ftrong rage they have for the purfuit of deer, if on the journey they crofs ‡ the fcent; when the mafter finds it very difficult to make them purfue their way.

The great traveller of the thirteenth century, *Marco Polo*, had knowlege of this fpecies of conveyance from the merchants who went far north to traffic for the precious furs. He defcribes the fledges; adds, that they were drawn by fix great dogs; and that they changed them and the fledges on the road, as we do at prefent in going poft ∥.

The *Kamtfchatkans* make ufe of the fkins of dogs for cloathing, and the long hair for ornament: fome nations are fond of them as a food; and reckon a fat dog a great delicacy §. Both the *Afiatic* and *American* Savages ufe thefe animals in facrifices to their gods ¶, to befpeak favor, or avert evil. When the *Koreki* dread any infection,

* *Hift. Kamtfchatka*, 107. 197.   † The fame, 107.

‡ The fame.—There is a variety of black wolves in the *Vekroturian* mountains. The fhe-wolves have been fuccefsfully coupled with dogs in fome noblemen's parks about *Mofcow*.

∥ In *Bergeron*, 160.   § *Hift. Kamtfchatka*, 231. The *Americans* do the fame, *Drage*, i. 216.   ¶ *Hift. Kamtfch.* 226. *Drage*, ii. 41.

they

they kill a dog, wind the inteftines round two poles, and pafs between them.

The *Greenlanders* are not better mafters. They leave their dogs
to feed on muffels or berries; unlefs in a great capture of feals,
when they treat them with the blood and garbage. Thefe people
alfo fometimes eat their dogs: ufe the fkins for coverlets, for cloathing, or to border and feam their habits: and their beft thread is
made of the guts.

The Dogs in general are large; and, in the frigid parts at left,
have the appearance of Wolves: are ufually white, with a black face;
fometimes varied with black and white, fometimes all white; rarely
brown, or all black: have fharp nofes, thick hair, and fhort ears: and
feldom bark; but fet up a fort of growl, or favage howl. They fleep
abroad; and make a lodge in the fnow, lying with only their nofes
out. They fwim moft excellently: and will hunt, in packs, the
ptarmigan, arctic fox, polar bear, and feals lying on the ice. The
natives fometimes ufe them in the chafe of the bear. They are exceffively fierce; and, like wolves, inftantly fly on the few domeftic
animals introduced into *Greenland*. They will fight among themfelves, even to death. Canine madnefs is unknown in *Greenland* *.
They are to the natives in the place of horfes: the *Greenlanders* faften
to their fledges from four to ten; and thus make their vifits in
favage ftate, or bring home the animals they have killed. *Egede*
fays that they will travel over the ice fifteen *German* miles in a day,
or fixty *Englifh*, with fledges loaden with their mafters and five or fix
large feals †.

Thofe of the neighboring ifland of *Iceland* have a great refemblance to them. As to thofe of *Newfoundland*, it is not certain that
there is any diftinct breed: moft of them are curs, with a crofs of the
maftiff: fome will, and others will not, take the water, abfolutely
refufing to go in. The country was found uninhabited, which
makes it more probable that they were introduced by the *Europeans*;

* *Faun. Grœnl.* p. 19.   † *Egede,* 63. *Crantz,* i. 74.

G

who

who ufe them, as the factory does in *Hudfon's Bay*, to draw firing from the woods to the forts.

The Savages who trade to *Hudfon's Bay* make ufe of the wolfifh kind to draw their furs.

It is fingular, that the race of *European* Dogs fhew as ftrong an antipathy to this *American* fpecies, as they do to the Wolf itfelf. They never meet with them, but they fhew all poffible figns of diflike, and will fall on and worry them; while the wolfifh breed, with every mark of timidity, puts its tail between its legs, and runs rom the rage of the others. This averfion to the Wolf is natural to all genuine Dogs: for it is well known that a whelp, which has never feen a wolf, will at firft fight tremble, and run to its mafter for protection: an old dog will inftantly attack it.

I fhall conclude this article with an abftract of a letter from Dr. *Pallas*, dated *October* 5th 1781; in which he gives the following confirmation of the mixed breed of thefe animals and Dogs.

" I have feen at *Mofcow* about twenty fpurious animals from dogs
" and black wolves. They are for the moft part like wolves, except
" that fome carry their tails higher, and have a kind of coarfe bark-
" ing. They multiply among themfelves: and fome of the whelps
" are greyifh, rufty, or even of the whitifh hue of the Arctic wolves:
" and one of thofe I faw, in fhape, tail, and hair, and even in bark-
" ing, fo like a cur, that, was it not for his head and ears, his ill-
" natured look, and fearfulnefs at the approach of man, I fhould
" hardly have believed that it was of the fame breed."

10. ARCTIC.   Arctic Fox, *Hift. Quad.* N°   —LEV. MUS.

DOG. With a fharp nofe: ears almoft hid in the fur, fhort and rounded: hair long, foft, and filky: legs fhort: toes covered above and below with very thick and foft fur: tail fhorter than that of the common Fox, and more bufhy.

Inferior in fize to the common Fox: color a blueifh-grey, and

                                                              fometimes

ſometimes white. The young, before they come to maturity, duſky. The hair, as uſual in cold regions, grows much thicker and longer in winter than ſummer.

Theſe animals are found only in the Arctic regions, a few degrees within and without the Polar circle. They inhabit *Spitzbergen*, *Greenland*, and *Iceland* *: are only migratory in *Hudſon's Bay*, once in four or five years †: are found again in *Bering's* and ‡ *Copper Iſle*, next to it; but in none beyond: in *Kamtſchatka*, and all the countries bordering on the frozen ſea, which ſeems their great reſidence; comprehending a woodleſs tract of heath land, generally from 70 to 65 degrees lat. They abound in *Nova Zembla* ‖: are found in *Cherry* iſland, midway between *Finmark* and *Spitzbergen* §, to which they muſt have been brought on iſlands of ice; for it lies above four degrees north of the firſt, and three ſouth of the laſt: and laſtly, in the bare mountains between *Lapland* and *Norway*.

They are the hardieſt of animals, and even in *Spitzbergen* and *Nova Zembla* prowl out for prey during the ſeverity of winter. They live on the young wild geeſe, and all kind of water fowl; on their eggs; on hares, or any leſſer animals; and in *Greenland*, (through neceſſity) on berries, ſhell-fiſh, or whatſoever the ſea throws up. But in the north of *Aſia*, and in *Lapland*, their principal food is the *Lemings* ¶. The Arctic foxes of thoſe countries are as migratory as thoſe little animals; and when the laſt make their great migrations, the latter purſue them in vaſt troops. But ſuch removals are not only uncertain, but long: dependent on thoſe of the *Leming*. The Foxes will at times deſert their native countries for three or four years, probably as long as they can find any prey. The people of *Jeniſea* imagine, that the wanderers from their parts go to the banks of the *Oby*.

---

* *Egede*, 62. *Marten's Spitzb.* 100. *Horrebow's Iceland*, 43.　　† *Mr. Graham.*
‡ *Muller's Col. voy.* 53.　　‖ *Heemſkirk's voy.* 34.　　§ *Purchas*, iii. 559.
¶ Of which I apprehend there are two ſpecies—the *Lapland*, *Hiſt Quad.* N° 317, and the *Mus Migratorius* of *Pallas*, or *Yaik Rat*, *Hiſt. Quad.* N° 326. which inhabits the country near the *Yaik*.

Thoſe

Those found on *Bering's* and *Copper* isles were probably brought from the *Asiatic* side on floating ice : *Steller* having seen in the remoter islands only the black and brown foxes : and the same only on the continent of *America*. They burrow in the earth, and form holes many feet in length ; strewing the bottom with moss. But in *Spitzbergen* and *Greenland,* where the ground is eternally frozen, they live in the cliffs of rocks : two or three inhabit the same hole. They swim well, and often cross from island to island in search of prey. They are in heat about *Lady-day* ; and during that time continue in the open air : after that, retreat to their earths. Like dogs, continue united in copulation : bark like them : for which reason the *Russians* call them *Peszti*. They couple in *Greenland* in *March,* and again in *May* ; and bring forth in *April* and in *June* *.

They are tame and inoffensive animals ; and so simple, that there are instances of their standing by when the trap was baiting, and instantly after putting their heads into it. They are killed for the sake of their skins, both in *Asia* and *Hudson's Bay :* the fur is light and warm, but not durable. Mr. *Graham* informed me, that they have appeared in such numbers about the fort, that he has taken, in different ways, four hundred from *December* to *March*. He likewise assured me, that the tips of their tails are always black ; those of the common foxes always white : and that he never could trace the breeding-places of the former..

The *Greenlanders* take them either in pitfalls dug in the snow, and baited with the Capelin fish ; or in springs made with whale-bone, laid over a hole made in the snow, strewed over at bottom with the same kind of fish ; or in traps made like little huts, with flat stones, with a broad one by way of door, which falls down (by means of a string baited on the inside with a piece of flesh) whenever the fox enters and pulls at it †. The *Greenlanders* preserve the skins for traffic ; and in cases of necessity eat the flesh. They also make

---

* *Faun. Groenl. 20.*   † *Crantz. i. 72.*

buttons

buttons of the fkins: and fplit the tendons, and make ufe of them inftead of thread. The blue furs are much more efteemed than the
white.

European Fox, *Hift. Quad.* N° 139.—*Smellie*, iv. 214.—LEV. MUS.

DOG. With a pointed nofe: pointed erect ears: body of a
tawny red mixed with afh-color: fore part of the legs black:
tail long and bufhy, tipt with white.

Inhabits the northern parts of *North America* from *Hudfon's Bay*,
probably acrofs the continent to the iflands intermediate between *America* and *Kamtfchatka*. Captain *Bering* faw there five quite tame,
being unufed to the fight of man.

This fpecies gradually decreafes to the fouthward, in numbers and
in fize: none are found lower than *Penfylvania*. They are fuppofed
not to have been originally natives of that country. The *Indians*
believe they came from the north of *Europe* in an exceffive hard
winter, when the feafon was frozen. The truth feems to be, that
they were driven in fome fevere feafon from the north of their own
country, and have continued there ever fince. They abound about
Hudfon's Bay, the *Labrador* country, and in *Newfoundland* and *Canada*; and are found in *Iceland**. They burrow as the *European* foxes
do; and in *Hudfon's Bay*, during winter, run about the woods in
fearch of prey, feeding on birds and leffer animals, particularly
mice.

New England is faid to have been early ftocked with foxes by a
gentleman who imported them from *England*, for the pleafure of the
chafe†; and that the prefent breed fprung from the occafion. This
fpecies is reckoned among the pernicious animals, and, being very
deftructive to lambs, are profcribed at the rate of two fhillings a
head.

The variety of *Britifh* fox, with a black tip to the tail, feems unknown in *America*.

* *Olaffen*, i. 31. † *Kalm*, i. 283.

2 The

The skins are a great article of commerce: abundance are im ported annually from *Hudson's Bay* and *Newfoundland*. The natives o *Hudson's Bay* eat the flesh, rank as is it is.

This species abounds in *Kamtschatka*, and is the finest red fur of any known: grows scarce within the Arctic circle of the *Afiatic* regions and is found there often white.

α BLACK. THIS variety is found very often entirely black, with a white tip to the tail; and is far inferior in value and beauty to those of *Kamtfchatka* and *Sibiria*, where a single skin sells for four hundred rubles.

The best in *North America* are found on the *Labrador* side of *Hudson's Bay*. They are also very common on the islands opposite to *Kamtfchatka*. The *American* black foxes, which I have examined, are frequently of a mixed color: from the hind part of the head to the middle of the back is a broad black line: the tail, legs, and belly, black: the hairs on the face, sides, and lower part of the back, cinereous; their upper ends; black the tip white.

β CROSS. FOX. With a bed of black running along the top of the back, crossed by another passing down each shoulder; from whence it took the name. The belly is black: the color of the rest of the body varies in different skins; but in all is a mixture of black, cinereous, and yellow: the fur in all very soft: and the tail very bushy and full of hair; for nature, in the rigorous climate of the North, is ever careful to guard the extremities against the injury of cold.

This is likewise a very valuable variety. It is remarked, that the more defireable the fur is, the more cunning and difficult to be taken is the fox which owns it[*]. The *Coffacks* quartered in *Kamtfchatka* have attempted for two winters to catch a single black fox. The Crofs-fox, *vulpes crucigera* of *Gefner*, and *Kors-raef* of the *Swedes* [†], is found in all the Polar countries.

[*] *Hift. Kamtfchàtka,* 95.   [†] *Gefner Quad.* 967.   *Faun. Suec.* N° 4.

In

In the new-difcovered *Fox iflands* thefe animals abound: one in three or four are found entirely black, and larger than any in *Sibiria*: the tail alfo is tipt with white. But as they live among the .rocks, there being no woods in thofe iflands, their hair is almoft as coarfe as that of the Wolf, and of little value compared to the *Sibirian*.

Brant Fox, Hift. Quad. p. 235.

γ BRANT.

FOX. With a very fharp and black nofe: fpace round the ears ferruginous: forehead, back, fhoulders, fides, and thighs, red, cinereous, and black: the afh-color predominates, which gives it a hoary look: belly yellowifh: tail black above, cinereous on the fides, red beneath.

About half the fize of the common fox. Defcribed from one Mr. *Brooks* received from *Penfylvania*, under the name of *Brandt-fox*; but it had not that bright rednefs to merit the name of either *Brandt-fuchfe*, or *Brand-raef*, given by *Gefner* and *Linnæus*.

Corfak Fox, Hift. Quad. p. 236.

δ CORSAK.

FOX. With upright ears: yellowifh-green irides: throat white: color, in fummer, pale tawny; in winter, cinereous: middle of the tail cinereous; bafe and tip black; the whole very full of hair: the fur is coarfer and fhorter than that of the common fox.

I difcovered this fpecies among the drawings of the late *Taylor White*, Efq; who informed me that it came from *North America*. I imagine, from *Hudfon's Bay*.

This fpecies is very common in the hilly and temperate parts of *Tartary*, from the *Don* to the *Amur*; but never is found in woody places: it burrows deep beneath the furface. It is alfo faid to inhabit the banks of the rivers *Indigifky* and *Anadyr*, where the hills grow bare. In the reft of *Sibiria* it is only known beyond lake *Baikal*; and from fkins brought -by the *Kirghifian* and *Bucharian* traders. In *Ruffia* it is found in the defarts towards *Crimea* and *Aftra-can*, and alfo on the fouthern end of the *Urallian* mountains.

Grey

12. GREY.   Grey Fox, *Hift. Quad.* N° 142.

FOX. With a fharp nofe: long fharp upright ears: long legs: color entirely grey, except a little rednefs about the ears.

Inhabits from *New England* to the fouthern end of *North America*; but are far more numerous in the fouthern colonies. They have not the rank fmell of the red foxes. They are alfo lefs active, and grow very fat*. They breed in hollow trees: give no diverfion to the fportfmen, for after a mile's chafe they run up a tree†. They feed on birds; are deftructive to poultry; but never deftroy lambs‡. The fkins are ufed to line clothes: the fur is in great requeft among the hatters. The greafe is reckoned efficacious in rheumatic diforders.

13. SILVERY.   Silvery Fox, *Hift. Quad.* N° 143.

FOX. With a fine and thick coat of a deep brown color, ove fpread with long filvery hairs of a moft elegant appearance.

Inhabits *Louifiana*, where their holes are feen in great abundance on the woody heights. As they live in forefts, which abound in game; they never moleft the poultry, fo are fuffered to run at large§.

They differ fpecifically from the former, more by their nature in burrowing, than in colors.

* *Lawfon*, 125.   † *Catefby*, ii. 78. *Joffelyn*, 82.   ‡ *Kalm*, i. 282.
§ *Du Pratz*, ii. 64. *Charlevoix*, v. 196.

HIST. QUAD. Genus XIX.

Hist. Quad. N° 160.—*Smellie,* v. 197. 200.—Lev. Mus.

CAT. With a small head: large eyes: ears a little pointed: chin white: back, neck, sides, and rump, of a pale brownish red, mixed with dusky hairs: breast, belly, and inside of the legs, cinereous: tail a mixture of dusky and ferruginous, the tip black.

The teeth of a vast size: claws whitish; the outmost claw of the fore feet much larger than the rest: the body very long: the legs high and strong. The length of that I examined was five feet three from head to tail; of the tail, two feet eight.

Inhabits the continent of *North America,* from *Canada* to *Florida*; and the species is continued from thence low into *South America,* through *Mexico, Guiana, Brasil,* and the province of *Quito,* in *Peru,* where it is called *Puma,* and by the *Europeans* mistaken for a Lion: it is, by reason of its fierceness, the scourge of the country. The different climate of *North America* seems to have subdued its rage, and rendered it very fearful of mankind: the left cur, in company with his master, will make it run up a tree*, which is the opportunity of shooting it. It proves, if not killed outright, a dangerous enemy; for it will descend, and attack either man or beast. The flesh is white, and reckoned very good. The *Indians* use the skin for winter habits; and when dressed is made into shoes for women, and gloves for men †.

It is called in *North America* the Panther, and is the most pernicious animal of that continent. Lives in the forests. Sometimes purs, at other times makes a great howling. Is extremely destructive to do-

* *Catesby, App.* xxv.   † *Lawson,* 118.

H   mestic

meftic animals, particularly to hogs. It preys alfo upon the Moofe,
and other deer; falling on them from the tree it lurks in, and never
quits its hold .   The deer has no other way of faving itfelf, but by
plunging into the water, if there happens to be any near; for the
Panther, like the Cat, detefts that element. It will feed even on
beafts of prey.   I have feen the fkin of one which was fhot, juft as it
had killed a wolf.   When it has fatisfied itfelf with eating, it carefully conceals the reft of the carcafe, covering it with leaves. If any
other animal touches the reliques, it never touches them again.

15. Lynx.   *Hift. Quad.* Nº 170.—*Smellie,* v. 207. 217.—Lev. Mus.

CAT. With pale yellow eyes : ears erect, tufted with black long
hair : body covered with foft and long fur, cinereous tinged
with tawny, and marked with dufky fpots, more or lefs vifible in different fubjects, dependent on the age, or feafon in which the animal
is killed : the legs ftrong and thick : the claws large. About three
times the fize of a common Cat : the tail only four inches long,
tipt with black.

Inhabits the vaft forefts of *North America :* is called in *Canada, Le
Chat, ou Le Loup-cervier* †, on account of its being fo deftructive to
deer ; which it drops on from the trees, like the former, and, fixing
on the jugular vein, never quits its hold till the exhaufted animal
falls through lofs of blood ‡.

The *Englifh* call it a Wild Cat. It is very deftructive to their young
pigs, poultry, and all kind of game. The fkins are in high efteem
for the foftnefs and warmnefs of the fur; and great numbers are annually imported into *Europe.*

* *Charlevoix,* v. 189, who by miftake calls it *Carcajou,* and *Kincajou ;* two very different animals.   † *Charlevoix,* v. 195.   ‡ *Lawfon,* 118. *Catefby, App.* xxv.

Bay Lynx. *Hiſt. Quad.* Nᵒ 171.

CAT. With yellow irides : ears like the former : color of the head, body, and outſide of the legs and thighs, a bright bay, obſcurely marked with duſky ſpots : the forehead marked with black ſtripes from the head to the noſe : cheeks white, varied with three or four incurvated lines of black : the upper and under lip, belly, and inſides of the legs and thighs, white : the inſide of the upper part of the fore legs croſſed with two black bars : the tail ſhort ; the upper part marked with duſky bars, and near the end with one of black ; the under ſide white. In ſize, about twice that of a common Cat ; the fur ſhorter and ſmoother than that of the former.

This ſpecies is found in the internal parts of the province of *New York*. I ſaw one living a few years ago in *London*. The black bars on the legs and tail are ſpecific marks.

Hiſt. Quad. Nᵒ 168.
Cat-a-mountain ? *Lawſon*, 118. *Du Pratz*, ii. 64.

CAT. With upright pointed ears, marked with two brown bars : head and upper part of the body of a reddiſh brown, with long narrow ſtripes of black : the ſides and legs with ſmall round ſpots : chin and throat of a clear white : belly of a dull white : tail eight inches long, barred with black. Length from noſe to tail two feet and a half.

Inhabits *North America*. Is ſaid to be a gentle animal, and to grow very fat. Deſcribed originally in the *Memoires de l'Academie* ; ſince which an account of another, taken in *Carolina*, was communicated by the late Mr. *Collinſon* to the Count *de Buffon* *. The only difference is in ſize ; for the laſt was only nineteen inches long : the tail four ; but the ſame characteriſtic ſtripes, ſpots, and bars, on the tail, were ſimilar in both.

* Supplem. iii. 227.

H 2

There

OBSCURE SPECIES. There still remain undescribed some animals of the Feline race, which are found in *North America*, but too obscurely mentioned by travellers to be ascertained. Such is the beast which *Lawson* saw to the westward of *Carolina*, and calls a Tiger. He says it was larger than the Panther, i. e. *Puma*, and that it differed from the Tiger of *Asia* and *Africa**. It possibly may be the *Brasilian* Panther, *Hist. Quad.* Nº 158, which may extend further north than we imagine. It may likewise be the Cat-a-mount of *Du Pratz*†; which, he says, is as high as the Tiger, i. e. *Puma*, and the skin extremely beautiful.

The *Pijoux* of *Louisiana*, mentioned by *Charlevoix*‡, are also obscure animals. He says they are very like our Wild Cats, but larger: that some have shorter tails, and others longer. The first may be referred to one of the three last species: the last may be our *Cayenne* Cat, Nº 163.

Domestic Cats are kept in *Iceland* and *Norway* §. Some of them escape and relapse to a savage state. In *Iceland* those are called *Urdakelter*, because they live under rocks and loose stones, where they hide themselves. They prey on small birds. The most valuable of their skins are sold for twelve *Danish* skillings, or six pence a-piece. *Linnæus*, speaking of the cats of *Sweden*, says, they are of exotic origin ‖. They are not found wild either in that kingdom, or any part of the *Russian* dominions. Unknown in *America*.

* *Hist. Carolina*, 119.

† ii. 64. I wish to suppress the synonym of *Cat-a-mount*, as applied to the *Cayenne Cat*, as it seems applicable to a much larger species.

‡ *Hist. de la Nouv. France*, vi. 158. § *Oluf. Iceland*, i. Paragr. 80. *Pontop.* ii. 8. ‖ *Faun. Suec.* Nº 9.

HIST.

HIST. QUAD. Genus XX.

Hiſt. Quad. N° 175.—Lev. Mus.

BEAR. With a long narrow head and neck: tip of the noſe black: teeth of a tremendous magnitude: hair of a great length, ſoft, and white, and in part tinged with yellow: limbs very thick and ſtrong: ears ſhort and rounded.

Travellers vary about their ſize. *De Buffon* quotes the authority of *Gerard le Ver** for the length of one of the ſkins, which, he ſays, was twenty-three feet. This ſeems to be extremely miſrepreſented; for *Gerard*, who was a companion of the famous *Barentz*, and *Heemſkirk*, a voyager of the firſt credit, killed ſeveral on *Nova Zembla*, the largeſt of which did not exceed thirteen feet in length†. They ſeem ſmaller on *Spitzbergen*: one meaſured by order of a noble and able navigator ‡, in his late voyage towards the Pole, was as follows: I give all the meaſurements to aſcertain the proportions.

	Feet.	Inches.
Length from ſnout to tail -	7	1
from ſnout to ſhoulder-bone -	2	3
Height at the ſhoulder -	4	3
Circumference near the fore legs ..	7	0
of the neck near the ear	2	1
Breadth of the fore-paw -	0	7
Weight of the carcaſe without the head, ſkin, or entrails - - -	610 lb.	

This ſpecies, like the Rein and Arctic Fox, almoſt entirely ſurrounds the neighborhood of the Polar circle. It is found within it,

---

* *De Buffon, Suppl.* iii. 200.   † See *Le Ver*, p. 14. ed. 1606. *Amſteld.*
‡ The Honorable *Conſtantine John Phipps*, now Lord *Mulgrave*.

far

far as navigators have penetrated; in the ifland of *Spitzbergen*, and within *Baffin's Bay*; in *Greenland* and *Hudfon's Bay*; in *Terra di Labrador* *; and, by accident, wafted from *Greenland*, on iflands of ice, to *Iceland* and *Newfoundland*.   It perhaps attends the courfe of the Arctic circle along the vaft regions of *America*; but it is unknown in the groupes of iflands between that continent and *Afia*; neither is it found on the *Tchuktki Nofs*, or the Great Cape, which juts into the fea north of *Kamtfchatka* †.   None are ever feen in that country.   But they are frequent on all the coafts of the Frozen Ocean, from the mouth of the *Ob* ‡, eaftward; and abound moft about the eftuaries of the *Jenefei* and *Lena*.   They appear about thofe favage tracts, and abound in the unfrequented iflands of *Nova Zembla*, *Cherry*, and *Spitzbergen*, where they find winter quarters undifturbed by mankind.   The fpecies is happily unknown along the fhores of the White fea, and thofe of *Lapland* and *Norway*.   Poffibly even thofe rigorous climates may be too mild for animals that affect the utmoft feverity of the Arctic zone.   They never are feen farther fouth in *Sibiria* than *Mangafea*, nor wander into the wooly parts, unlefs by accident in great mifts.

They are fometimes brought alive into *England*.   One which I faw was always in motion, reftlefs, and furious, roaring in a loud and hoarfe tone; and fo impatient of warmth, that the keeper was obliged to pour on it frequently pailfuls of water.   In a ftate of nature, and in places little vifited by mankind, they are of dreadful ferocity.   In *Spitzbergen*, and the other places annually frequented by the human race, they dread its power, having experienced its fuperiority, and fhun the conflict: yet even in thofe countries prove tremendous enemies, if attacked or provoked.

Barentz, in his voyages in fearch of a north-eaft paffage to *China*, had fatal proofs of their rage and intrepidity on the ifland of *Nova Zembla*: his feamen were frequently attacked, and fome of them

---

* *Phil. Tranf.* lxiv. 377.   † *Muller*, Pref. xxv.   ‡ *Purchas's Pilgrims,* iii. 805.

killed.

killed. Those whom they seized on they took in their mouths, ran away with the utmost ease, tore to pieces, and devoured at their leisure, even in sight of the surviving comrades. One of these animals was shot preying on the mangled corpse, yet would not quit its hold; but continued staggering away with the body in its mouth, till dispatched with many wounds *.

They will attack, and attempt to board, armed vessels far distant from shore; and have been with great difficulty repelled †. They seem to give a preference to human blood; and will greedily dis-inter the graves of the buried, to devour the cadaverous contents ‡.

Their usual food is fish, seals, and the carcases of whales. On land, they prey on deer ‖, hares, young birds, and eggs, and often on whortleberries and crowberries. They are at constant enmity with the Walrus, or Morse: the last, by reason of its vast tusks, has generally the superiority; but frequently both the combatants perish in the conflict §.

FOOD.

They are frequently seen in *Greenland*, in lat. 76, in great droves; where, allured by the scent of the flesh of seals, they will surround the habitations of the natives, and attempt to break in ¶; but are soon driven away by the smell of burnt feathers **. If one of them is by any accident killed, the survivors will immediately eat it ††.

They grow excessively fat; a hundred pounds of fat has been taken out of a single beast. Their flesh is coarse, but is eaten by the seamen: it is white, and they fancy it tastes like mutton. The liver is very unwholesome, as three of *Heemskirk*'s sailors experienced, who fell dangerously ill on eating some of it boiled ‡‡. The skin is an article of commerce: many are imported, and used chiefly for covers to coach-boxes. The *Greenlanders* feed on the flesh and fat; use the skins to sit on, and make of it boots, shoes, and gloves; and split the tendons into thread for sewing.

---

* *Heemskirk's voy.* 14. † The same, 18. ‡ *Martin's Spitzb.* 102.
‖ *Faun. Groenl.* p. 23. § *Egede*, 83. ¶ The same, 60. ** *Faun. Groenl.* p. 23.
†† *Heemskirk*, 51. ‡‡ The same. 45.

During

During summer they reside chiefly on islands of ice, and pass frequently from one to the other. They swim most excellently, and sometimes dive, but continue only a small space under water. They have been seen on islands of ice eighty miles from any land, preying and feeding as they float along. They lodge in dens formed in the vast masses of ice, which are piled in a stupendous manner, leaving great caverns beneath: here they breed, and bring one or two at a time, and sometimes, but very rarely, three. Great is the affection between parent and young; they will sooner die than desert one another *. They also follow their dams a very long time, and are grown to a very large size before they quit them.

During winter they retire, and bed themselves deep beneath, forming spacious dens in the snow, supported by pillars of the same, or to the fixed ice beneath some eminence; where they pass torpid the long and dismal night †, appearing only with the return of the sun ‡. At their appearance the *Arctic* Foxes retire to other haunts ‖.

The Polar Bear became part of the royal menagery as early as the reign of *Henry* III. Mr. *Walpole* has proved how great a patron that despised prince was of the Arts. It is not less evident that he extended his protection to Natural History. We find he had procured a White Bear from *Norway*, from whence it probably was imported from *Greenland*, the *Norwegians* having possessed that country for some centuries before that period. There are two writs extant from that monarch, directing the sheriffs of *London* to furnish six pence a day to support *our* White Bear in our Tower of *London*; and to provide a muzzle and iron chain to hold him when out of the water; and a long and strong rope to hold him, when he was fishing in the *Thames* §. Fit provision was made at the same time for the king's Elephant.

* *Marten's Spitzb.* 102.

† *Egede*, 60. *Martens* says, that the fat is used in pains of the limbs, and that it assists parturition.

‡ *Heemskirk's voy.* in *Purchas*, iii. 500, 501.   ‖ The same, 499.   § *Madox's Antiquities of the Exchequer*, i. 376.

The

The skins of this species, in old times, were offered by the hunters to the high altars of cathedrals, or other churches, that the priest might stand on them, and not catch cold when he was celebrating high mass in extreme cold weather. Many such were annually offered at the cathedral at *Drontheim* in *Norway* ; and also the skins of wolves, which were sold to purchase wax lights to burn in honor of the saints *.

Hist. Quad. N° 174.—*Smellie*, v. 19.

19. BLACK.

B EAR. With a long pointed nose, and narrow forehead : the cheeks and throat of a yellowish brown color : hair over the whole body and limbs of a glossy black, smoother and shorter than that of the *European* kind.

They are usually smaller than those of the old world ; yet Mr *Bartram* gives an instance of an old he-bear killed in *Florida* which was seven feet long, and, as he guessed, weighed four hundred pounds †.

These animals are found in all parts of *North America*, from *Hudson's Bay* to the southern extremity ; but in *Louisiana* and the southern parts they appear only in the winter, migrating from the north in search of food. They spread across the northern part of the *American* continent to the *Kamtschatkan* sea. They are found again in the opposite country ‡, and in the *Kurilski* islands, which intervene between *Kamtschatka* and *Japan* ‖, *Jeso Masima*, which lies north of *Japan* §, and probably *Japan* itself ; for *Kæmpfer* says, that a few small bears are found in the northern provinces ¶.

It is very certain that this species of bear feeds on vegetables. *Du Pratz*, who is a faithful as well as intelligent writer, relates, that

FOOD.

* *Olaus Magnus*, lib. xviii. c. 20. † Journal of his travels into *East Florida*, 26. ‡ *Hist. Kamtsch.* iii. 385. ‖ The same, 287. § *Voy. au Nord.* iv. 5. ¶ *Hist. Japan*, i. 126.

I

in

in one severe winter, when these animals were forced in multitudes
from the woods, where there was abundance of animal food, they re-
jected that, notwithstanding they were ready to perish with hunger;
and, migrating into the lower *Louisiana,* would often break into the
courts of houses.   They never touched the butchers meat which lay
in their way, but fed voraciously on the corn or roots they met with *.

Necessity alone sometimes compels them to attack and feed on
the swine they meet in the woods: but flesh is to them an unnatural diet.   They live on berries, fruits, and pulse of all kinds;
are remarkably fond of potatoes, which they very readily dig up with
their great paws; make great havock in the fields of maize; and are
great lovers of milk and honey.   They feed much on herrings, which
they catch in the season when those fish come in shoals up the creeks,
which gives their flesh a disagreeable taste; and the same effect is
observed when they eat the bitter berries of the *Tupelo.*

They are equally inoffensive to mankind, provided they are not
irritated, but if wounded, they will turn on their assailant with
great fury, and, in case they can lay hold, never fail of hugging him
to death; for it has been observed they never make use, in their rage,
of either their teeth or claws.   If they meet a man in a path they
will not go out of his way; but will not attack him.   They never
seek combat.   A small dog will make them run up a tree.

The bears of *Kamtschatka* resemble those of *America:* they are
neither large nor fierce.   They also wander from the hills to the
lower lands in summer, and feed on berries and fish.   They reject
carnivorous food, nor ever attack the inhabitants, unless they find
them asleep, when, through wantonness, they bite them severely,
and sometimes tear a piece of flesh away; yet, notwithstanding they
get a taste of human blood, are never known to devour mankind.
People thus injured are called *Dranki* †, or the *flayed.*

The *American* bears do not lodge in caves or clefts of rocks, like
those of *Europe.*   The bears of *Hudson's Bay* form their dens beneath

---

* *Du Pratz,* ii. 57.   † *Hist. Kamtschatka,* iii. 386.

the

the fnow, and fuffer fome to drop at the mouth, to conceal their retreat.

The naturalift's poet, with great truth and beauty, defcribes the
retreat of this animal in the frozen climate of the north :

> There through the piny foreft half abforpt,
> Rough tenant of thofe fhades, the fhapelefs BEAR,
> With dangling ice all horrid, ftalks forlorn ;
> Slow pac'd, and fourer as the ftorms increafe,
> He makes his bed beneath th' inclement drift,
> And with ftern patience, fcorning weak complaint,
> Hardens his heart againft affailing want.

Thofe of the fouthern parts dwell in the hollows of antient trees.
The hunter difcovers them by ftriking with an ax the tree he fufpects
they are lodged in, then fuddenly conceals himfelf. The Bear is
immediately rouzed, looks out of the hollow to learn the caufe of
the alarm ; feeing none, finks again into repofe *. The hunter then
forces him out, by flinging in fired reeds ; and fhoots him while he
defcends the body of the tree, which, notwithftanding his aukward
appearance, he does with great agility ; nor is he lefs nimble in
afcending the tops of the higheft trees in fearch of berries and
fruits.

The long time which thefe animals fubfift without food is amazing. They will continue in their retreat for fix weeks without the left
provifion, remaining either afleep or totally inactive. It is pretended that they live by fucking their paws ; but that is a vulgar error.
The fact is, they retire immediately after autumn, when they have
fattened themfelves to an exceffive degree by the abundance of
the fruits which they find at that feafon. This enables animals, which perfpire very little in a ftate of reft, to endure an abftinence of uncommon length. But when this internal fupport is exhaufted, and they begin to feel the call of hunger, on the approach of
the fevere feafon, they quit their dens in fearch of food. Multitudes

---

* *Du Pratz,* ii. 61.

then migrate into the lower parts of *Louifiana* : they arrive very lean ; but foon fatten with the vegetables of that milder climate * They never wander far from the banks of the *Miffifipi,* and in their march form a beaten path like the track of men.

Lawfon and *Catefby* † relate a very furprizing thing in refpect to this animal, which is, that neither *European* or *Indian* ever killed a Bear with young. In one winter were killed in *Virginia* five hundred bears, and among them only two females ; and thofe not pregnant. The caufe is, that the male has the fame unnatural diflike to its offspring as fome other animals have : they will kill and devour the cubs. The females therefore retire, before the time of parturition, into the depth of woods and rocks, to elude the fearch of their favage mates. It is faid that they do not make their appearance with their young till *March* ‡.

All who have tafted the flefh of this animal fay, that it is moft delicious eating : a young Bear, fattened with the autumnal fruits, is a difh fit for the niceft epicure. It is wholefome and nourifhing, and refembles pork more than any other meat. The tongue and the paws are efteemed the moft exquifite morfels ; the hams are alfo excellent, but apt to ruft, if not very well preferved.

Four inches depth of fat has been found on a fingle Bear, and fifteen or fixteen gallons of pure oil melted from it §. The fat is of a pure white, and has the fingular quality of never lying heavy on the ftomach, notwithftanding a perfon drank a quart of it ‖. The *Americans* make great ufe of it for frying their fifh. It is befides ufed medicinally, and has been found very efficacious in rheumatic complaints, achs, and ftrains.

The *Indians* of *Louifiana* prepare it thus :—As foon as they have killed the Bear, they fhoot a Deer ; cut off the head, and draw the fkin entire to the legs, which they cut off : they then ftop up every orifice, except that on the neck, into which they pour the melted fat

---

* *Du Pratz,* ii. 60.   † *Lawfon,* 117. *Catefby, App.* xxvi.   ‡ *Joffelyn's voy.* 91.   § *Bartram's journ. E. Florida,* 26.   ‖ *Lawfon,* 116.

of the Bear; which is prepared by boiling the fat and flesh together. This they call a *Deer of oil*, and sell to the *French* for a gun, or something of equal value*.

Bears greafe is in great repute in *Europe* for its suppofed quality of making the hair to grow on the human head. A great chymist in the *Haymarket* in *London* used to fatten annually two or three Bears for the fake of their fat.

The skin is in ufe for all purpofes which the coarfer forts of furs are applied to: it ferves in *America*, in diftant journies, for coverlets; and the finer parts have been in fome places ufed in the hat manufacture †.

The *Indians* of *Canada* daub their hands and face with the greafe, to preferve them from the bite of mufketoes: they alfo fmear their bodies with the oil after exceffive exercife ‡. They think, like the *Romans* of old, that oil fupples their joints, and preferves them in full activity.

Black Bear, *Hift. Quad.* Nº 174.— *Smellie*, v. 19.

20. BROWN.

BEAR. With long fhaggy hair, ufually dufky or black, with brown points; liable to vary, perhaps according to their age, or fome accident, which does not create a fpecific difference.

α.   A variety of a pale brown color, whofe skins I have feen imported from *Hudfon's Bay*. The fame kind, I believe, is alfo found in *Europe*. The cubs are of a jetty black, and their necks often encircled with white.

β.   Bears fpotted with white.

γ.   Land Bears, entirely white. Such fometimes fally from the lofty mountains which border on *Sibiria*, and appear in a wandering manner in the lower parts of the country ‖. *Marco Polo* relates, that they were frequent in his time in the north of *Tartary*, and of a very great fize.

* *Du Pratz*, ii. 62.   † *Lawfon*, 117.   ‡ *Kalm*, iii. 13.   ‖ *Doctor Pallas*.

Grizzly

Grizzly Bears. Thefe are called by the *Germans Silber-bar*, or the *Silver-bear*, from the mixture of white hairs. Thefe are found in *Europe*, and the very northern parts of *North America*, as high as lat. 70; where a hill is called after them, *Grizzle Bear Hill*, and where they breed in caverns *. The ground in this neighborhood is in all parts turned by them in fearch of the hoards formed by the Ground Squirrels for winter provifion.

All thefe varieties form but one fpecies. They are granivorous and carnivorous, both in *Europe* and *America*; and I believe, according to their refpective palates or habits, one may be deemed a variety which prefers the vegetable food; another may be diftinguifhed from its preference of animal food. Mr. *Graham* affures me, that the brown Bears, in the inland parts of *Hudfon's Bay*, make great havock among the Buffaloes: are very large, and very dangerous when they are attacked and wounded.

BEARS VENERAT-
ED IN AMERICA.

In all favage nations the Bear has been an object of veneration. Among the *Americans* a feaft is made in honor of each that is killed. The head of the beaft is painted with all colors, and placed on an elevated place, where it receives the refpects of all the guefts, who celebrate in fongs the praifes of the Bear. They cut the body in pieces, and regale on it, and conclude the ceremony †.

CHASE.

The chafe of thefe animals is a matter of the firft importance, and never undertaken without abundance of ceremony. A principal warrior firft gives a general invitation to all the hunters. This is followed by a moft ferious faft of eight days, a total abftinence from all kinds of food; notwithftanding which, they pafs the day in continual fong. This they do to invoke the fpirits of the woods to direct them to the place where there are abundance of bears. They even cut the flefh in divers parts of their bodies, to render the fpirits more propitious. They alfo addrefs themfelves to the *manes* of the beafts flain in preceding chafes, as if it were to direct them in their dreams to plenty of game. One dreamer alone cannot determine

* Mr. *Samuel Hearne*.   † *Charlevoix, Nouv. Fr.* v. 443.

the

the place of chase, numbers must concur; but, as they tell each other their dreams, they never fail to agree: whether that may arise from complaisance, or by a real agreement in the dreams from their thoughts being perpetually turned on the same thing.

The chief of the hunt now gives a great feast, at which no one dares to appear without first bathing. At this entertainment they eat with great moderation, contrary to their usual custom. The master of the feast alone touches nothing; but is employed in relating to the guests antient tales of the wonderful feats in former chases: and fresh invocations to the manes of the deceased bears conclude the whole. They then sally forth amidst the acclamations of the village, equipped as if for war, and painted black. Every able hunter is on a level with a great warrior; but he must have killed his dozen great beasts before his character is established: after which his alliance is as much courted as that of the most valiant captain.

They now proceed on their way in a direct line: neither rivers, marshes, or any other impediments, stop their course; driving before them all the beasts which they find in their way. When they arrive in the hunting-ground, they surround as large a space as their company will admit, and then contract their circle; searching, as they contract, every hollow tree, and every place fit for the retreat of the bear, and continue the same practice till the time of the chase is expired.

As soon as a bear is killed, a hunter puts into its mouth a lighted pipe of tobacco, and, blowing into it, fills the throat with the smoke, conjuring the spirit of the animal not to resent what they are going to do to its body; nor to render their future chases unsuccessful. As the beast makes no reply, they cut out the string of the tongue, and throw it into the fire: if it crackles and runs in (which it is almost sure to do) they accept it as a good omen; if not, they consider that the spirit of the beast is not appeased, and that the chase of the next year will be unfortunate.

The hunters live well during the chase, on provisions which they bring with them. They return home with great pride and self-sufficiency; for to kill a bear forms the character of a complete man. They again give a great entertainment, and now make a point to leave nothing. The feast is dedicated to a certain genius, perhaps that of Gluttony, whose resentment they dread, if they do not eat every morsel, and even sup up the very melted grease in which the meat was dressed. They sometimes eat till they burst, or bring on themselves some violent disorders. The first course is the greatest bear they have killed, without even taking out the entrails, or taking off the skin, contenting themselves with singeing the skin, as is practised with hogs *.

The *Kamtschatkans*, before their conversion to Christianity, had almost similar superstitions respecting bears and other wild beasts: they entreated the bears and wolves not to hurt them in the chase, and whales and marine animals not to overturn their boats. They never call the two former by their proper name, but by that of *Sipang*, or *ill-luck*.

At present the *Kamtschatkans* kill the bear and other wild beasts with guns: formerly they had variety of inventions; such as filling the entrance of its den with logs, and then digging down upon the animal and destroying it with spears †. In *Sibiria* it is taken by making a trap-fall of a great piece of timber, which drops and crushes it to death: or by forming a noose in a rope fastened to a great log; the bear runs its head into the noose, and, finding itself engaged, grows furious, and either falls down some precipice and kills itself, or wearies itself to death by its agitations.

The killing of a bear in fair battle is reckoned as great a piece of heroism by the *Kamtschatkans* as it is with the *Americans*. The victor makes a feast on the occasion, and feasts his neighbors with the beast; then hangs the head and thighs about his tent by way of trophies.

These people use the skins to lie on, and for coverlets; for bonnets, gloves, collars for their dogs, soles for their shoes, to prevent them

* *Charlevoix*, v. 169 to 174.   † *Hist. Kamtschatka*, Fr. iii. 73.

from

from flipping on the ice. Of the shoulder-blades they make inftruments to cut the grafs; of the inteftines, covers for their faces, to
protect them from the fun during fpring; and the *Coffacks* extend
them over their windows inftead of glafs. The flefh and fat is among
the chief dainties of the country *.

Superftitions, relative to this animal, did not confine themfelves to
America and *Afia*, but fpread equally over the north of *Europe*. The
Laplanders held it in the greateft veneration: they called it the *Dog*
of GOD, becaufe they efteemed it to have the ftrength of ten men,
and the fenfe of twelve †. They never prefume to call it by its
proper name of *Guouzhja*, leaft it revenge the infult on their flocks;
but ftyle it *Moedda-aigia*, or the *old man in a furred cloak* ‡.

The killing of a Bear was reckoned as great an exploit in *Lapland*
as it was in *America*, and the hero was held in the higheft efteem by
both fexes; and, by a fingular cuftom, was forbid all commerce with
his wife for three days. The *Laplanders* bring home the flain beafts
in great triumph. They erect a new tent near their former dwelling,
but never enter it till they have flung off the drefs of the chafe.
They continue in it three entire days; and the women keep at home
the fame fpace. The men drefs the flefh of the Bear in the new
tent, and make their repaft, giving part to the females; but take
great care never to beftow on them a bit of the rump. Neither will
they deliver to them the meat through the common entrance of the
hut, but through a hole in another part. In fign of victory, the men
fprinkle themfelves with the blood of the beaft.

After they have finifhed eating the flefh, they bury the bones with
great folemnity, and place every bone in its proper place, from a
firm perfuafion that the Bear will be reftored, and re-animate a new
body.

At the pulling off the fkin, and cutting the body into pieces, they
were ufed to fing a fong, but without meaning or rhyme ‖; but the

* *Hift. Kamtfchatka, Fr.* iii. 390.　　　† *Leems Lapmark, Suppl.* 64.　　　‡ The
fame, 502.　　　‖ The fame, *Suppl.*

K

antient

antient *Fins* had a fong, which, if not highly embellifhed by the
tranflator, is far from inelegant.

> Beaft ! of all foreft beafts fubdued and flain,
> Health to our huts and prey a hundred-fold
> Reftore ; and o'er us keep a conftant guard !
> I thank the Gods who gave fo noble prey !
> When the great day-ftar hides beyond the *alps,*
> I hie me home ; and joy, all clad in flowers,
> For three long nights fhall reign throughout my hut.
> With tranfport fhall I climb the mountain's fide.
> Joy op'd this day, joy fhall attend its clofe.
> Thee I revere, from thee expect my prey :
> Nor e'er forgot my carol to the BEAR *.

21. WOLVERENE.   *Hift. Quad.* N° 176, 177. *Syn. Quad.*
                      Gulo, *Pallas Spicil. Zool. Fafc.* xiv. 25. tab. ii.—LEV. MUS.

BEAR. With fhort rounded ears, almoft concealed by the fur :
face fharp, black, and pointed : back broad, and, while the
animal is in motion, much elevated, or arched; and the head carried
low : the legs fhort and ftrong : claws long and fharp, white at their
ends.

The length from nofe to tail twenty-eight inches; of the trunk of
the tail feven inches. It is covered with thick long hairs, reddifh at
the bottom, black at the end; fome reach fix inches beyond the
tip.

The hairs on the head, back, and belly, are of the fame colors,
but much finer and fofter. Before they are examined, the animal appears wholly black. The throat whitifh, marked with black. Along
the fides, from the fhoulders to the tail, is a broad band of a ferruginous color : in feveral of the fkins, brought from *Hudfon's Bay,* I ob-

* *Nicholf's Ruffian Nations,* i. 50.

ferved

ferved this band to be white. The legs are black; the feet covered with hair on the bottom. On the fore feet of that which I examined were fome white fpots. On each foot were five toes, not greatly divided.

It hath much the action of a Bear; not only in the form of its back, and the hanging down of its head, but alfo in refting on the hind part of the firft joint of its legs.

This is one of the local animals of *America*. I trace it as far north as the *Copper* river, and to the countries on the weft and fouth of *Hudfon's Bay*, *Canada*, and the tract as far as the ftraits of *Michillmakinac*, between the lakes *Huron* and *Superior*.

I have reafon to think that the Glutton of the old writers is the fame with this animal; and that in my Hiftory of Quadrupeds I unneceffarily feparated them. Since I have received the late publication of Dr. *Pallas*, I am fatisfied that it is common to the north of *America*, *Europe*, and *Afia*, even to *Kamtfchatka*; inhabiting the vaft forefts of the north, even within the *Polar* circle. The *Kamtfchatkans* value them fo highly as to fay, that the heavenly beings wear no other furs. The fkins are the greateft prefent they can make their miftreffes; and the women ornament their heads with the parts of the white banded variety. The *Ruffians* call thefe animals *Roffomak*; the *Kamtfchatkans*, *Tymi*, or *Tummi*.

It is a beaft of uncommon fiercenefs, the terror of the Wolf and Bear; the former, which will devour any carrion, will not touch the carcafe of this animal, which fmells more fetid than that of a Polecat. It has great ftrength, and makes vaft refiftance when taken; will tear the traps often to pieces; or if wounded, will fnap the ftock from the barrel of the mufket; and often do more damage in the capture than the fur is worth.

It preys indifferently on all animals which it can mafter. It feeds by night, and, being flow of foot, follows the track of wolves and foxes in the fnow, in order to come in for fhare of their prey. It will dig up the carcafes of animals, and the provifions concealed by the

huntfmen

huntfmen deep in the fnow, which it will carry away to other places to devour. About the *Lena* it will attack horfes, on whofe backs are often feen the marks of its teeth and claws. By a wonderful fagacity it will afcend a tree, and fling from the boughs a fpecies of mofs which Elks and Reins are very fond of; and when thofe animals come beneath to feed on it, will fall on them and deftroy them: or, like the Lynx, it afcends to the boughs of trees, and falls on the Deer which cafually pafs beneath, and adheres till they fall down with fatigue. It is a great enemy to the Beaver *, and is on that account fometimes called the *Beaver-eater*. It watches at the mouth of their holes, and catches them as they come out. It fearches the traps laid for taking other beafts, and devours thofe which it finds taken. It breaks into the magazines of the natives, and robs them of the provifions; whether they are covered with logs, brufhwood, and built high between two or three ftanding trees †.

It lodges in clefts of rocks, or in hollows of trees, and in *Sibiria* often in the deferted holes of Badgers; never digging its own den, nor having any certain habitation. It breeds once a year, bringing from two to four at a litter ‡. Its fur is much ufed for muffs. Notwithftanding its great fiercenefs when wounded, or firft feizure, it is capable of being made very tame ‖.

FUR.

The fkins are frequently brought from *Hudfon's Bay*, and commonly ufed for muffs. In *Sibiria* the fkin is moft valued which is black, and has left of the ferruginous band. Thefe are chiefly found in the mountanous forefts of *Jakutfk*, and ufed by the natives to adorn their caps. Few of the *Sibirian* fkins are fent into *Ruffia*, but are chiefly fold to the *Mongals* and *Chinefe*.

The relations of the exceffive gluttony of this animal; that it eats till it is ready to burft, and that it is obliged to unload itfelf by fqueezing its body between two trees; are totally fabulous: like other animals, they eat till they are fatisfied, and then leave off §.

* *Dobbs*, 40.   † *Mr. Graham*.   ‡ The fame.   ‖ *Edw.* ii. 103.
§ *Hift. Kamtfch.* 385.

Hift.

Hiſt. Quad. N° 178.—*Smellie,* v. 46.—Lev. Mus.

BEAR. With upper jaw larger than the lower: face ſharp-pointed, and fox-like: ears ſhort and rounded: eyes large, of a yellowiſh green; the ſpace round them black: a duſky line extends from the forehead to the noſe; the reſt of the face, the cheeks, and the throat, white: the hair univerſally long and ſoft; that on the back tipt with black, white in the middle, and cinereous at the roots: tail annulated with black and white, and very full of hair: toes black, and quite divided: the fore-feet ſerve the purpoſes of a hand.

These animals vary in color. I have ſeen ſome of a pale brown, others white. Their uſual length, from noſe to tail, is two feet: near the tail about one.

Raccoons inhabit only the temperate parts of *North America,* from *New England* * to *Florida* †. They probably are continued in the ſame latitudes acroſs the continent, being, according to *Dampier,* found in the iſles of *Maria,* in the South Sea, between the ſouth point of *California* and *Cape Corientes.* It is alſo an inhabitant of *Mexico,* where it is called *Mapach* ‖.

PLACE.

It lives in hollow trees, and is very expert at climbing. Like other beaſts of prey, keeps much within during day, except it proves dark and cloudy. In ſnowy and ſtormy weather it confines itſelf to its hole for a week together. It feeds indifferently on fruits or fleſh; is extremely deſtructive to fields of mayz, and very injurious to all kinds of fruits; loves ſtrong liquors, and will get exceſſively drunk. It makes great havock among poultry, and is very fond of eggs. Is itſelf often the prey of Snakes §.

MANNERS.

• *Joſſelyn's voy.* 85.　† *Account of Florida,* 50.　‖ *Fernand. Nov. Hiſp.* i.
§ *Kalm,* i. 97. ii. 63.

Thoſe

Thofe which inhabit places near the fhore live much on fhell-fifh, particularly oyfters. They will watch the opening of the fhell, dextroufly put in its paw, and tear out the contents; fometimes the oyfter fuddenly clofes, catches the thief, and detains it till drowned by the return of the tide. They likewife feed on crabs, both fea and land. It has all the cunning of the Fox. *Lawfon* * fays, that it will ftand on the fide of a fwamp, and hang its tail over into the water: the crabs will lay hold, miftaking it for a bait; which, as foon as the Raccoon feels, it pulls out with a fudden jerk, and makes a prey of the cheated crabs.

It is made tame with great eafe, fo as to follow its mafter along the ftreets; but never can be broke from its habit of ftealing, or killing of poultry †. It is fo fond of fugar, or any fweet things, as to do infinite mifchief in a houfe, if care is not taken ‡.

It has many of the actions of a Monkey; fuch as feeding itfelf with its fore feet, fitting up to eat, being always in motion, being very inquifitive, and examining every thing it fees with its paws. Notwithftanding it is not fond of water, it dips into it all forts of dry food which is given to it; and will wafh it's face with its feet, like a Cat.

It is fought after on account of the fur. Some people eat it, and efteem it as very good meat. The *Swedes* call it *Siup*, and *Efpan*; the *Dutch*, *Hefpan*; and the *Iroquefe*, *Affigbro*. The hair makes the beft hats, next to that of the Beaver. The tail is worn round the neck in winter, by way of prefervative againft the cold ‖.

---

* 121.   † *Kalm*, i. 208.   ‡ The fame.   ‖ *Kalm*, ii. 97.

HIST.

HIST. QUAD. Genus XXI.   BADGER.

American Badger, *Hift. Quad.* p. 298. β.—*Smellie*, iv. 226.—Lev. Mus.   23. COMMON.

BADGER. With rounded ears: forehead, and middle of the cheeks, marked with a white line, extending to the beginning of the back, bounded on each fide by another of black: cheeks white: fpace round the ears dufky: body covered with long coarfe hair, cinereous and white.

The legs were wanting in the fkin which I faw; but I fupply that defect from *M. de Buffon*'s defcription. They were dufky, and the toes furnifhed with claws, like the *European* kind. *M. de Buffon* obferved only four toes on the hind feet; but then he fufpected that one was torn off from the dried fkin he faw.

Thefe animals are rather fcarce in *America*. They are found in   PLACE.
the neighborhood of *Hudfon's Bay*, and in *Terra di Labrador*; and perhaps as low as *Penfylvannia*, where they are called Ground Hogs *. They do not differ fpecifically from the *European* kind; but are fometimes found white in *America* †.

I do not difcover them in northern *Afia*, nearer than the banks of the *Yaik* ‡. They are common in *China*, where they are frequently brought to the fhambles, being an efteemed food ‖. In northern *Europe*, they are found in *Norway* and *Sweden* §.

Le Comte de Buffon imagines this animal ¶ to be the *Carcajou* of the *Americans*, and not the Wolveren. The matter is uncertain:

---

* *Kalm*, i. 189.   † *Briffon Quad.* 185.   ‡ *Pallas.*   ‖ *Bell's*
travels, ii. 83.   § *Pontoppidan*, ii. 28. *Faun. Suec.* N° 20.   ¶ *Suppl.*
tom. iii. 242.

yet

yet I find that name beftowed on the latter by *La Hontan*; by *Dobbs*, who makes it fynonymous; and by *Charlevoix*, though the laft miftakes the animal, yet not the manners of that which he afcribes it to.   On the other hand, Mr. *Graham* and Mr. *Edwards* omit that title, and call it only Wolveren, or Queequehatch.

HIST.

HIST. QUAD. Genus XXII.

Hist. Quad. N° 181.

OPOSSUM. With ten cutting teeth above, eight beneath: 24. Virginian eyes black, fmall, and lively: ears large, naked, membranace-ous, and rounded: face long and pointed; whifkers on each fide of the nofe, and tufts of long hairs over the eyes: legs are fhort; the thumb on the hind feet has a flat nail, the reft of the toes have on them fharp talons: the body is fhort, round, and thick: the tail long; the bafe is covered with hair for three inches, the reft is co-vered with fmall fcales, and has the difgufting refemblance of a Snake.

On the lower part of the belly of the female is a large pouch, in which the teats are placed, and in which the young lodge as foon as they are born. The body is cloathed with very long foft hairs, ly-ing ufually uneven: the color appears of a dirty white; the lower parts of the hairs dufky: and above each eye is a whitifh fpot: the belly tinged with yellow.

The length of one I examined was feventeen inches, of the tail fourteen.

This fpecies is found as far north as *Canada* *, where the *French* call it *Le Rat de bois*; from thence it extends fouthward, even to the *Brafils* and *Peru*. The fingularity of the ventral pouch of the fe-male, and the manner of its bringing up its young, places it among the moft wonderful animals of the new continent.

Place.

As foon as the female finds herfelf near the time of bringing forth, fhe prepares a neft of coarfe grafs, covered with long pieces of fticks, near four feet high and five in diameter, confufedly put to-

* *Charlevoix*, v. 197.

L

gether.

gether *. She brings forth from four to fix at a time. As foon as they come into the world they retreat into the falfe belly, blind, naked, and exactly refembling little fœtufes. They faften clofely to the teats, as if they grew to them; which has given caufe to the vulgar error, that they were created fo. There they adhere as if they were inanimate, till they arrive at a degree of perfection in fhape, and attain fight, ftrength, and hair: after which they undergo a fort of fecond birth. From that time they run into the pouch as an afylum from danger. The female carries them about with the utmoft affection, and would rather be killed than permit this receptacle to be opened; for fhe has the power of contracting or dilating the orifice by the help of fome very ftrong mufcles. If they are furprifed, and have not time to retreat into the pouch, they will adhere to the tail of the parent, and efcape with her †.

The Opoffum is both carnivorous and frugivorous. It is a great enemy to poultry; and will fuck the blood and leave the flefh untouched ‡. It climbs trees very expertly, feeding on wild fruits, and alfo on various roots. Its tail has the fame prehenfile quality as that of fome fpecies of Monkies. It will hang from the branches by it, and by fwinging its body, fling itfelf among the boughs of the adjacent trees. It is a very fluggifh animal; has a very flow pace, and makes fcarcely any efforts to efcape. When it finds itfelf on the point of being taken, it counterfeits death; hardly any torture will make it give figns of life §. If the perfon retires, it will put itfelf in motion, and creep into fome neighboring bufh. It is more tenacious of life than a Cat, and will fuffer great violence before it is killed ‖.

The old animals are efteemed as delicate eating as a fucking pig; yet the fkin is very fœtid. The *Indian* women of *Louifiana* dye the hair, and weave it into girdles and garters ¶.

* *Bartram's journal E. Florida,* 30.   † The fame.   ‡ *Du Pratz,* ii. 65.
§ The fame, 56.   ‖ *Lawfon,* 120.   ¶ *Du Pratz,* ii. 66.

HIST.

HIST. QUAD. Genus XXIII.

Hift. Quad. N° 192.—*Smellie,* iv. 257.—Lev. Mus.

25. Common.

WEESEL. With fmall rounded ears: beneath each corner of the mouth is a white fpot: breaft and belly white; reft of the body of a pale tawny brown. Its length, from the tip of the nofe to the tail, is about feven inches; the tail two and a half.

Inhabits the country about *Hudfon's Bay, Newfoundland,* and as far fouth as *Carolina**. Mr. *Graham* fent fome over, both in their fummer coat, and others almoft entirely white, the color they affume in winter. We meet with them again in *Kamtfchatka,* and all over *Ruffia* and *Sibiria*; and in thofe northern regions they regularly turn white during winter. One, which was brought from *Natka Sound* in *North America,* had between the ears and nofe a bed of gloffy black, which probably was its univerfal color before its change. Dr. *Irving* faw on *Moffen* ifland, north of *Spitzbergen,* lat. 80. an animal, perhaps of this kind, fpotted black and white †.

Place.

Hift. Quad. N° 193.—*Smellie,* iv. 262.—Lev. Mus.

26. Stoat.

WEESEL. With fhort ears, edged with white: head, back, fides, and legs, of a pale tawny brown: under fide of the body white: lower part of the tail brown, the end black.

In northern countries, changes in winter to a fnowy whitenefs, the end of the tail excepted, which retains its black color: in this ftate is called an *Ermine.*

* *Catefby, App.* † *Phipps's voy.* 58.

L 2 Length,

Length, from nofe to tail, ten inches ; the tail is five and a half.

Inhabits only *Hudfon's Bay*, *Canada*, and the northern parts of *North America*. In *Newfoundland* it is fo bold as to commit its thefts in open view. Feeds on eggs, the young of birds, and on the mice with which thofe countries abound. They alfo prey on Rabbits, and the White Grous. The fkins are exported from *Canada* among what the *French* call *la menuë pelleterie*, or fmall furs.*.

It is found again in plenty in *Kamtfchatka* †, the *Kurili* iflands, *Sibiria*, and in all the northern extremities of *Europe*. It is fcarce in *Kamtfchatka* ; and its chafe is not attended to, amidft the quantity of fuperior furs. But in *Sibiria* and *Norway* they are a confiderable ar-

ticle of commerce. In the former, they are taken in traps, baited with a bit of flefh ‡ ; in the latter, either fhot with blunt arrows, or taken, as garden mice are in *England*, by a flat ftone propped by a baited ftick, which falls down on the leaft touch, and crufhes them to death §. They are found in *Sibiria* in great plenty in woods of birch, yet are never feen in thofe of fir. Their fkins are fold there on the fpot from two to three pounds fterling per hundred ‖.

They are not found on the *Arctic* flats. The inhabitants of the *Ifchuktfchi Nofs* get them in exchange from the *Americans*, where they are of a larger fize than any in the *Ruffian* dominions.

The exceffive cold of certain winters has obliged even thefe hardy animals to migrate, as was evident in the year 1730, and 1744 ¶.

Hift. Quad. N° 200.—*Br. Zool.* i. N° 16.—*Smellie*, iv. 245.—LEV. MUS.

WEESEL. With white cheeks and tips of ears ; yellow throat and breaft ; reft of the fur of a fine deep chefnut-color in the male, paler in the female : tail bufhy, and of a deeper color than the body.

* *Charlevoix*, v. 197.   † *Hift. Kamtfchatka*, 99.   ‡ *Bell's travels*, i. 199.
§ *Pontoppidan*, ii. 25.   ‖ *Gmelin Ruff. Samlung*, 516.   ¶ *Nov. Sp. an.* 188.

Thefe

Thefe animals inhabit, in great abundance, the northern parts of *America*; but I believe the fpecies ceafes before it arrives at the temperate provinces.. They appear again in the north of *Europe*, extend acrofs the *Urallian* chain, but do not reach the *Oby*.

They inhabit forefts, particularly thofe of fir and pine, and make their nefts in the trees. Breed once a year, and bring from two to foar at a litter. They feed principally upon mice; but deftroy alfo all kinds of birds which they can mafter. They are taken by the natives of *Hudfon's Bay* in fmall log-traps, baited, which fall on and kill them. The natives eat the flefh.

Their fkins are among the more valuable furs, and make a moft important article of commerce. I obferved, that in one of the *Hudfon's Bay* Company's annual fales, not fewer than 12,370 good fkins, and 2360 damaged, were fold; and in that year (1743) 30,325 were imported by the *French* from *Canada* into the port of *Rochelle*. They are found in great numbers in the midft of the woods of *Canada*; and once in two or three years come out in great multitudes, as if their retreats were overftocked: this the hunters look on as a forerunner of great fnows, and a feafon favorable to the chafe *.

It is remarkable, that notwithftanding this fpecies extends acrofs the continent of *America*, from *Hudfon's Bay* to the oppofite fide, yet it is loft on the *Afiatic* fide of the ftraits of *Tfchuktfchi*; nor is it recovered till you reach *Catherinebourg*, a diftrict of *Sibiria* weft of *Tobolfk*, and twenty-five degrees weft longitude diftant from *America*. The fineft in the known world are taken about *Ufa*, and in the mountains of *Caucafus* †. It is known that the *Tfchuktfchi* ‡ procure the fkins for cloathing themfelves from the *Americans*; their country being deftitute of trees, and confequently of the animals, inhabitants of forefts, furnifhing thofe ufeful articles.

The Houfe Martin, *Hift. Quad.* N° 199, is found neither in *America*, or the *Arctic* countries.

---

* *Charlevoix*, v. 197.   † *Doctor Pallas.*   ‡ *Muller*, Pref. xxix.

Hift.

Hiſt. Quad. Nº 204.—*Smellie,* vii. 307.—Lɛv. Mus.

WEESEL. With ears a little pointed: body and head covered with hair of a mixture of grey, cheſnut, and black, and be-neath protected by a cinereous down: the lower jaw encircled with white: legs and tail black: on the breaſt, between the fore-legs, a ſpot of white, and another on the belly between the hind-legs: toes covered above and below with fur.

I ſaw this and the following animal at *Paris,* in the cabinet of M. *Aubry, Curè de St. Louis en L'Iſle.* They were in glaſs caſes, ſo I could get only an imperfect view of them. According to M. *de Buffon,* the length of this was a foot and a half *French* meaſure; the tail ten inches *. The fur is fine; and the ſkins were often imported by the *French* from *Canada.*

This ſeems to me to be very nearly allied to the *European* Martin, Nº 15. *Br. Zool.* vol. i. It agrees very much in dimenſions, and in the white marks. It is alſo the animal which Mr. *Graham* ſent to the Royal Society from *Hudſon's Bay,* under the name of *Jackaſh,* which he ſays harbours about creeks, and lives on fiſh. Brings from two to four young at a time. Is caught by the natives, who eat the fleſh and barter the ſkins.

Hiſt. Quad. Nº 205.—*Smellie,* vii. 307.

WEESEL. With a long neck and body: ſhort legs: head and body brown tinged with tawny: tail black: the down of a bright aſh-color.

* *Le Pekan,* tom. xiii. 304. tab. xlii. xliii.

Length

Length from head to tail one foot four inches, *French*; tail seven inches, or to the end of the hairs nine.

Inhabits *Canada*.

Hift. Quad. N° 201, and p. 328.—*Smellie*, vii. 309.

Muftela Zibellina, *Pallas Sp. Zool. fafc.* xiv. 54. tab.

30. SABLE.

WEESEL. With head and ears whitifh: the ears broad, inclining to a triangular form, and rounded at top, in the *Afiatic* fpecimens; in the *American*, rather pointed: whole body of a light tawny: feet very large, hairy above and below: claws white.

Length, from nofe to tail, twenty inches; of the trunk of the tail, four inches; from the bafe to the end of the hairs eight: of a dufky color.

This defcription is taken from a fkin fent from *Canada*: but it extends acrofs the whole continent, being frequently found among the furs which the *Americans* traffic with among the inhabitants of the *Tfchuktfchi Nofs* *. The *American* fpecimen, which I had opportunity of examining, was of the bleached, or worft kind; probably others may equal in value thofe of *Afia*.

PLACE.

The great refidence of thefe animals is in *Afia*, beginning at the *Urallian* chain, and growing more and more plentiful as they advance eaftward, and more valuable as they advance more north. None are found to the north-eaft of the *Anadir*, nor in any parts deftitute of trees. They love vaft forefts, efpecially thofe of fir, in which thofe of moft exquifite beauty are found. They are frequent in *Kamtfchatka*, and are met with in the *Kuril* ifles †. They extend from about lat. 50 to lat. 58.

They are very eafily made tame: will attach themfelves fo to their mafter, as to wander a confiderable way, and return again to their home.

---

* *Doftor Pallas.*   † *Defcr. Kamtfchatka*, 275.

3

They

They abhor water : therefore the notion of their being the *Satherion* of *Ariſtotle* is erroneous.

CAPTURE.

Another way of taking them, beſides thoſe which I before mentioned, is by placing a piece of timber from tree to tree horizontally; near one end of this is placed a bait : over the lower piece of wood is placed another, ſuſpended obliquely, and reſting at one end on a poſt very ſlightly : a rod extends from it to a nooſe to which the bait is faſtened. As ſoon as the Sable ſeizes the meat, the upper timber falls, and kills the precious animal *. The hunting-ſeaſon always begins with the firſt ſnows : but they are now become ſo very ſcarce, as to be confined to the vaſt foreſts of the extreme parts of *Sibiria*, and to the diſtant *Kamtſchatka*. Such has been the rage of luxury !

FURS WHEN FIRST USED AS A LUXURY.

It was not till the later ages that the furs of beaſts became an article of luxury. The more refined nations of antient times never made uſe of them : thoſe alone whom the former ſtigmatized as barbarians, were cloathed in the ſkins of animals. *Strabo* deſcribes the *Indians* covered with the ſkins of Lions, Panthers, and Bears †; and *Seneca* ‡, the *Scythians* cloathed with the ſkins of Foxes, and the leſſer quadrupeds. *Virgil* exhibits a picture of the ſavage *Hyperboreans*, ſimilar to that which our late circumnavigators can witneſs to in the cloathing of the wild *Americans*, unſeen before by any poliſhed people.

Gens effræna virum *Riphæo* tunditur Euro ;
Et pecudum fulvis velantur corpora ſetis.

Moſt part of *Europe* was at this time in ſimilar circumſtances. *Cæſar* might be as much amazed with the ſkin-dreſſed heroes of *Britain*, as our celebrated *Cook* was at thoſe of his new-diſcovered regions. What time hath done to us, time, under humane conquerors, may effect for them. Civilization may take place, and thoſe ſpoils of animals, which are at preſent eſſential for cloathing, become the mere objects of ornament and luxury.

---

* *Decouvertes dans le Ruſſe, &c.* iv. 237. tab. vi. vii.   † *Strabo*, lib. xvii.
p. 1184.   ‡ *Epiſt.* Ep. xc.

I can—

I cannot find that the *Greeks* or old *Romans* ever made use of furs. It originated in those regions where they most abounded, and where the severity of the climate required that species of cloathing. At first it consisted of the skins only, almost in the state in which they were torn from the body of the beast; but as soon as civilization took place, and manufactures were introduced, furs became the lining of the dress, and often the elegant facing of the robes. It is probable, that the northern conquerors introduced the fashion into *Europe*. We find, that about the year 522, when *Totila*, king of the *Visigoths*, reigned in *Italy*, that the *Suethons* (a people of modern *Sweden*) found means, by help of the commerce of numberless intervening people, to transmit, for the use of the *Romans*, *saphilinas pelles*, the precious skins of the Sables *. As luxury advanced, furs, even of the most valuable species, were used by princes as lining for their tents: thus *Marco Polo*, in 1252, found those of the *Cham* of *Tartary* lined with Ermines and Sables †. He calls the last *Zibelines*, and *Zambolines*. He says that those, and other precious furs, were brought from countries far north; from the *land of Darkness*, and regions almost inaccessible, by reason of morasses and ice ‡. The *Welsh* set a high value on furs, as early as the time of *Howel Dda* ‖, who began his reign about 940. In the next age, furs became the fashionable magnificence of *Europe*. When *Godfrey* of *Boulogne*, and his followers, appeared before the emperor *Alexis Comnene*, on their way to the *Holy Land*, he was struck with the richness of their dresses, *tam ex oftro quam aurifrigio et niveo opere harmelino et ex mardrino grifioque et vario*. How different was the advance of luxury in *France*, from the time of their great monarch *Charlemagne*, who contented himself with the plain fur of the Otter! *Henry* I. wore furs; yet in his distress was obliged to change them for warm *Welsh* flannel §. But in the year 1337 the luxury had got to such a head, that *Edward* III. enacted, that all persons who could not spend a hundred a year, should absolutely be prohibited the use of this species of finery.

* *Jornandes de Rebus Geticis.*   † *In Bergeron's Coll. 70. Purchas*, iii. 86.
‡ 160, 161, 162.   ‖ *Leges Wallicæ*.   § *Barrington on the Statutes*, 4th ed. 243.

M   These,

Thefe, from their great expence, muft have been foreign furs, obtained from the *Italian* commercial ftates, whofe traffic was at this
period boundlefs. How ftrange is the revolution in the fur trade !
The north of *Afia*, at that time, fupplied us with every valuable kind ;
at prefent we fend, by means of the poffeffion of *Hudfon's Bay*, furs,
to immenfe amount, even to *Turkey* and the diftant *China*.

31. FISHER.   *Hift. Quad.* N° 202.—*Smellie*, v. 297.—LEV. MUS.

WEESEL. With ears broad, round, and dufky, edged with
white : head and fides of the neck pale brown mixed with
afh-color and black : hairs on the back, belly, legs, and tail, brown
at the bafe, and black at their ends : fides of the body brown.

The feet very large and broad, covered above and below thickly
with hair : on each foot are five toes, with white claws, fharp, ftrong,
and crooked : the fore legs fhorter than thofe behind : the tail is full
and bufhy, fmalleft at the end. Length, from nofe to tail, is twenty-
eight inches ; of the tail feventeen.

This animal inhabits *Hudfon's Bay*, and is found in *New England*,
and as low as *Penfylvania*. About *Hudfon's Bay* they are called *We-
jacks*, and *Woodfhocks*. They harbour about creeks, feed upon fifh, and
probably birds. They breed once a year, and have from two to four
at a birth. The natives catch them, and difpofe of the fkins, which
are fold in *England* for four or fix fhillings apiece. Such is the account I received from Mr. *Graham*.

The late worthy Mr. *Peter Collinfon* tranfmitted to me the following relation, which he received from Mr. *Bartram*:—" They are
" found in *Penfylvania* ; and, notwithftanding they are not amphibi-
" ous, are called *Fifhers*, and live on all kinds of leffer quadrupeds."
I do not know how to reconcile thefe accounts of the fame animal
(for fuch it is) unlefs it preys indifferently on fifh and land animals,
as is often the cafe with rapacious beafts, and that both Mr. *Graham*
and *Bartram* may have overlooked that circumftance.

Hift.

WEESEL. With fmall and rounded ears: the ground color of the whole animal black, marked on the back and fides with five long parallel lines of white; one extending from the head along the top of the back to the bafe of the tail; with two others on each fide, the higheft of which reaches a little way up the tail: the tail is long, and very bufhy towards the end.

This fpecies varies in the difpofition of the ftripes, and I fufpect the male is entirely black, as defcribed by *M. Du Pratz* *; who fays, that the female has rings of white intermixed. If that is the cafe, the *Coafe,* which *M. de Buffon* † received from *Virginia,* is of this kind. It is of an uniform color; but what is a ftronger proof of their differing only in fex, is the agreement in number of toes in the fore feet, there being four on each; an exception to the character of this Genus.

In fize it is equal to an *European* Pole-cat, but carries its back more elevated.

Thefe animals are found from *Penfylvania* as far as *Louifiana,* where they are known by the name of the Pole-cat ‡ or Skunk; which is given indifferently to both of thefe foetid beafts.

Nature hath furnifhed this and the following a fpecies of defence fuperior to the force either of teeth or claws. The *French* moft juftly call thefe animals *enfans du Diable,* or children of the Devil, and *Bêtes puantes,* or the ftinking beafts; as the *Swedes* beftow on them that of *Fifkatta.* The peftiferous vapour which it emits from behind, when it is either attacked, purfued, or frightened, is fo fuffocating and foetid, as at once to make the boldeft affailant retire with precipita-

---

* Vol. ii. 67. xiii. *Coafe,* p. 288. *Le Conepute* (the female) ibid. tab. xxxviii. xl. ‡ *Catefby,* ii. tab. 62.

tion. A fmall fpace is often no means of fecurity; the animal either
will turn its tail, and by a frequent crepitus prevent all repetition of
attempts on its liberty; or elfe ejaculate its ftifting urine to the diftance of eighteen feet *.   Its enemy is ftupified with the abominable
ftench; or perhaps experiences a temporary blindnefs, fhould any of
the liquid fall on his eyes.   No wafhing will free his cloaths from
the fmell: they muft even be buried in frefh foil, in order to be effectually purified.

Perfons who have juft undergone this misfortune, naturally run to
the next houfe to try to free themfelves from it; but the rights of
hofpitality are denied to them: the owner, dreading the infection, is
fure to fhut the door againft them.

Profeffor *Kalm* ran the danger of being fuffocated by the ftench of
one, which was purfued into a houfe where he was.

A maid-fervant, who deftroyed another in a room where meat was
kept, was fo affected by the vapour as to continue ill for feveral days;
and the provifions were fo infected, that the mafter of the houfe was
forced to fling them away †.

Travellers are often obliged, even in the midft of forefts, to hold
their nofes, to prevent the effects of its ftench.

The brute creation are in like dread of its effluvia.   Cattle will
roar with agony; and none but true-bred dogs will attack it: even
thofe are often obliged to run their nofes into the ground before they
can return to complete its deftruction.   The fmell of the dogs, after
a combat of this nature, remains for feveral days intolerable.

Notwithftanding this horrible quality, the flefh is eaten, and is
efteemed as fweet as that of a Pig.   The bladder muft be taken out,
and the fkin flayed off, as foon as the animal is killed ‡.

I fhould think it a very difagreeable companion: yet it is often
tamed fo as to follow its mafter like a Dog; for it never emits its

---

* *Kalm*, i. 273.   † The fame, 277.   ‡ *Lawfon*, 119. *Kalm*, i. 278.

vapour

vapour unlefs terrified *. It furely ought to be treated with the higheft attention.

The fkin is neglected by the *Europeans*, by reafon of the coarfenefs of the hair. The *Indians* make ufe of it for tobacco pouches, which they carry before them like the *Highlanders*.

It climbs trees with great agility. It feeds on fruits † and infects. Is a great enemy to birds, deftroying both their eggs and young. It will alfo break into hen-roofts, and deftroy all the poultry ‡. It breeds in holes in the ground, and hollow trees, where it leaves its young, while it is rambling in queft of prey.

Hift. Quad. N° 218.—*Smellie*, v. 297.—LEV. MUS.

33. SKUNK.

WEESEL. With fhort rounded ears: fides of the face white: from the nofe to the back extends a bed of white; along the top of the back, to the bafe of the tail, is another broad one of black, bounded on each fide by a white ftripe: the belly, feet, and tail, black. But the colors vary: that which is figured by *M. de Buffon* has a white tail: the claws on all the feet very long, like thofe of a Badger: the tail very full of hair.

This inhabits the continent of *America*, from *Hudfon's Bay* § to *Peru* ‖. In the laft it is called *Chinche*. It burrows like the former, and has all the fame qualities. It is alfo found in *Mexico*, where it is called *Conepatl*, or *Boy's little Fox* ¶.

* *Kalm*, i. 278. † *Catefby*, ii. tab. 62. ‡ *Kalm*, i. 274. § Sent from thence by Mr. *Graham*. ‖ *Feuillée Obf. Peru*, 1714, p. 272. ¶ *Hernandez*, *Mex.* 382.

HIST.

OTTER. *HIST. QUAD.* Genus XXIV.

34. Common. *Hift. Quad.* N° 226.—*Br. Zool.* i. N° 19.—*Smellie,* iv. 232.—Lev. Mus.

OTTER. With fhort rounded ears: head flat and broad: long whifkers: aperture of the mouth fmall: lips very mufcular, defigned to clofe the mouth firmly while in the action of diving: eyes fmall, and placed nearly above the corners of the mouth: neck fhort: body long: legs fhort, broad, and thick: five toes on each foot, each furnifhed with a ftrong membrane or web: tail depreffed, and tapering to a point.

The fur fine; of a deep brown color, with exception of a white fpot on each fide of the nofe, and another under the chin.

Thefe animals inhabit as far north as *Hudfon's Bay, Terra di Labrador,* and *Canada,* and as low fouth as *Carolina* and *Louifiana* *; but in the latter provinces are very fcarce. The fpecies ceafes farther fouth. *Lawfon* fays that they are fometimes found, to the weftward of *Carolina,* of a white color, inclining to yellow. Thofe of *North America* are larger than the *European,* and the furs of fuch which inhabit the colder parts are very valuable. Their food is commonly fifh; but they will alfo attack and devour the Beaver †.

They are found again in *Kamtfchatka,* and in moft parts of northern *Europe* and *Afia,* but not on the *Arctic* flats: are grown very fcarce in *Ruffia.* The *Kamtfchatkans* ufe their furs to face their garments, or to lap round the fkins of Sables, which are preferved better in Otter fkins than any other way. They ufually hunt them with dogs, in time of deep fnow, when the Otters wander too far from the banks of rivers ‡.

* *Lawfon,* 119, and *Du Pratz,* ii. 69. † *Dobbs,* 40. ‡ *Hift. Kamtf.* 115, 116.

The

The *Americans* round *Hudson's Bay* shoot or trap them for the sake of the skins, which are sent to *Europe*. They also use the skins for pouches, ornamented with bits of horn; and eat the flesh.

Otters are probably continued along the *Arctic* parts of *America*, westward; being found on the most eastern, or the greater *Fox Islands*, which are supposed to be pretty near to that continent.

Lesser Otter, *Hist. Quad.* N° 228.

35. MINX.

OTTER. With a white chin: rounded ears: top of the head in some hoary, in others tawny: the body covered with short tawny hairs, and longer of a dusky color: the feet broad, webbed, and covered with hair: the tail dusky, ending in a point. This animal is of the shape of the common Otter, but much smaller: its length being only twenty inches from head to tail; of the tail only four.

DESCRIPTION

It inhabits the middle provinces of *North America*, from *New Jersey* to the *Carolinas*. I did not discover it among the skins sent by Mr. *Graham* from *Hudson's Bay*; the animal described as one of this species differing from the many I have seen from the more southern colonies: yet possibly it may be found in a more northern latitude than that which I have given it, if the *Foutereaux*, an amphibious sort of little Polecats mentioned by *La Hontan*, be the same *.

AMERICA.

It frequents the banks of rivers, inhabiting hollow trees, or holes which it forms near the water †. It has, like the Skunks, when provoked, a most excessively fœtid smell. It lives much upon fish, frogs, and aquatic insects; dives admirably, and will continue longer under water than the Musk-beaver ‡: yet at times it will desert its watery haunts, and make great havoke in the poultry yards, biting off the heads of the fowls, and sucking the blood. At times it lurks amidst

* i. 62.   † *Kalm*, ii. 62.   ‡ Letter from *Mr. Peter Collinson.*

the

the docks and bridges of towns, where it proves a useful enemy to rats *.

It is besides very destructive to the Tortoise; whose eggs it scrapes out of the sand and devours: and eats the fresh-water muscles; whose shells are found in great abundance at the mouth of their holes. It is capable of being made tame, and domesticated †.

ASIA.

The species is spread in *Asia*, along the banks of the *Yaik*, in the *Orenburg* government ‡. None are seen in *Sibiria*; but appear again near the rivers which run into the *Amur*. Its fur is in those parts very valuable, and esteemed as next in beauty to the Sable. It

EUROPE.

is either hunted with dogs or taken in traps. In *Europe* it is found in *Poland* and *Lithuania*, where it is named *Nurek*; and the *Germans* call it *Nurtz*. It is also an inhabitant of *Finland*: the natives call it *Tichuri*; the *Swedes*, *Mænk* §, a name carried into *America* by some *Swedish* colonist, and with a slight variation is still retained.

36. SEA.

Hist. Quad. N° 230.
Lutra Marina, Kalan. *Nov. Com. Petrop.* ii. 367. tab. xvi.
Castor Marin, *Hist. Kamtschatka*, 444.
Sea Otter, *Muller*, 57, 58 ‖.—LEV. MUS.

DESCRIPTION.

OTTER. With hazel irides: upper jaw long, and broader than the lower: nose black: ears erect, conic, small: whiskers long and white: in the upper jaw six, in the lower four, cutting teeth: grinders broad: fore legs thick; on each four toes, covered with hair, and webbed: the hind feet resemble exactly those of a Seal: the toes divided by a strong shagreened membrane, with a skin skirting the external side of the outmost toe, in the manner of some water fowl.

* *Kalm*, ii. 61.   † *Lawson*, 122.   ‡ *Dr. Pallas.*   § *Fauna Suec.* N° 13.
‖ I here insert the synonyms; for in the Synopsis of Quadrupeds, following *Linnæus* and *Brisson*, I confound the *Brasilian* Otter of *Marcgrave* with this animal.

The

The skin is extremely thick, covered closely with long hair, remarkably black and glossy; and beneath that is a soft down. The hair sometimes varies to silvery. The hair of the young is soft and brown.

The length, from nose to tail, is about three feet; that of the tail thirteen inches and a half. The tail is depressed, full of hair in the middle, and sharp-pointed. The weight of the biggest, seventy or eighty pounds.

SIZE.

These are the most local animals of any we are acquainted with, being entirely confined between lat. 44 and 60, and west longitude 126 to 150 east from *London*, in the coast and seas on the north-east parts of *America*; and again only between the *Kamtschatkan* shores and the isles which intervene between them and *America*. They land also on the *Kuril* islands; but never are seen in the channel between the north-east part of *Sibiria* and *America*.

PLACE.

They are most extremely harmless, and most singularly affectionate to their young. They will never desert them, and will even pine to death on being robbed of them, and strive to breathe their last on the spot where they experienced the misfortune.

MANNERS.

It is supposed that they bring but one at a time. They go between eight and nine months with young, and suckle it almost the whole year. The young never quits its dam till it takes a mate. They are monogamous, and very constant.

They bring forth on land: often carry the young between their teeth, fondle them, and frequently fling them up and catch them again in their paws. Before the young can swim, the old ones will take them in their fore feet, and swim about upon their backs.

They run very swiftly: swim sometimes on their sides, on their backs, and often in a perpendicular direction. They are very sportive, embrace each other, and kiss.

They never make any resistance; but endeavour, when attacked, to save themselves by flight: when they have escaped to some distance, they will turn back, and hold one of their fore feet over their

N eyes,

eyes, to gaze, as men do their hands to fee more clearly in a funny day; for they are very dull-fighted, but remarkably quick-fcented.

They are fond of thofe parts of the fea which abound moft with weeds, where they feed on fifh, fepiæ, lobfters, and fhell-fifh, which they comminute with their flat grinders.

CAPTURE. They are taken different ways: in the fummer, by placing nets among the fea-plants, where thefe animals retire in the frequent ftorms of this tempeftuous coaft.

They are killed with clubs or fpears, either while they lie afleep on the rocks, or in the fea floating on their backs.

Thirdly, they are purfued by two boats till they are tired, for they cannot endure to be long at a time under water.

During winter they are brought in great numbers to the *Kurilian* iflands, by the eaftern winds, from the *American* fhore.

The hunter goes with a dog, who points them. He knocks it on the head, and flays it, while the dog is beating about for another.

They are called in the *Kamtfchatkan* tongue *Kalan,* in the plural *Kalani.*

Their flefh is preferred to that of Seals by the natives; but the unfortunate crew who were fhipwrecked in the expedition in 1741, under Captain *Bering,* found it to be infipid, hard, and tough as leather; fo that they were obliged to cut it in fmall pieces before they could eat it. Others pretend, that the flefh of the young is very delicate, and fcarcely to be diftinguifhed from young lamb.

FUR. But the valuable part of them is their fkin. Few are brought into *Europe*; but great quantities are fold to the *Chinefe,* at vaft prices, from feventy to a hundred rubles apiece, or 14 or 25 l. fterling each. What a profitable trade might not a colony carry on, was it poffible to penetrate to thefe parts of *North America* by means of the rivers and lakes! The accefs to *Pekin* would be then eafy, by failing up the gulph of *Petcheli.* At prefent, thefe valuable furs are carried by land above three thoufand miles to the frontiers of *China,* where they are delivered to the merchants.

I   Thefe

Thefe animals partake very much of the nature of Seals, in their almoft conftant refidence in the water, their manner of fwimming, fin-like legs, and number of fore teeth. In their ears they greatly refemble the *little Seal* of my Hiftory of Quadrupeds, N° 386, and feem the animals which connect the genera of Otters and Seals.

They are feen very remote from land, fometimes even at the diftance of a hundred leagues.

DIV. II. Sect. II.
DIGITATED QUADRUPEDS.

Without CANINE TEETH: and with two CUTTING TEETH
only in each jaw.

D I V. II. Sᴇᴄᴛ. II. Digitated Quadrupeds.

Without Cᴀɴɪɴᴇ Tᴇᴇᴛʜ : and with two Cᴜᴛᴛɪɴɢ Tᴇᴇᴛʜ only in each jaw.

Generally Herbivorous, or Frugivorous.

Hᴀʀᴇ.   *H I S T.   Q U A D.*   Gᴇɴᴜs XXVI.

37. Vᴀʀʏɪɴɢ.   *Hiſt. Quad.* Nº 242.—Alpine Hare, *Br. Zool.* i. Nº 21.—Lᴇv. Mᴜs.

HARE. With the edges of the ears and tips black : the colors, in ſummer, cinereous, mixed with black and tawny : tail always white.

Mr. *Graham* ſays, that thoſe of *Hudſon's Bay* are of the ſame ſize with the common ; but thoſe which I have examined in *Scotland* are much leſs, weighing only ſix pounds and a half : the common Hare weighs upwards of eight.

This ſpecies inhabits *Greenland*, where alone they continue white throughout the year * ; and are very numerous amidſt the ſnowy mountains. They are uſually fat ; and feed on graſs, and the white moſs of the country. They are found about the rocks at *Churchill*, and the ſtreights of *Hudſon's Bay* ; but are not common. They breed once a year, and bring two at a time †. They change their color to white at approach of winter. They are met with in *Canada* and *Newfoundland*; after which the ſpecies ceaſes to the ſouthward, or at leſt I have no authority for its being continued ; the Hare of *New England* ſeeming, by *Joſſelyn's* account, to be the following ſpecies.

* *Crantz*, i. 70. *Egede*, 62.   † Mr. *Graham*.

The

The *Greenlanders* eat the flesh dressed, and the contents of the stomach raw. They use the excrements for wick for their lamps; and cloath their children with the soft and warm skins.

This species abounds from *Livonia* to the north-east part of *Sibiria* and *Kamtschatka*; and from *Archangel* to *Saratof*, on the banks of the *Wolga*, in east lat. 49. 52, and even farther into the *Orenburg* government. In *Sibiria* they quit the lofty mountains, the southern boundaries of that country, and, collecting in flocks innumerable, at approach of winter migrate to the plains, and northern wooded parts, where vegetation and food abound. Mr. *Bell* met with them daily in their progress *. Multitudes of them are taken in toils by the country people, not for the sake of the flesh, but the skins; which are sent to *Petersburg*, and from thence exported to various parts.

American Hare, *Ph. Transf.* lxii. 4. 376. *Hift. Quad.* Nº 243.   38. AMERICAN.

HARE. With ears tipt with grey: neck and body rusty, cinereous, and black: legs pale rust color: belly white: tail black above, white beneath.

The distinctions between this and the common Hares and varying Hares are these:—They are less, weighing only from three pounds eight ounces to four pounds and a half: the length to the setting-on of   SIZE. the tail only nineteen inches. The hind legs are longer in proportion than those of the common Hare or *varying* Hare; the length of this, from the nose to the tip of the hind legs, extended, being two feet five: of a *varying* Hare, measured at the same time, in *Hudson's Bay*,   OF A VARYING. HARE. only two feet seven and a half; but from the nose to the tail was two feet: its weight seven pounds six ounces.

* *Travels*, octavo ed. i. 246.

Thefe

These animals are found from *Hudson's Bay* to the extremity of *North America*; but swarm in countries bordering on the former. In the time of *M. Jeremie*, who resided in *Hudson's Bay* from 1708 to 1714, twenty-five thousand were taken in one season *. At present they are a principal winter food to our residents there. They are taken in wire snares, placed at certain intervals in small openings made in a long extent of low hedging formed for that purpose; the animals never attempting to jump over, but always seek the gaps. These hedges are removed, on the falling of the snows, to other places, when the Rabbets seek new tracks †. Their flesh is very good; but almost brown, like that of the *English* Hare.

From *Hudson's Bay*, as low as *New England*, these animals, at approach of winter, receive a new coat, which consists of a multitude of **long** white hairs, twice as long as the summer fur, which still remains beneath. About the middle of *April* they begin to shed their winter covering.

From *New England* southward they retain their brown color the whole year. In both warm and cold climates they retain the same nature of never burrowing; but lodge in the hollow of some decayed tree, to which they run in case they are pursued. In the cultivated parts of *America*, they make great havoke among the fields of cabbage, or turnips ‡. In *Carolina*, they frequent meadows and marshy places; and are very subject to have maggots breed in the skin §. In that province they breed very often, and even in the winter months, and bring from two to six at a time; but usually two or four ‖.

I know of no use that is made of the skins, excepting that the natives of *Hudson's Bay* wrap them round the limbs of their children, to preserve them against the cold.

---

* *Voyages au Nord*, iii. 344.   † *Drage*, i. 176.   ‡ *Kalm*, ii. 46.
§ *Lawson*, 122.   ‖ *Doctor Garden*.

** Without a tail.

Hiſt. Quad. N° 248.—*Blackb. Muſ.*

HARE. With ſhort, broad, rounded ears: long head, and whiſkers: fur duſky at the roots; of a bright bay near the ends; tips white: intermixed are divers long duſky hairs.

Length nine inches.

Found from the *Altaic* chain to lake *Baikal*, and from thence to *Kamtſchatka*. They dwell amidſt the ſnows of the loftieſt and moſt dreadful rocky mountains, and never deſcend to the plains. They alſo are ſaid to inhabit the fartheſt Fox or *Aleutian* iſlands: therefore poſſibly may be met with in *America*.

The manners are ſo amply deſcribed in my Hiſtory of Quadrupeds, that I ſhall not repeat an account of them.

BEAVER.　　　*HIST QUAD.*　Genus XXVII.

40. Castor.　　　*Hiſt. Quad.* N° 251.—*Smellie,* v. 21.—Lev. Mus.

Description.　　BEAVER. With a blunt noſe: ears ſhort, rounded, and hid in the fur: eyes ſmall: very ſtrong cutting teeth: hair of a deep cheſnut brown: fore feet ſmall, and the toes divided: hind feet large, and the toes webbed: the tail eleven inches long, and three broad; almoſt oval, flat, and covered with thin ſcales.

The uſual length, from noſe to tail, is about two feet four; but I have meaſured the ſkin of one, which was near three feet long.

Beavers vary in color. They are ſometimes found of a deep black, eſpecially in the north. In Sir *Aſhton Lever's muſeum* is a ſpecimen quite white. As they advance ſouthward, the beauty of their fur decreaſes. Among the *Illinois* they are tawny, and even as pale as ſtraw color *

Place.　　In *North America* theſe animals are found in great plenty all round *Hudſon's Bay,* and as low as *Carolina* † and *Louiſiana* ‡. They are not known in *Eaſt Florida* §. The ſpecies alſo ceaſes before it arrives in *South America.* To ſpeak with preciſion, it commences in lat. 60, or about the river of Seals, in *Hudſon's Bay*; and is loſt in lat. 30, in *Louiſiana.*

From *Hudſon's Bay* and *Canada,* I can trace them weſtward to 120 degrees of longitude, as far as the tract weſt of *Lac Rouge,* or the Red lake ‖. The want of diſcovery prevents us from knowing whether they are continued to the weſtern extremity of this great continent oppoſite to *Aſia*: probably they are, for the *Ruſſian* adventurers got ſome of their ſkins on the iſle of *Kadjak,* which the natives muſt

* *Charlevoix,* v. 140.　　† *Lawſon.*　　‡ *Du Pratz,* ii. 69.　　§ *Account of Eaſt Florida,* 50.　　‖ *Dobbs,* 35.

have

have had from *America*.   They certainly are not found in the iſlands
of the new Archipelago; nor yet in *Kamtſchatka* *, by reaſon of the
interruption of woods, beyond the river *Konyma*.   From thence I
doubt whether they are met with aſſociated, or in a civilized ſtate,
nearer than the banks of the river *Jeneſei*, or the *Konda*, and other
rivers which run into the *Oby* : but they are found ſcattered in the
woody parts of independent *Tartary*; alſo in *Caſan*, and about the
Yaik, in the *Orenburg* government.   In the ſame unſociable ſtate they
inhabit *Europe*, and are found in *Ruſſia*, in *Lapland*, *Norway*, and
Sweden.

Theſe are the moſt ſagacious and induſtrious of animals.   They   SAGACITY.
live in ſociety, and unite in their labors, for the good of the com-
monwealth they form.   They erect edifices, ſuperior in contrivance
to the human beings.   They uſually live near, and ſhew a dexterity
in their œconomy unequalled by the four-footed race.

In order to form a habitation, they ſelect a level piece of ground,   DWELLINGS.
with a ſmall rivulet running in the midſt.   To effect their works, a
community of two or three hundred aſſembles : every individual
bears his ſhare in the laborious preparation.   Some fall trees of great   LABORS.
ſize, by gnawing them aſunder with their teeth, in order to form
beams or piles ; others are employed in rolling the pieces to the water ; others dive, and ſcrape holes with their feet in order to fix them ;
and another ſet exert their efforts to rear them in their proper places.
A fifth party is buſied in collecting twigs to wattle the piles.   A ſixth,
in collecting earth, ſtones, and clay; others carry it on their broad
tails to proper places, and with their feet beat and temper the earth
into mortar, or ram it between the piles, or plaiſter the inſide of the
houſes.

All theſe preparations are to form their dwellings within an arti-   DAMS.
ficial piece' of water or pond, which they make by raiſing a dam
acroſs the level ſpot they had pitched on.   This is done, firſt by driv-

* The Sea Beaver (as it is called) Sp. of this work, muſt not be confounded with
this.

ing into the ground ftakes, five or fix feet long, placed in rows, and fecuring each row by wattling it with twigs, and filling the inter-ftices with clay, ramming it down clofe. The fide next to the water is floped, the other perpendicular. The bottom is from ten to twelve feet thick ; the thicknefs gradually diminifhes to the top, which is about two or three. The centre of the dam forms a fegment of a circle ; from which extends, on each fide, a ftrait wing : in the midft of the centre is ufually a gutter left for the wafte water to dif-charge itfelf. Thefe dams are often a hundred feet long, and nicely covered with turf.

Houses.   The houfes thefe wife animals make, are placed in the water col-lected by means of the dam, and are feated near the fhore. They are built upon piles, and are fometimes round, fometimes oval ; the tops are vaulted, fo that their infide refembles an oven, their outfide a dome. The walls are made of earth, ftones, and fticks, and ufually two feet thick. They are commonly about eight feet high above the furface of the water *, and are very neatly and clofely plaiftered on the infide. The floor is a foot higher than the water. The houfe, fometimes, has only one floor, which is ftrewed with leaves or mofs, on which each Beaver lies in its proper place ; at other times there are three apartments ; one to lodge, another to eat in, and a third to dung in † : for they are very cleanly, and inftantly caufe the filth to be carried off by the inferior Beavers.

M. Du Pratz ‡ fays, that thofe of *Louifiana* form numbers of cells, and that each animal, or more probably each pair, poffefs one. He fays, that he has feen no lefs than fifteen of thefe cells furrounding the centre of one houfe. He alfo acquaints us, that the Beavers of *Louifiana* are a third lefs than the brown fort ; are covered with a ci-nereous down, which is covered with long filvery hairs.

In each houfe are two openings ; one towards the land, the other is within, and communicates with the water, for the conveniency of

* *Clerk*, i. 142.   † *Mr. Graham*. *Catefby*, *App*. xxx.   ‡ i. 241.

getting

getting to their magazine of provifion in frofty weather. This ori-fice is formed fo as to be beyond the thicknefs of the ice; for they lodge their provifions under the water, and dive and bring it into their houfe according as they want it.

Their food is laid in before winter by the tenants of each houfe; it confifts of the bark and boughs of trees. *Lawfon* fays that they are fondeft of the faffafras, afh, and fweet gum. In fummer they live on leaves, fruits, and fometimes crabs and cray-fifh; but they are not fond of fifh.

The numb r of houfes in each pond is from ten to twenty-five: the number of animals in each, from two to thirty. They are fup-pofed to affociate in pairs; are therefore monogamous: another proof of their advances towards civilization.

I think I have heard that every family confifts of an even number. Sometimes the community, within the precinct of a dam, confifts of four hundred; but I prefume this muft be in places little frequented by mankind.

They begin to build their houfes, when they form a new fettle-ment, in the fummer; and it cofts them a whole feafon to finifh their work, and lay in their provifions.

They are very attentive to their fafety; and on any noife take to the water for their further fecurity. They form vaults or burrows in the banks of the creek formed by the dam, into which they retreat in cafe of imminent danger.

They feem to be among quadrupeds, what Bees are among infects. They have a chief, or fuperintendant, in their works, who directs the whole. The utmoft attention is paid to him by the whole commu-nity. Every individual has his tafk allotted, which they undertake with the utmoft alacrity. The overfeer gives a fignal by a certain number of fmart flaps with his tail, expreffive of his orders. The moment the artificers hear it, they haften to the place thus pointed out, and perform the allotted labor, whether it is to carry wood, or draw the clay, or repair any accidental breach.

They

CENTINELS.

They have also their centinels, who, by the same kind of signal, give notice of any apprehended danger.

SLAVES.

They are said to have a sort of slavish Beaver among them (analogous to the Drone) which they employ in servile works, and the domestic drudgery *.

THEIR WOOD HOW CUT.

I have mentioned before their sagacity in laying in the winter provision. They cut the wood they prefer into certain lengths; pile them in heaps beneath the water, to keep them moist; and, when they want food, bite the wood into small pieces, and bring it into their houses. The *Indians* observe the quantity which the Beavers lay in their magazine at approach of winter. It is the Almanack of the Savages; who judge from the greater or less stock, of the mildness or severity of the approaching season †.

The expedition with which they cut down trees, for the forming their dams, is amazing. A number surrounds the body, and will in a few minutes gnaw through a tree of three feet in circumference; and always contrive to make it fall towards the spot they wish ‡.

Beavers have in *America* variety of lakes and waters in which they might fix their seats; but their sagacity informs them of the precarious tenure of such dwellings, which are liable to be overthrown by every flood. This induces them to undertake their mighty and marvellous labors. They therefore select places where no such inconveniences can be felt. They form a dam to support a reservoir, fed only by a small rill; and provide for the overflow of the waste water by a suitable channel in the middle of their bank. They have nothing to fear but from land floods, or the sudden melting of the snows. These sometimes make breaches, or damage their houses; but the defects are instantly repaired.

During the winter they never stir out, except to their magazines of provision; and in that season grow excessively fat.

They are strongly attached to certain haunts, there being places which they will not quit, notwithstanding they are frequently dis-

* *Mr. Graham.*   † *Charlevoix, v. 151.*   ‡ *Catesby, App. 30.*

turbed.

turbed. There is, fays *Charlevoix*, a ftrong inftance on the road between *Montreal* and lake *Huron*, which travellers, through wantonnefs,
annually moleft; yet is always repaired by the induftrious inhabitants.

In violent inundations they are fometimes overpowered in their
attempts to divert the fury of the water. In thofe cafes they fly into
the woods : the females return as foon as the water abates ; the males
continue abfent till *July*, when they come back to repair the ravages
made in their houfes *.

Beavers breed once a year, and bring forth the latter end of winter;
and have two or three young at a birth.

There is a variety of the Beaver kind, which wants either the fagacity or the induftry of the others, in forming dams and houfes.
Thefe are called *Terriers*. They burrow in the banks of rivers, and
make their holes beneath the freezing depth of the water, and work
upwards for a great number of feet. Thefe alfo form their winter
magazines of provifion. Beavers which efcape the deftruction of a
community, are fuppofed often to become *Terriers*.

Strange animal feen by Mr. *Phipps* and others in *Newfoundland*, of
a fhining black : bigger than a Fox : fhaped like an *Italian* grehound :
legs long : tail long and taper. One gentleman faw five fitting on a
rock with their young, at the mouth of a river; often leapt in and
dived, and brought up trouts, which they gave to their young. When
he fhewed himfelf, they all leapt into the water, and fwam a little
way from fhore, put their heads out of the water and looked at him.
An old furrier faid, he remembered a fkin of one fold for five guineas.
The *French* often fee them in *Hare Bay*.

Beavers have, befides man, two enemies ; the Otter, and the Wolverene ; which watch their appearance, and deftroy them. The laft
is on tha account called, in fome parts of *America*, the *Beaver-eater*.
They are very eafily overcome; for they make no refiftance : and have
no fecurity but in flight.

TERRIER
BEAVERS.

THEIR ENEMIES

* *Charlevoix*, v. 151.

2

It

It is not wonderful that such sociable animals should be very affectionate. Two young Beavers, which were taken alive and brought to a neighboring factory in *Hudson's Bay*, were preserved for some time; and throve very fast, till one of them was killed by an accident. The survivor instantly felt the loss, began to moan, and abstained from food till it died*.

They are taken several ways: sometimes in log-traps, baited with poplar sticks, laid in a path near the water. The *Indians* always wash their hands before they bait the traps, otherwise the sagacious animal is sure to shun the snare.

Sometimes they are shot, either while they are at work, or at food, or in swimming across the rivers. But these methods are used only in summer, and not much practised; for the skins in that season are far less valuable than in the winter. At that time they are taken in nets placed above and below their houses, across the creeks, on stakes. If the water is frozen, the ice is cut from shore to shore, in order to put down the stakes. When the net is set, the *Indians* send their women to the Beaver-houses to disturb the animals; who dart into the water, and are usually taken in the net, which is instantly hauled up; and put down again with all expedition. If the Beaver misses the net, it sometimes returns to its house, but oftener into the vaults on the sides of the banks; but the poor creature seldom escapes, being pursued into all his retreats, the houses being broke open, and the vaults searched by digging along the shores.

The value of the fur of these animals, in the manufacture of hats, is well known. It began to be in use in *England* in the reign of *Charles* I.†, when the manufacture was regulated, in 1638, by proclamation; in which is an express prohibition of using any materials except *Beaver stuff*, or *Beaver wool*; and the hats called demi-castors were forbidden to be made, unless for exportation.

This caused a vast encrease of demand for the skins of the Beavers. The *Indians*, on the discovery of *America*, seem to have paid very

* *Drage's voy.* i. 151.   † *Rymer's Fœdera*, xx. 230.

little

little attention to them, amidst the vast variety of beasts they at that time possessed, both for food and cloathing. But about the period of the fashion of hats, they became an article of commerce, and object of chase. The southern colonies soon became exhausted of their Beavers; and of later years the traffic has been much confined to *Canada* and *Hudson's Bay*. The importance of this trade, and the ravages made among the animal creation in those parts, will appear by the following state of the imports into the ports of *London* and *Rochelle* in 1743. I take that year, as I have no other comparative state:

Hudson's Bay company sale, begun *November* 17th 1743.

26,750 Beaver skins.
14,730 Martins.
590 Otters.
1,110 Cats, i. e. Lynx.
320 Fox.
600 Wolverenes.
320 Black Bears.
1,850 Wolves.
40 Woodshocks, or Fishers.
10 Minx.
5 Raccoon.
120 Squirrels.
130 Elks, i. e. Stags.
440 Deer.

Imported into *Rochelle* in the same year.

127,080 Beavers.
16,512 Bears.
110,000 Raccoon.
30,325 Martins.
12,428 Otters and Fishers.
1,700 Minx.
1,220 Cats.

P 1,267 Wolves.

1,267 Wolves.

92 Wolverenes.

10,280 Grey Foxes and Cats.

451 Red Foxes.

This great balance in favor of the *French* arifes not only from their fuperior honefty in their dealings with the ignorant *Indians*, but the advantageous fituation of *Canada* for the fur trade.   They had both fides of the river St. *Lawrence*; the country round the five great lakes; and the countries bordering on the rivers flowing into them; and finally, the fine fur countries bordering on the *Hudfon's Bay* company, many of whofe waters falling into the St. *Lawrence*, gave an eafy conveyance of thofe commodities to *Montreal*; where a fair is annually kept, with all the favage circumftances attendant on *Indian* concourfe.

The traffic carried on in *Hudfon's Bay* is chiefly brought from the chain of lakes and rivers that empty themfelves into the bay at *Nelfon*'s river, running foutherly from lat. 56 to lat. 45.   Lake *Pachegoia* is the moft northerly: there the *Indians* rendezvous in *March*, to make their canoes for the tranfportation of the furs; for at that feafon the bark of the birch-tree feparates very eafily from the wood.

41. MUSK.   *Hift. Quad.* Nº 252.—*Smellie*, v. 260.

BEAVER.   With a thick nofe, blunt at the end: ears fhort, hid in the fur: eyes large: body thick, and in form quite refembles that of the Beaver; its color, and that of the head, a reddifh brown: breaft and belly cinereous, tinged with ruft-color: the fur is very foft and fine.

The toes on every foot are diftinct and divided: thofe of the hind feet fringed on both fides with ftiff hairs or briftles, clofely fet together: tail compreffed, and thin at the edges, covered with fmall fcales, with a few hairs intermixed.

Length,

Length, from nofe to tail, one foot: of the tail nine inches.

Thefe animals are in fome parts of *America* called the Little Beaver, on account of its form, and fome parts of its œconomy. From its fcent it is ftyled the Mufk Rat, and Mufquafh. The *Hurons* call it *Ondathra*; from which *M. de Buffon* gives it the name of *Ondatra* *.

It is found from *Hudfon's Bay* to as low at left as *Carolina* †. Like the Beaver, it forms its houfe of a round fhape, covered with a dome, and conftructed with herbs and reeds cemented with clay. At the bottom and fides are feveral pipes, through which they pafs in fearch of food; for they do not lay in a ftock of provifion, like the former. They alfo form fubterraneous paffages, into which they retreat whenever their houfes are attacked.

Thefe houfes are only intended for winter habitations; are deferted, and rebuilt annually. During fummer, they live in pairs, and bring forth their young from three to fix at a time. At approach of winter, they conftruct their houfes, and retire into them, in order to be protected from the inclemency of the feafon. Several families occupy the fame dwelling, which is oft-times covered many feet with fnow and ice; but they creep out and feed on the roots which lie beneath. They are very fond of the *Acorus Verus*, or *Calamus Aromaticus* ‡. This perhaps gives them that ftrong mufky fmell thefe animals are fo remarkable for; which they lofe during winter, probably when this fpecies of plant is not to be got. They alfo feed on the frefh-water Muffels. They feed too on fruit; for *Kalm* fays, that apples are the baits ufed for them in traps. We may add, that in winter they eat the roots of nettles, and in fummer, ftrawberries and rafberries ‖, during which time it is rare to fee the male and female feparate.

The flefh is fometimes eaten. The fur is made ufe of in the manufacture of hats. The Mufk-bag is fometimes put among cloths, to preferve them from worms or infects

Thefe animals, as well as the Beaver, feem to have their *Terriers*, or fome which do not give themfelves the trouble of building houfes,

---

* x. 12.　　† *Lawfon*, 120.　　‡ *Lawfon*, 120.　　‖ *Charlevoix*, v. 158.

　　　　　　　　　　but

but burrow, like Water-rats, in banks adjacent to lakes, rivers, and ditches *, and often do much damage, by admitting the water through the embankments of meadows. They continue in their holes, except when they are in the water in fearch of food. They make their nefts with fticks, placing a lining of fome foft materials within †. *Charlevoix* ‡ adds, that they fometimes make ufe of a hollow tree for their refidence.

When taken young, they are capable of being tamed; are very playful and inoffenfive, and never bite.

---

* *Kalm*, ii. 56, and *Charlevoix*.   † *Kalm*, ii. 58.   ‡ v. 158.

HIST.

HIST. QUAD. Genus XXVIII.

Hiſt. Quad. Nº 257.—Lev. Mus.   42. Canada.

PORCUPINE. With ſhort ears, hid in the fur: hair on the head, body, legs, and upper part of the tail, long, ſoft, and of a dark brown color; but ſometimes found white: on the upper part of the head, body, and tail, are numbers of ſtrong ſharp quils; the longeſt, which are thoſe on the back, are three inches long; the ſhorteſt are towards the head and on the ſides, and concealed in the hair; mixed with them are certain ſtiff ſtraggling hairs, at leſt three inches longer than the others, tipt with dirty white: the under ſide of the tail is white.

On each fore foot are four toes; on the hind five; all armed with long claws, hollowed on the under ſide.

The ſize of one, which Sir Joseph Banks brought from *Newfoundland*, was about that of a Hare, but more compactly made; the back arched; and the whole form reſembling that of the Beaver: the tail is ſix inches long, which, in walking, is carried a little bent upwards.

This ſpecies inhabits *America*, from *Hudſon's Bay* to *Canada* *, *Newfoundland*, *New England*, and, but rarely, as far ſouth as *Virginia* †. *Lawſon* makes no mention of it among the animals of *Carolina*. *Du Pratz* ‡ ſays, it loves the cold, and is found on the banks of the *Illinois*. It may therefore be ranked among the local northern animals.

They are found in great plenty about *Hudſon's Bay*, where the trading *Indians* depend much on them for food. They are reckoned excellent eating, even by the *Europeans*, taſting, when roaſted, like ſuck-

* *Charlevoix*, v. 198.   † *Cateſby, App.* xxx.   ‡ ii. 68.

5   ing-

ing-pig.   The bones, during winter, are of a greenish yellow, owing, as is supposed, to their feeding during that season on the bark of the pine.   It is observed, that the bones of animals sometimes take a tincture from their diet; for example, those of beasts which feed on madder become red *.   They are also very fond of the bark of juniper.   In summer, they live on the wild fruits, and lap water, but will not go into it.   In winter, take snow by way of beverage.

They nestle under the roots of great trees, and will also, in quest of fruits, ascend the boughs.   When the *Indians* discover one in a tree, they cut it down, and kill the animal by a blow on the nose.

They defend themselves with their quills.   They fly from their pursuer; but when they cannot escape, will sidle towards their enemy, in order to push their quils into him : they are but weak instruments of offence; for a small stroke with the hand against the grain will bring them from the skin, leaving them sticking slightly in the flesh.   The *Indians* use them to make holes in their noses and ears, for the placing their nose and ear-rings, and other finery †.   They also trim the edges of their deer-skin habits with fringes of dyed quils ‡, or make pretty linings with them for the bark boxes.

They are very indolent animals, sleep much, and seldom travel a mile from their haunts ‖.

M. *de Buffon* gives two figures of this beast, under the name of *Le Coendou* and *L'Urson*.   The first he makes an inhabitant of *Brasil*: the last, of *Hudson's Bay*: but the *Coendou* is a very different animal §.   The two figures he has exhibited are of our Porcupine in the winter and summer dress, the hair growing thinner as the warm season approaches ¶.   His *Coendou* shews it in the first state, his *Urson* in the second **.

They are said to copulate in *September*, and to bring only one young, the first week in *April*; another, which it brings forth, being always dead ††.

* *Phil Trans.* lxii. 374.   † *Drage's voy.* i. 177.   ‡ The same, 191.
‖ *Mr. Graham,*   § See *Syn. Quad.* No   ¶ *Edw. Hist. Birds,* i. 52.
** *Hist. Nat.* xii. tab. liv. lv.   †† *Mr. Graham.*

                                                                    HIST.

HIST. QUAD. Genus XXIX.   MARMOT.

Hiſt. Quad. Nº 259.   43. Quebec.
Mus empetra, Pallas *Nov. Sp. Quad. faſc.* i. 75.

MARMOT. With ſhort rounded ears: blunt noſe: cheeks
ſwelled, and of a cinereous color: end of the noſe black:
top of the head cheſnut: the hair on the back grey at the bottom,
black in the middle, and the tips whitiſh: the belly and legs of a
deep orange, or a bright ferruginous color.

Toes black, naked, and quite divided: four toes, with the rudi
ments of another, on the fore feet, five on the hind feet: tail ſhort,
duſky, and full of hair.

The ſpecimen which I ſaw formerly at Mr. *Brook*'s, alive, appeared
larger than a Rabbet; but the ſpecimen in the Royal Society's Muſeum* was only eleven inches long from noſe to the tail, and the
tail three inches. This probably was a young one.

Hiſt. Quad. Nº 260.—*Smellie,* iv. 346.

                                                                  44. Maryland.

MARMOT. With prominent dark eyes: ſhort rounded ears:
noſe ſharper-pointed than that of the laſt, and of a cinereous
colour: head and body of a brown color, which is lighter on the
ſides, and ſtill more ſo on the belly: the legs and feet duſky: toes
long, and divided: claws long, and ſharp: tail duſky, and buſhy;
half the length of the body: a ſpecific diſtinction from the other
kinds.

Size of a Rabbet.

* *Phil. Tranſ.* lxii. 378.

                                                                  Inhabits

Inhabits the temperate and warm parts of *North America*, from *Penſylvania* to the *Bahama Iſlands*. It feeds on fruits, berries, and vegetables. In the provinces it inhabits the hollows of trees, or burrows under ground, ſleeping for a month together. The *European* ſpecies continues dormant half the year: whether it takes a long ſleep in the warm climate of the *Bahamas* I am uncertain. It dwells there among the rocks, and makes its retreat into the holes on the approach of the hunters. In thoſe iſlands it is very fond of the berries of the *Ehretia Bourreria*, called there *Strong Back*. The fleſh is reckoned very good, but reſembles more that of a Pig than a Rabbet *.

It is called there the *Bahama* Coney. By Mr. *Edwards*, who figures one from *Maryland*, the *Monax*, or Marmot of *America* †.

45. HOARY.   *Hiſt. Quad.* Nº 261.—LEV. MUS.

MARMOT. With the tip of the noſe black: ears ſhort, and oval: cheeks whitiſh: crown duſky and tawny: hair in all parts rude and long; on the back, ſides, and belly, cinereous at the bottoms, black in the middle, and tipped with white, ſo as to ſpread a hoarineſs over the whole: legs black: claws duſky: tail full of hair, black and ferruginous.

Size of the preceding.

Inhabits the northern parts of *North America*.

46. TAIL-LESS.   *Hiſt. Quad.* Nº   —LEV. MUS.

MARMOT. With ſhort ears: color of the head and body a cinereous brown: the extremities of the hairs white: two cutting teeth above, four below: no tail.

About the ſize of the common Marmot.

Inhabits *Hudſon's Bay*.

* *Cateſby Carol.* ii. 79. *App.* xxviii.   † *Hiſt. Birds,* ii. 104.

Hiſt.

MARMOT. Without ears: face cinereous: back, and hind part of the head, of a light yellowifh brown; fometimes fpotted diftinctly with white, at others undulated with grey: belly and legs of a yellowifh white: tail about four inches and a half long. Length, from nofe to tail, about nine and a half. But there is a pygmy variety wholly yellow, and with a fhort tail, frequent near the falt lakes, between the mouths of the *Yaik* and the *Jemba.*

Inhabits *Bohemia, Auftria,* and *Hungary,* and in the *Ruffian* empire; begins to be common about the *Occa,* eaft of *Mofcow;* extends over all the temperate and open parts of *Sibiria,* and about *Jakutfk,* and in *Kamtfchatka.* It is alfo on the ifland of *Kadjak,* and was feen in great numbers by *Steller* on *Schamagin*'s ifles, almoft clofe on the fhore of *North America,* which give it place in this part of the work.

They burrow, and fink the pipes to their retreats obliquely, and then winding; and at the end is an arched oblong chamber, a foot in diameter, ftrewed with dried grafs. The entrances, or pipes, of the males are of greater, and thofe of the females of leffer, diameters. Towards winter they make a new pipe to their neft, but that only reaches to the turf; and with the earth which is taken out they fill up the fummer pipe.

They live entirely in a ftate of folitude, unlefs in the amorous feafon, when the females are found in the fame burrows with the males; but they bring forth in their own burrows, and by that means prevent the males from deftroying the young, as they cannot enter by reafon of the narrownefs of the pipes, the males being fuperior in fize to their mates. They fleep all night; but in the morning quit their holes, efpecially in fine weather, and feed and fport till approach of night. If the males approach one another, they fight fharply. The females often fet up a very fharp whiftle; the males are, for the moft part, filent. At the fight of a man, they

Q   inftantly

inftantly run into their burrows; and are often feen ftanding upright, and looking about them, as if on the watch : and if they fpy any body, give a loud whiftle, and difappear.

They are very eafily tamed, and become very fportive and amufing; and are very fond of being ftroked and cherifhed. In this ftate they will eat grain, and many forts of herbs. In a wild ftate they prey on mice, and fmall birds, as well as vegetables. *Gmelin* fays, that in *Sibiria* they inhabit granaries; but I do not find it confirmed by Doctor *Pallas*. *Gmelin* adds, that thofe who frequent granaries, feek for prey during the whole winter * : as to the others, they certainly remain torpid all the fevere feafon, and revive on the melting of the fnows.

They bring forth from three to eight at a time. The young grow very quick, and defert the maternal burrows in the fummer.

Their enemies are all forts of Weefels, which dig them out of their holes. More males than females perifh, as the latter are fiercer, and defend themfelves much better. During day they are fnatched up by hawks and hungry crows.

In fome places they are taken in fnares, for the fake of their fkins, which are ufually fent to *China*. The *Kamtfchatkans* make moft elegant garments and hoods of them; fpecimens of the latter are preferved in the *Leverian Mufeum*. In *Sibiria* their flefh is efteemed a great delicacy, efpecially in autumn, when they are a lump of fat.

The *Ruffians* call them *Suflik*; the *Sibirians*, *Jevrafcha*, and *Jemuranka*; the *Kamtfchatkans*, *Syræth*.

* *Voyage en Sibirie,* i. 378,

A. Bobak, *Hist. Quad.* N° 262.—*Smellie*, vii. 198.

MARMOT. With short oval thick ears: small eyes: upper part of the body greyish, mixed with long black and dusky hairs, tipt with grey: throat rust-colored: rest of the body and inside of the limbs yellowish rust: four claws on the fore feet, and a short thumb furnished with a strong claw: five toes behind: tail short, slender, and full of hair. Length sixteen inches: of the tail five.

Inhabits *Poland*, the *Ukraine*, *Tartary*, *Sibiria*, and even in plenty in *Kamtschatka*.

Its manners most amply described in the History of Quadrupeds.

SQUIRREL.   *HIST. QUAD.*   G e n u s XXX.

N. B. The ears of the *American* Squirrels have no tufts.

48. HUDSON.   Hudfon's Bay Squirrel, *Hiſt. Quad.* Nº 274.—LEV. MUS.

SQUIRREL. Of a ferruginous colour, marked along the top
of the back with a line of a deeper hue: belly of a pale aſh-color, mottled with black, and divided from the ſides by a duſky line:
tail ſhorter and leſs buſhy than that of the *European* kind ; of a ruſt-
color, barred, and ſometimes edged with black.

Inhabits the pine-forefts of *Hudſon's Bay* and *Labrador :* live upon
the cones: keep in their nefts the whole winter. Are found as high
as the Copper river ; yet do not change their colors by the ſeverity
of the winter, like the *Petits gris* of northern *Europe* and *Aſia*, from
which they form a diſtinct ſpecies.   I know of only one exception in
change of color in thoſe of *America*, Sir *Aſhton Lever* being poſſeſſed
of a ſpecimen of a milky whiteneſs ; but he did not know from
what part of the continent it came.

Carolina *. With the head, back, and ſides, grey, white, and ferruginous, intermixed: belly white: the color divided from that of
the ſides by a ruſt-coloured line: lower part of the legs red: tail
brown, mixed with black, and edged with white.

Both theſe are rather leſs than the *European* Squirrels.

49. GREY.   *Hiſt. Quad.* Nº 272.—*Smellie*, v. 321.—LEV. MUS.

SQUIRREL. With hair of a dull grey color, intermixed with
black, and frequently tinged with dull yellow: belly white. But
they vary, the body being ſometimes of a fine uniform grey.

* Leſſer Grey Squirrel, *Hiſt. Quad.* p.

This

This is the largeſt of the genus, and grows to half the ſize of a Rabbet.

In *America* I do not diſcover this animal farther north than *New England* [*] ; from whence they are found in vaſt numbers as far ſouth as *Louiſiana* [†]. Theſe, and the other ſpecies of Squirrels, are the greateſt peſts to the farmers of *North America*. They ſwarm in ſeveral of the provinces, and often deſcend in troops from the back ſettlements, and join the reſt in their ravages on the plantations of mayz, and the various nuts and maſt which that fertile country produces.

Thoſe which migrate from the mountains generally arrive in autumn ; inſtantly clear the ground of the fallen acorns, nuts, and maſt, and form with them magazines for their winter proviſions, in holes which they dig under ground for that purpoſe. They are often robbed of their hoards ; for the coloniſts take great pains to find them out ; and oft-times the hogs, which rove about the woods, root up and devour their magazines. It is from theſe that they ſupply themſelves, from time to time, with proviſions, quitting their neſts, and returning with a ſufficient ſtock to laſt them for ſome ſpace ; it being obſerved, that during winter they do not care to quit their warm retreat, unleſs on a viſit to their ſtorehouſes ; therefore, whenever they are obſerved to run about the woods in greater numbers than uſual, it is a certain ſign of the near approach of ſevere cold ; for inſtinct directs them to lay in a greater ſtock than uſual, leaſt the inclemency of the weather ſhould deprive them of acceſs to their ſubterraneous magazines.

The damage which they do to the poor planters, by deſtroying the mayz, is incredible. They come by hundreds into the fields, climb up the ſtalks, and eat the ſweet corn wrapt up in the heads, and will deſtroy a whole plantation in a night. For this reaſon they were proſcribed. In ſome places the inhabitants were, each, obliged annually to bring in four Squirrels heads. In others, a ſum was given,

[*] *Joſſelyn's voy.* 86.   [†] *Boſſu*, i, 361.

about

about three pence, for every one that was killed. This proved such
an encouragement, as to set all the idle people in the province in
pursuit of them. *Pensylvania* paid, from *January* 1749 to *January*
1750, 8000l. currency : but on complaint being made by the depu-
ties, that their treasuries were exhausted by these rewards, they were
reduced to one half. How improved must the state of the *Americans*
then be, in thirty-five years, to wage an expensive and successful war
against its parent country, which before could not bear the charges
of clearing the provinces from the ravages of these insignificant animals !

It has been observed, that the Squirrels are greatly multiplied
within these few years, and that in proportion to the encrease of
the fields of mayz, which attract them from all parts; I mean not
only the grey species, but all the others.

They are eaten by some people, and are esteemed very delicate.
Their skins, in *America*, are used for ladies shoes ; and are often imported into *England*, for lining or facing for cloaks.

They make their nests in hollow trees, with moss, straw, wool, and
other warm materials. They chiefly inhabit trees of the deciduous
kind; but sometimes in pines, whose cones are an article of their
provision. They keep their nests for several days together, seldom
stirring out, except for a fresh supply of food. Should a deep snow
prevent them from getting to their storehouses, multitudes perish
with hunger.

When they are sitting on a bough, and perceive a man, they instantly move their tails backward and forward, and gnash their teeth
with a very considerable noise. This makes them detested by the
sportsmen, who lose their game by the alarm they give. The GREY
Squirrel is a difficult animal to kill : it sits on the highest trees, and
often between the boughs, and changes its place with such expedition that the quickest marksman can scarcely find time to level his
piece ; and if it can once get into a hole, or into any old nest, nothing
                                                                    can

can provoke it to get out of its afylum. They run up and down the bodies of trees, but very rarely leap from one to the other.

They are eafily made tame; will even be brought to play with cats and dogs, which in a ftate of domefticity will not hurt them. They will alfo attach themfelves fo far as to follow children to and from the woods.

They agree in their manner of feeding with the *European* kinds; and have all the fame fort of attitudes.

SQUIRREL. With coarfe fur, mixed with dirty white and black: the throat, and infide of the legs and thighs, black: the tail is much fhorter than is ufual with Squirrels, and of a dull yellow, mixed with black: in fize equal to that of the Grey.

β. CAT.

Inhabits *Virginia*. Mr. *Knaphan*, in whofe collection I found it, informed me, that the planters called it the *Cat* Squirrel.

I fufpect that this animal is only a variety. *Lawfon* fays, that he has feen the Grey fpecies pied, reddifh, and black; but this point muft be determined by natives of the countries which they inhabit, who, from obfervation, may decide by their manners, or their colors, in different feafons, or periods of life.

Hift. Quad. N° 273.—*Brown's Zoology*, tab. xlvii.—LEV. MUS.

50. BLACK.

SQUIRREL. With white ears, nofe, and feet: the body totally black: the tail black, tipt with white: in fize equal to the former.

Thefe fometimes vary: there being examples of individuals which are wholly deftitute of any white marks. The beautiful figure of one of thefe animals from *Eaft Florida*, in Mr. *Brown's* Zoology, has ears edged with white, and a much longer tail than ufual.

* *Hift. Carol.* 124.

Q 4

Inhabits

Inhabits neither *Hudson's Bay* nor *Canada*, but is found in moſt other parts of *America*, as far as *Mexico* *. It is equally numerous, and as deſtructive to the mayz as the Grey Squirrel, but breeds and aſſociates in troops ſeparate from that ſpecies †; yet makes its neſt in the ſame manner, and like it forms magazines of proviſion againſt the ſevere ſeaſon.

In *Mexico*, and probably in other parts of *America*, they eat the cones of pine-trees; and lodge in the hollows of the trees.

A. With membranes from leg to leg.

51. FLYING.   *Hiſt. Quad.* Nº 283.—*Smellie*, v. 307.—LEV. MUS.

SQUIRREL. With large black eyes: round and naked ears: a membrane, covered with hair, extending from the fore to the hind legs: the hairs on the tail diſpoſed flatways on each ſide: are long in the middle, ſhort towards the end, which tapers to a point: that and the body of a browniſh cinereous: the belly white, tinged with yellow.

Inhabits all parts of *North America*, and as low as *Mexico*, where it is called *Quimichpatlan* ‡. The natives of *Virginia* named it *Aſſa-panic* ‖.

They live in hollow trees. Like the *Dormouſe*, they ſleep the whole day; but towards night creep out, and are very lively and active. They are gregarious, numbers being found in one tree. By means of the lateral membranes, they take prodigious leaps, im-properly called flying; and can ſpring ten yards at an effort. When they would leap, they extend the hind legs, and ſtretch

* Is the *Quauhtechallotliltic*, or *Tlilacotequillin*, of the *Mexicans*. *Fernandez*, 8.
† *Catesby*, ii. 73.   ‡ *Fernand. Nov. Hiſp.* 8.   ‖ *Smith's Virginia*, 27.

2

out the intervening fkin, which producing a larger furface, makes the animals fpecifically lighter than they would otherwife be: even with all this advantage, they cannot keep in a ftrait line, but are urged downward with their weight. Senfible of this, they mount the tree in proportion to the diftance of the leap they propofe to take, leaft they fhould fall to the ground before they had reached a place of fecurity.

They never willingly quit the trees, or run upon the ground being conftant refidents of the branches. They go in troops of ten or twelve, and feem in their leaps, to people unaccuftomed to them, like leaves blown off by the wind.

They bring three or four young at a time. They ufe the fame food, and form their hoards like other Squirrels.

They are very eafily tamed, and foon grow very familiar. They feem of a tender nature, and to love warmth, being very fond of creeping to the fleeve or pocket of the owner. If they are flung down, they fhew their diflike to the ground, by inftantly running up and fheltering themfelves in his cloaths.

Hift. Quad. Nº 284.

52. HOODED.

SQUIRREL. With the lateral fkins beginning at the ears, uniting under the chin, and extending, like thofe of the former, from fore leg to hind leg: the ears naked, and rather long: the hairs on the tail difpofed horizontally.

The color of the upper part of the body reddifh: the lower part cinereous, tinged with yellow.

This fpecies, according to *Seba*, who is the only perfon who has defcribed or figured it, came from *Virginia* *. *Linnæus* is very confufed in his fynonyms of this and the former kind; that of Mr. *Edwards* refers to the other fpecies; and that of *Seba*, in his article of *Sciurus Volans*, to both †.

* *Seb.Muf.* i. tab. xliv. p. 72.

† *Syft. Nat.* 85, where he calls it *Mus Volans*; and p. 88, where he ftyles it *Sciurus.*

R It

It is fingular that there fhould be only one fpecimen ever brought of this fpecies, from a country we have had fuch great intercourfe with. It may perhaps be a monftrous variety, by the extent of the fkin into a fort of hood.   As to color, that is an accidental difference, which happens to numbers of other animals.

53. SEVERN RIVER.

Hift. Quad. N° 282.
Greater Flying Squirrel, *Ph. Tr.* lxii. 379.

SQUIRREL.   With the hair on the body and fides of a deep cinereous color at the bottom ; the ends ferruginous : breaft and belly of a yellowifh white : the whole coat long and full : the tail thick of long hairs, difpofed in a lefs flatted manner than thofe of the *European* kind ; brown on the upper fide, darkeft at the end ; the lower part of the fame color with the belly : the lateral fkin, the inftrument of flight, difpofed from leg to leg, in the fame manner as in the firft fpecies, N° 51.

In fize it is far fuperior to the common Flying Squirrel, being at leaft equal to the *Englifh* kind.

This fpecies is found in the fouthern parts of *Hudfon's Bay*, in the forefts of the country bordering on *Severn* river in *James's* bay.

A. COMMON. *Hift. Quad.* N° 266.—*Smellie,* iv. 268.—LEV. MUS.

SQUIRREL.   With tufted ears : head, body, and legs, ferruginous : breaft and belly white : tail reddifh brown.

This fpecies inhabits the northern world, as high as *Lapmark* ; is

con-

continued through all the Arctic countries, wherefoever wood is found; abounds throughout *Sibiria*, except in the north-eaft parts, and in *Kamtfchatka*, where it ceafes, by reafon of the ceffation of forefts.

In all thefe countries they are red in fummer, but at approach of winter change to various and elegant greys. Their furs are of exquifite foftnefs, and are the forts known to us by the name of *Petit Gris*. In the more foutherly parts of thefe cold climates, they retain a tinge of the fummer red, and are lefs valuable. The change of color is effected gradually, as is its return in fpring to its ferruginous coat.

It is very fingular, that the alteration is not only performed in the feverity of the open air, but even in the warmth of a ftove. Dr. *Pallas* made the experiment on one which was brought to him on the 12th of *September*, and was at that time entirely red. About the 4th of *October* many parts of the body began to grow hoary; and at the time it happened to die, which was on the 4th of *November*, the whole body had attained a grey color, and the legs, and a fmall part of the face, had alone the reddifh tinge *.

The varieties are as follow:—A blackifh one, with the fur footy tipt with red, and full black gloffy tail, are common about lake *Baikal*, and the whole courfe of the *Lena*. Sir *Afhton Lever* is in poffeffion of one of a jetty blacknefs, with a white belly: its ears, as well as thofe of all the *Petit Gris*, are adorned with very long tufts. Thefe change in winter to a lead-color, and are taken in the thick *Alpine* forefts, where the *Pinus Cembra*, or Stone Pines, abound. The fkins of thefe are neglected by the *Chinefe*, but greatly efteemed in *Europe*, efpecially the tails, for facings of dreffes.

This variety is obferved fometimes to migrate in amazing numbers from their lofty alpine abodes, compelled to it whenever there happens to be a fcarcity of provifion †. Swarms have appeared even

* *Nov. Sp. Quad.* 373. † *Nov. Sp. an.* 188.

in the town of *Tomſk*, in deſerted houſes, and in the towers of the fortifications ; where numbers are taken alive, and of great ſize, by the children of the place.

A beautiful and large variety, about the *Baraba*, called the *Teleutian*, is in great eſteem for its beautiful grey color, like a Gull's back, with a ſilvery gloſs, and finely undulated. Their ſummer color is uſually duſky red, and the ſides and feet black. Theſe are highly eſteemed by the *Chineſe*, and ſell at the rate of 6 or 7*l.* ſterling per thouſand *

A ſmall variety of this, leſſer even than the common kind, is met with about the neighborhood of the *Kaſym* and *Iſet*.

A variety is alſo met with which change to a white color; and others again retain a white color both in winter and ſummer.

The late navigators to the Icy ſea brought home with them from *Pulo Condor*, a knot of iſlands in north lat. 8. 40. on the coaſt of *Cambodia*, a Squirrel totally black.

B.   EUROPEAN FLYING SQUIRREL, *Hiſt. Quad.* Nᵒ 285.—*Smellie,* v. 307.—LIV. MUS.

SQUIRREL. With naked ears: flying membranes extending from the middle of the hind legs to the baſe of the fore feet, and ſpreading there in a rounded ſail: tail full of hair, and round at the end: color of the upper part of the body a fine grey, like that of a Gull's back: the lower part white.

Length to the tail four inches and a quarter; of the tail, five.

Inhabits the birch-woods of *Finmark, Lapland, Finland, Lithuania,* and *Livonia*. Is found in *Aſia*, in the woods of the *Urallian* chain, and from thence to the river *Kolyma*. Neſtles in the hollows of trees remote from the ground, where it makes its neſt of the ſofteſt moſſes. Is always ſolitary, except in the breeding-ſeaſon, and never appears.

* *Mem. Ruſſ. Aſiat.* vii. 124.

în

in the day-time. Lives on the buds and catkins of the birch, and on the fhoots and buds of pines, which give its juices a ftrong refinous fmell; and its excrements will burn ftrongly, with a pitchy fcent. The laft are always found at the root of the tree, as if the animal defcended to eafe nature. It feldom comes out in bad weather; but certainly does not remain torpid during winter; for it is often taken in the traps laid for the Grey Squirrels. The fkins are often put up in the bundles with the latter, fo that the purchafer is defrauded, as their fur is of no value. They leap at vaft diftances from tree to tree, and never defcend but for the purpofe before mentioned. By reafon of fimilitude of color between them and the birch bark, they are feen with great difficulty, which preferves them from the attacks of rapacious birds.

They bring forth two, three, and rarely four, young at a time. When the parent goes out for food, fhe laps them carefully up in the mofs. They are very difficult to be preferved, and feldom can be kept alive, by reafon of want of proper food. They are born blind, and continue fo fourteen days. The mother pays them great attention; broods over them, and covers them with its flying membrane. The *Ruffians* call them *Ljetaga,* or the *Flying.*

HIST.

DORMOUSE.   *HIST. QUAD.*   Genus XXXI.

54. Striped.   Ground Squirrel, *Hiſt. Quad.* N° 286.—*Smellie*, v. 329.—Lev. Mus.

DORMOUSE. With naked rounded ears: the eyes full and black; about them a whitiſh ſpace: the head, body, and tail, of a reddiſh brown, deepeſt on the laſt: from neck to tail a black line extends along the top of the back: on each ſide run two others, parallel to the former, including between them another of a yellowiſh white: breaſt and belly white: the toes almoſt naked, and of a fleſh-color; long, ſlender, and very diſtinct; four, with the rudiment of a fifth, on the fore feet; five perfect toes on the hind.

Size.   The length is about five inches and a half; of the tail, to the end of the hairs, rather longer.

Inhabits all parts of *North America*, I think, from *Hudson's Bay* to *Louiſiana*; certainly from *Canada*, where the *French* call them *Les Suiſſes*, from their ſkins being rayed with black and white, like the breeches of the *Switzers* who form the Pope's guard *.

They are extremely numerous: live in woods, yet never run up trees, except when purſued, and find no other means of eſcape. They live under ground, burrow, and form their habitations with two entrances, that they may ſecure a retreat through the one, in caſe the other ſhould be ſtopped. Theſe little animals form their ſubterraneous dwellings with great ſkill, working them into the form of long galleries, with branches on each ſide, every one terminating in an enlarged apartment, in which they hoard their ſtock of winter proviſion †. Their acorns are lodged in one, in a ſecond the mayz, in a third the hickery-nuts, and in the laſt their moſt favorite food,

* *Charlevoix*, v. 198.   † *Kalm*, i. 322. 325.

the

the *Chinquaquin*, or chefnut. Nature has given to them, as to the Hamfter *, a fine conveniency for collecting its provifions, having furnifhed them with pouches within their cheeks, which they fill with mayz, and other articles of food, and fo convey them to their magazines.

Thofe of *Sibiria* live chiefly on feeds, and particularly on the kernels of the *Cembra*, or Stone Pine; and thefe they hoard up in fuch quantities, that ten or fifteen pounds of the moft choice have been found in a fingle magazine †.

They pafs the whole winter either in fleep or in eating. During the fevere feafon, they very rarely ftir out, at left as long as their provifions laft; but if by an unexpected continuance of bad weather their provifions fail, they will then fally out, and dig under ground in cellars where apples are kept, or into barns where *mayz* is ftored, and make great devaftations. They will even enter houfes, and eat undifmayed, before the inhabitants, any corn they chance to meet with ‡. The Cat makes great havock among them, being at all feafons as great an enemy to them as to domeftic Mice. It is hunger alone that tames them. They are naturally of a very wild nature, will bite moft feverely, and cannot by any means be rendered familiar.

They are remarkably nice in the choice of their food, when the variety of autumnal provifions gives opportunity. They have been obferved, after having ftuffed their pouches with the grains of rye, to fling it out when they meet with wheat, and to replace the rye with the more delicious corn.

Their fkins, form a trifling article of commerce, being brought over among *le menue pelliterie*, the fmall furs, and ufed for the lining of ladies cloaks.

In *Sibiria* they are killed with blunt arrows, or caught in fall-traps. About the *Lena*, the boys go out in the amorous feafon of thefe little animals, and, ftanding behind a tree, mimic the noife of the females, which brings the males within reach of their fticks, with which

---

* *Hift. Quad.* N° 324.   † *Pallas, Nov. Sp. an.* 379.   ‡ *Du Pratz*, ii. 68.

they

they kill them.   The ſkins are ſold to the *Chineſe* merchants. About the *Lena*, a thouſand of their ſkins are not valued at more than ſix or eight rubles *.

Theſe animals are found in great numbers in *Aſia*, beginning about the river *Kama* †, and from thence growing more and more frequent in the wooded parts of *Sibiria*; but theſe, and all the ſpecies of Squirrel, ceaſe towards the north-eaſt extremity of the country, by reaſon of the interruption of woods, which cuts them off from *Kamt-ſchatka*.

55. ENGLISH ?   Dormouſe, *Br. Zool.* i. N° 234.—*Hiſt. Quad.* N° 289.—*Smellie*, iv. 334.—Lev. Mus.

Mr. *Lawſon* ſays that the *Engliſh* Dormouſe is found in *Carolina*; but it has not as yet been tranſmitted to *Great Britain*.   In order to aſcertain the ſpecies, I add a brief deſcription.

DORMOUSE.   With full black eyes: broad, thin, ſemi-tranſparent ears: throat white: reſt of the body and the tail of a tawny red.   Size of the common Mouſe; but the body of a plumper form, and the noſe more blunt: tail two inches and a half long, covered on every ſide with hair.

In *Europe*, inhabits thickets; forms its neſt at the bottom of a tree or ſhrub; forms magazines of nuts for winter food; ſits up to eat, like the Squirrel; lies torpid moſt of the winter; in its retreat, rolled up into the ſhape of a ball; retires to its neſt at approach of cold weather.

* *Pallas, Nov. Sp. an.* 380.
† A river falling into the *Wolga* about forty miles below *Caſan*.

HIST.

HIST. QUAD. Genus XXXIII.

Br. Zool. i. N° 25.—*Hiſt. Quad.* N° 297.—*Smellie,* iv. 275.—Lev. Mus.

56. BLACK.

RAT. With head and body of a very deep iron grey, nearly black: belly aſh-colored: legs almoſt naked, and duſky: a claw inſtead of a fifth thumb on the fore feet. Length, from noſe to tail, ſeven inches; tail near eight.

Inhabits *North America,* not only the ſettled parts, but even the rocks of the *Blue* mountains *, remote from all human dwellings. There they live among the ſtones, and in the ſubterraneous grottos frequent in thoſe hills. They lie cloſe during day, but at night come out, and make a moſt horrible noiſe amidſt theſe *American* alps. In violent froſts they lie torpid; and in the inhabited parts of the country they are obſerved to redouble their ſcreaks before ſevere weather, as if they had ſome preſage from their conſtitutions.

By Mr. *Bartram's* obſervations it appears very certain, that theſe animals are natives of *America.* They are found even at preſent in the moſt deſolate places, as well as in the houſes and barns of the inhabited parts. It is unknown in *Europe,* that either the common Rat or Mouſe ever deſerted the haunts of mankind, for rocks and deſerts: they therefore have been there from the earlieſt times. It is likely, that if ever the *Blue* mountains become inhabited, the wild Rats will quit their rocks, and reſort to thoſe places where they find harveſted food, and will quickly become perniciouſly domeſtic.

We are poſitively told, that *South America* was free from theſe troubleſome animals, till they were introduced there from *Europe,* by the means of ſhips, in 1544 †.

* *Bartram,* as quoted by Mr. *Kalm,* ii. 47.   † *Garcilaſſo de la Vega,* 384.

S   We

We find none of the fpecies in *Kamtfchatka*, nor any where to the eaft of the *Urallian* chain. *America* muft therefore have been ftocked with them from the fide of *Europe*. They are very common in *Ruffia*. Towards *Aftracan* they are very fmall, but of the fame color with the others.

57. BROWN.   *Br. Zool.* i. N° 26.—*Hift. Quad.* N° 298.—*Smellie,* iv. 336.—LEV. MUS.

R AT. Above, light brown mixed with tawny, dufky, and cinereous: below, of a dirty white: four toes before, and a claw inftead of a fifth toe.

I have no authority for giving this fpecies a place here: but muft fuppofe that the new world could not poffibly efcape the peft, as it is univerfally become a moft deftructive inhabitant of *European* fhips.

58. AMERICAN.   *Hift. Quad.* N° 299.
              Characho, *Jike Cholqomac,* or Great Moufe of the *Monguls.*—LEV. MUS.

R AT. With the upper jaw extending very far beyond the lower: ears large and naked: tail rather fhorter, in proportion, than that of the Black, to which it is rather fuperior in fize: color a dufky brown.

The fpecimen, from which this defcription is taken, was fent from *North America* to Sir *Afhton Lever;* but I am not informed, whether it only frequented the deferts, or infefted houfes.

59. WATER.   *Br. Zool.* N°300.—*Hift. Quad.* N° 300.—*Smellie,* iv. 290.—LEV. MUS.

R AT. With fmall eyes: ears covered with the fur: teeth yellow: body covered with long hairs, black mixed with a few of a ruft-color: belly of a deep grey.

Length

Length of the head and body feven inches; tail five, covered with fhort black hairs. Weight nine ounces.

Inhabits *North America*, from *Canada* to *Carolina* *. In the firft, varies to tawny and to white †: feeds on the fry of fifh, infects, fhell-fifh, frogs, and roots; burrows on the banks of ponds and rivers; and dives and fwims as well as an Otter, notwithftanding it is not web-footed.

In northern *Europe* and *Afia* it is extremely common; from *Peterf-burgh* to *Kamtfchatka* in *Sibiria*, they are twice as large as in other places. They are found alfo from *Lapland* to the *Cafpian* fea, and alfo in *Perfia*; and are one of the animals which endure the cold of the Arctic circle.

Br. Zool. i. Nº 30.—*Hift. Quad.* Nº 301.—*Smellie*, iv. 282.—Lev. Mus.   60. Mouse.

THIS common animal needs no defcription. It is very abun-dant in the inhabited parts of *America* ‡, and is to be found from *Peterfburgh* perhaps as far as *Kamtfchatka*.

Kalm imagines them to be natives of *America*; for he affures us that he has killed them in the crevices of the rocks in defert places, far from the haunt of man ∥.

Hift. Quad. Nº 302. α. AMERICAN.—*Smellie*, iv. 285.—Lev. Mus.   61. Field.

RAT. With great, naked, and open ears: cheeks, fpace below the ears, and fides quite to the tail, orange-colored: back dufky and ruft-colored, marked along the top, from head to tail, with a dark line: throat, breaft, and belly, of a pure white: tail dufky above, white beneath: feet white: hind legs longer than thofe of the *Englifh* kind.

Length about four inches and a half; of the tail, four inches.

Inhabits *Hudfon's Bay* and *New York*.

---

* *Lawfon Carolina*, 122.   † *De Buffon*, xiv. 401.   ‡ *Kalm*, ii. 46.
∥ The fame, 47.

Hift.

62. VIRGINIAN.

Hift. Quad. N° 307.

RAT. With a black nofe: fur fhort, and in all parts white: limbs flender: tail very thick at the bafe, tapering to a point, and cloathed with long hair.

Seba alone, vol. i. p. 76. tab. xlvii. fig. 4, defcribes this fpecies.

63. LABRADOR.

Hift. Quad. N° 295.

RAT. With a blunt nofe: mouth placed far below: upper lip bifid: ears large, naked, rounded: fore legs fhort, furnifhed with four toes, and a tubercle inftead of a thumb: HIND LEGS long and naked, like fome of the *Jerboas*: toes long, flender, and diftinct; the exterior toe the fhorteft: thumb fhort.

SIZE.

The whole length of the animal is eight inches, of which the tail is four and three quarters.

COLORS.

Color above a deep brown, beneath white, feparated on each fide by a yellow line.

Inhabits *Hudfon's Bay* and *Labrador*. Sent over by Mr. *Graham*.

** With fhort tails.

64. HUDSON's.

Hift. Quad. N° 319.—LEV. MUS.

RAT. With foft long hair, dufky at the bottom, whitifh brown at the points: along the middle of the back, from head to tail, runs a dufky line: fides yellowifh: belly and infide of the thighs of a dirty white.

Legs very fhort: on the toes of the fore feet of the male only are four very large and fharp claws, tuberculated beneath; in the female fmaller and weaker: on the hind feet five toes with flender claws.

Tail

Tail not three quarters of an inch long, terminating with long ftiff hairs; it is fcarcely vifible, being almoft loft in the fur.

Defcribed from a fkin which Doctor *Pallas* favored me with, which he received from the *Labrador* coaft.

This is nearly a-kin to the *Lemmus.*

Short-tailed Field Moufe? *Br. Zool.* i. Nº 31.—*Hift. Quad.* Nº 322 ?—*Smellie,* iv. 293.—LEV. MUS.

RAT. With a blunt nofe: great head: prominent eyes: ears buried in the fur: head and upper part of the body of a ferru-ginous brown mixed with black: belly of a deep afh-color.

Length, from head to tail, fix inches; tail only one and a half, with a fmall tuft at the end.

Inhabits *Hudfon's Bay* and *Newfoundland,* in the laft very nume-rous, and does vaft damage in the gardens; refides under ground.

Hift. Quad. Nº 320.

RAT. With fmall and rounded ears: head broad; color dufky and tawny brown: the belly of a dirty white: a dufky line paffes from between the eyes, and extends obfcurely along the back. Larger than the common Moufe. Defcribed from fo muti-lated a fpecimen, fent to the Royal Society from *Hudfon's Bay* *, that it was impoffible to determine the fpecies; only, by the dark line along the back, it feemed likeft the HARE-TAILED, an inhabitant of *Sibiria,* whofe manners are defcribed in the Hiftory of Qua-drupeds.

* *Ph. Tr.* lxii. 379. Sp. 15.

A. Œconomic, *Hiſt. Quad.* N° 313.—*Deſcr. Kamtſchatka,* Fr. ed. 392.

R AT. With naked ears, uſually hid in the fur: ſmall eyes: teeth tawny: limbs ſtrong: color, an intermixture of black and yellow, darkeſt on the back: under ſide hoary.

Length four inches and a quarter, to the tail; the tail one inch

Inhabits in vaſt abundance *Sibiria,* from the eaſt ſide of the *Urallian* chain, even within the Arctic circle, and quite to *Kamtſchatka.* It is the noted *Tegultſchitch* of that country, diſtinguiſhed by its curious œconomy and by its vaſt migrations.

They make their burrows with the greateſt ſkill, immediately below the ſurface of the ſoft turfy ſoil. They form a chamber of a flattiſh arched form, of a ſmall height, and about a foot in diameter, to which they ſometimes add as many as thirty ſmall pipes or entrances. Near the chamber they often form other caverns, in which they lodge their winter ſtores: theſe conſiſt of various kinds of plants, even ſome of ſpecies poiſonous to mankind. They gather them in ſummer, harveſt them, and even at times bring them out of the cells to give them a more thorough drying in the ſun. The chief labor reſts on the females. The males, during ſummer, go about ſolitary, and inhabit ſome old neſts; and in that ſeaſon never touch their hoards, but live on berries. They are monogamous, and the male and female at other times found in the ſame neſt. The female brings two or three young at a time, and breeds often in the year.

No little animals are fo refpected by the *Kamtfchatkans* as thefe, for to them they owe a delicious food; and with great joy, about autumn, rob the hoards, and leave there many ridiculous prefents by way of amends: they alfo never take the whole of their provifions, and leave befides a little dried ovaries of fifh for their fupport.

But the migrations of thefe Mice, in certain years, is as extraordinary a fact as any in natural hiftory: I will only mention thofe of *Kamtfchatka*. The caufe is unknown. Doctor *Pallas* thinks it may arife from the fenfations of internal fire in that vulcanic tract, or a prefcience of fome unufual and bad feafon. They gather together in the fpring in amazing numbers, except the few that are converfant about villages, where they can pick up fome fubfiftance. This makes it probable that the country is over-ftocked, and they quit it for want of food. The mighty hoft proceeds in a direct courfe weftward, and with the utmoft intrepidity fwims over rivers, lakes, and even arms of the fea: many are drowned, many deftroyed by waterfowl, or rapacious fifh; thofe which efcape reft awhile, to bafk, dry their fur, and refrefh themfelves. If the inhabitants find them in that fituation, they treat them with the utmoft tendernefs, and endeavour to bring them to life and vigor. As foon as they have croffed the river *Penfchim*, at the head of the gulph of the fame name, they turn fouthward, and reach the rivers *Judoma* and *Ochot* by the middle of *July*. The fpace is moft furprifing, on confulting the map of the country. The flocks are alfo fo numerous, that an obferver has waited two hours to fee them all pafs. Their return into *Kamtfchatka*, in *October*, is attended with the utmoft feftivity and welcome. The natives confider it as a fure prognoftic of a fuccefsful chafe and fifhery: the firft is certain, as the Mice are always followed by multitudes of beafts of prey. They equally lament their migration, as the feafon is certainly filled with rains and tempefts.

MIGRATIONS.

RED,

B. RED, *Hiſt. Quad.* N° 314.

R AT. With briſtly noſe and face: ears oval, riſing above the
hair, naked, only tipt with fur: color, from forehead to rump,
a bright red: ſides light grey and yellow: belly whitiſh: tail duſky
above, light below.

Length not four inches; tail more than one.

Grow very common beyond the *Ob*, and live ſcattered over all
Sibiria, in woods and mountains, and about villages; extend even
to the Arctic circle. It is the *Tſchetanauſtſchu*, or *Red Mouſe* of the
Kamtſchatkans. It is a ſort of drone: makes no proviſion for itſelf, but
robs the hoards of the laſt ſpecies * Lives under logs of trees; frequents houſes; dares the ſevereſt weather, and is abroad amidſt the
ſnows; feeds on any thing, and is often caught in the traps ſet for
Ermines, in attempting to devour the bait.

C. LEMMUS, *Hiſt. Quad.* N° 317.—Godde Saeppan, *Leems*, 224.

R AT. With ſmall eyes and mouth: upper lip divided: ears ſmall,
placed far backwards: four ſlender toes on the fore feet, and a
ſharp claw, like a cock's ſpur, in place of a thumb: ſkin very thin.
Color of the head black and tawny, of the belly yellow.

Length of thoſe of *Scandinavian Lapland*, above five inches; thoſe
of the *Ruſſian* dominions not four.

The manners and wonderful migrations of the *Lemmi* of *Europe*,
have been fully treated of in my Hiſtory of Quadrupeds.

They abound in the countries from the *White Sea* to the gulph of
the *Oby*, and in the northern end of the *Urallian* chain; but differ
in ſize and color from thoſe of *Europe*. Like them, they migrate at
certain periods; and tend from the *Urallian* mountains, ſometimes
towards *Jeneſei*, ſometimes towards *Petzorah*, and at thoſe times re-

* *Deſcr. Kamtſchatka*, 392.

joice

joice the *Samoieds* with a rich chafe of the animals which purfue the wanderers. The *Samoieds* affert, that the Rein-Deer will greedily devour them; perhaps they take them medicinally, as Sheep are known as greedily to feek and fwallow Spiders.

D. LENA, Mus *Gmelini, Pallas, Nov. Sp. an.* 195.

RAT. With fhort round ears: white whifkers: thick broad body, in all parts nearly of equal breadth: tail fhort, thickly covered with rude hairs: five toes on the fore feet, with claws very ftrong and white: four on the hind feet, with claws much weaker: the fur pretty long; three parts of its length, from the roots, cinereous, the reft white; fo that the animal appears entirely white, except the cheeks, which are afh-colored, and the chin, which is dufky.

The length is three inches one-fifth, the tail four-fifths of an inch.

They are feen in great numbers in autumn, on the borders of the Icy Sea, and about the parts of the *Lena* that fall into it. They appear fuddenly, and depart as expeditioufly. They feed on the roots of moffes, and are themfelves the food of *Arctic* Foxes. Perhaps they extend to the *Jenefei:* for it is faid that there are two forts of Mice found there; one wholly white; the other black, yellow, and white, which perhaps is the *Lemmus* *.

E. RINGED, *Hift. Quad.* Nº 205.

RAT. With a blunt nofe: ears hid in the fur: hair very fine: claws ftrong and hooked: color of the upper part, fometimes ferruginous, fometimes light grey undulated with deep ruft-color: a crefcent of white extends on each, from the hind part of the head towards the throat, bounded on each fide by a bed of ruft-color.

* *Nov. Sp. an.* 197.

T Length

Length to the tail little more than three inches; tail one, terminated by a briftly tuft.

Found in the *Arctic* neighborhood of the *Oby*. Makes its neft
with rein-deer and fnowy liver-worts, juft beneath the turfy furface.
Are faid to migrate, like the *Lemmus*.

F. TCHELAG, *Defcr. Kamtfchatka,* 392,

THE author of the defcription of that great peninfula fays no
more than that it is a very fmall fpecies; frequents houfes;
and will go out and eat boldly any thing it has ftolen. The natives
call it *Tchelagatchitch*.

HIST QUAD.   Genus XXXIX.   SHREW.

Br. Zool. i. N° 32.—*Hiſt. Quad.* N° 341.—*Smellie,* iv. 305.   67. FŒTID ?

SHREW. With the head and upper part of the body duſky: ſides of a browniſh ruſt-color: eyes very ſmall, almoſt hid in the fur: ears ſhort: noſe very long and ſlender: upper mandible extends far beyond the lower.

Inhabits *Hudſon's Bay,* and probably *Carolina,* as *Lawſon* mentions a Mouſe found there which poiſons Cats * if they eat it. It is a notion in *England* that they are venomous: it is notorious that our Cats will kill, but not feed on them ; probably thoſe of *America* have the ſame inſtinct: ſo that their deaths in the new world muſt ariſe from ſome other cauſe, and be falſely attributed to theſe animals.

Mr. *Graham* ſent over two other ſpecimens, beſides that deſcribed. They were of a duſky grey above, and of a yellowiſh white beneath : their ſize, rather leſs than the *Engliſh* kind ; one being only two inches and a quarter long, the other only two inches ; but they ſeemed not to differ ſpecifically from the other.

The common Shrew is found in *Ruſſia* ; in all parts of *Sibiria,* ever in the *Arctic* flats ; and in *Kamtſchatka.*

* *Hiſt. Carolina,* 125.

T 2   *HIST.*

MOLE.   *HIST. QUAD.*   G e n u s   XXXV.

68. L o n g   t a i l e d.   *Hift. Quad.* N° 352.—L e v. M u s.

MOLE. With two cutting teeth in each jaw, and two fharp
flender canine: the grinders fmall and fharp: nofe long, the
end radiated with fhort tendrils: fore feet not fo broad as thofe of
the *Englifh* Mole, furnifhed with very long white claws: toes on the
hind feet quite feparated: body not fo thick and full as that of
the common fpecies: hair long, foft, and of a rufty brown: tail covered with fhort hair.

Length of the body four inches two-tenths; of the tail, two and
a half.

Inhabits *North America.* Received from *New York.*

69. R a d i a t e d.   *Hift. Quad.* N° 351.—*Smellie,* iv. 316.—L e v. M u s.

MOLE. With a long nofe, radiated like the former: the
body fhorter, and more full: hair dufky, very long, fine, and
compact: fore feet refembling thofe of the preceding; but the toes
of the hind feet are clofely connected.

Length to the tail three inches three quarters: the tail flender,
round, and taper, one inch three-tenths long.

Received from *New York.*

M a n n e r s.   This fpecies forms fubterraneous paffages in the fields, running in
various directions, and very fhallow. Their courfe may be traced
by the elevation of the earth on the furface, in form of a little bank,
two inches high, and as broad as a man's hand. Thefe holes are
unable to fupport any weight, fo that walkers find it very trouble-

fome

fome to go over places where thefe animals inhabit, the ground perpetually breaking under their feet *.

Thefe Moles have all the ftrength in their legs as thofe of *Europe*,
and work in the fame manner. They feed on roots, are very iraf-
cible, and will bite very feverely.

Hift. Quad, N° 353.—Lev. Mus.

70. Brown.

MOLE. With a long and very flender nofe : two broad cutting
teeth in the upper, four fharp and flender in the lower, jaw ;
the two middlemoft fhort : the grinders very numerous, ftrong, fharp,
and feparate : the fore feet very broad ; thofe and the hind feet exactly like thofe of the *European* kind.

Length about fix inches ; tail one.

I received two fpecimens of this animal from *New York*. The
hair in both foft, filky, and gloffy : the hair in each dufky at the
bottom ; but in one, the ends were of a yellowifh brown ; in the
other, brown : the feet and tail of both were white. I fufpect that
they were varieties of the kind defcribed by *Seba* †, which he got
from *Virginia* : it was totally black, gloffed over with a moft refplendent purple. I may here note, the Tail-lefs Mole, figured by
Seba in the fame plate, is not a native of *Sibiria*, as he makes it ; but
is an inhabitant of the *Cape of Good Hope*.

PLACE.

Thefe three fpecies agree pretty nearly with the Shrew in the fore
teeth ; for which reafon *Linnæus* claffes the two he defcribes among the
Sorices. I call them Moles from their fhape, which differs not from
the *European* kind ; but thofe who chufe to be very fyftematic, may
divide the genus of Shrews, and ftyle thefe *Sorices Talpæ-formes*.

* *Kalm,* i. 190.   † P. 51. tab. xxxii.

A. European, *Hift. Quad.* ii. N°   *Br. Zool.* i.

MOLE. With fix cutting teeth in the upper; eight in the lower jaw; and two canine teeth in each: color of the fur black.

PLACE.   Inhabits *Sweden*; but does not extend farther than the fouth of *Norway*, where it is called *Vond*. Is frequent in the temperate parts of *Ruffia*, and even in *Sibiria*, as far as the *Lena*. In *Sibiria* it is twice as big as thofe of *Europe*. Is found there milk white, but more ufually fo in the *Verchoturian* mountains.

H E D G E - H O G, *Hift. Quad.* Genus XXXVI.

B. Common, *Hift. Quad.* ii. N° 355.—*Br. Zool.* i. N°

HEDGE-HOG. With noftrils bounded on each fide by a loofe flap: ears rounded: back covered with prickles, white, barred with black: face, fides, and rump, with ftrong coarfe hair: tail an inch long.

PLACE.   Is found in *Sweden*, in the diocefe of *Aggerhuys*; and in that of *Bergen*, in *Norway**. It is called, in the *Norwegian* tongue, *Buftedyvel*. Is common in *Ruffia*, except in the extreme northern and fouthern parts. None in *Sibiria*, or very fcarce at left.

* *Leems,* 229. *Pontoppidan,* ii. 28.

D I V. III.

PINNATED QUADRUPEDS;

Or, with FIN-LIKE FEET.

D I V. III.   Pinnated Quadrupeds;
Or, with FIN-LIKE FEET.

WALRUS.   *HIST. QUAD.*   GENUS XLI.

71. ARCTIC.   *Hiſt. Quad.* Nº 373.—*Phipps's voy.* 184.
                     Roſmarus, *Zimmerman,* 330.
                     Le Tricheque, *Schreber,* ii. 82. tab. lxxix.
                     Cheval Marin, *Hiſt. Kamtſchatka,* 427.—*Smellie,* vii. 354.—LEV. MUS.

DESCRIPTION.   WALRUS.   With a round head; ſhort neck; ſmall and fiery eyes, ſunk a finger's depth in the ſockets, and retractile from external injuries *: mouth very ſmall; lips very thick, beſet above and below with great whiſkers, compoſed of briſtles, tranſparent, and thick as a ſtraw: inſtead of ears are two minute orifices, placed in the moſt diſtant part of the head.

Body is very thick in the middle, leſſening gradually towards the tail.   The ſkin in general is an inch thick, and two about the neck †, and much wrinkled about the joints: it is covered with ſhort hair, of a mouſe-color; ſome with reddiſh, others with grey; others are almoſt bare, as if they were mangy, and full of ſcars ‡.

The legs are very ſhort; on each foot are five toes, connected by webs, with a ſmall blunt nail to each.   The hind feet, like thoſe of Seals, are very broad: the tail is very ſhort: the penis two feet long, and of a bony ſubſtance.

* *Crantz,* i. 126.   † *Crantz,* i. 125.   ‡ *Marten's Spitzberg.*

In

In the upper jaw are two very long tufks, bending downwards. **TEETH.**
No cutting teeth; but in each jaw, above and below, four grinders,
flat at top, and the furfaces of thofe which I examined much worn.
The length of the largeft tufk I have heard of, was two feet three
inches, *Englifh* meafure, the circumference at the lower end, eight
and a half; the greateft weight of a fingle tufk twenty pounds:
but fuch are rarely found, and only on the coafts of the *Icy* fea, where
they are feldom molefted, and of courfe permitted to attain their full
growth *.

The Walrus is fometimes found of the length of eighteen feet, **SIZE.**
and the circumference, in the thickeft part, ten or twelve. The
weight from fifteen hundred to two thoufand pounds.

Inhabits, in prefent times, the coafts of the *Magdalene* iflands, in **PLACE.**
the gulph of St. *Laurence*, between latitude 47 and 48, their moft **AMERICA.**
foutherly refidence in any part of the globe. They are not found
on the feas of *Labradore*. The *Efkimaux* purchafe the teeth, for
the heading their Seal-darts, from the *Indians* of *Nuckvank*, about lat.
60; who fay, that they are annually vifited in the winter by multitudes
of thefe animals †. They are found in *Davis's Streights*, and within *Hudfon's Bay* ‡, in lat. 62. They alfo inhabit the coaft of *Greenland*. I am uncertain whether they frequent *Iceland*; but they are
found in great numbers near the iflands of *Spitzbergen*, and on all the **SPITZBERGEN.**
floating ice from thence to *Cherry Ifle*, a folitary fpot intermediate **CHERRY ISLE.**
between the laft and the moft northerly point of *Norway*. In 1608,
they were found there in fuch numbers, huddled on one another,
like hogs, that a fhip's crew killed above nine hundred in feven
hours time §.

If they are found in the feas of *Norway*, it is very rare ‖ in thefe **NORWAY.**
days. *Leems*, p. 310, fays that they fometimes frequent the fea
about *Finmark*; but about the year 980, they feemed to have been
fo numerous in the northern parts, as to become objects of chafe and

* *Hift Kamtfchatka*, 120.   † *Ph. Tranf.* lxiv. 378.   ‡ *Ellis's voy.* 80.
§ *Martens Spitzberg.* 182.   ‖ *Pontoppidan*, ii. 157.

U   commerce

commerce. The famous *Octher* the *Norwegian*, a native of *Helge-land* in the diocefe of *Drontheim*, incited by a moſt laudable curioſity and thirſt of diſcovery, ſailed to the north of his country, doubled the *North Cape*, and in three days from his departure arrived at the fartheſt place, frequented by the *Horſe-whale* fiſhers. From thence he proceeded a voyage of three days more, and perhaps got into the White Sea. On his return he viſited *England*, probably incited by the fame of King *Alfred's* abilities, and the great encouregement he gave to men of diſtinguiſhed character in every profeſſion. The traveller, as a proof of the authenticity of his relation, preſented the *Saxon* monarch with ſome of the teeth of theſe animals, then a ſub-ſtitute of ivory, and valued at a high price. In his account of his voyage, he alſo added that their ſkins were uſed in the ſhips inſtead of ropes *.

NOVA ZEMBLA, AND ICY SEA.

They are found again on the coaſts of *Nova Zembla*, and on the headlands which ſtretch moſt towards the north Pole; and as far as the *Tſchutki* point, and the iſles off that promontory. They ſcarcely extend lower than the neighborhood of the country of the *Anadyr*, but are ſeen in great abundance about cape *Newnham*, on the coaſt of *America*. The natives of the iſlands off the *Tchutki Noſs* ornament themſelves with pieces of the Walrus ſtuck through their lips or noſes; for which reaſon they are called by their neighbors *Zoobatee*, or *large-teethed* †. The natives about *Unalaſcha*, *Sandwich Sound*, and *Turn-again* river, obſerve the ſame faſhion. I entertain doubts whether theſe animals are of the ſame ſpecies with thoſe of the Gulph of St. *Laurence*. The tuſks of thoſe of the Frozen Sea are much longer, more ſlender, and have a twiſt and inward curvature.

MANNERS.

They are gregarious, and ſometimes have been found together in thouſands; are very ſhy, and avoid the haunts of mankind. They uſually are ſeen on the floating ice, preferring that for their reſidence, as their bodies require cooling, by reaſon of the heat which ariſes from their exceſſive fatneſs ‡.

* *Hackluyt*, i. 5.　　† *Hiſt. Kamtſchatka*, 47.　　‡ *Nov. Com. Petrop.* ii. 291.

They

They are monogamous; couple in *June*, and bring forth in the earlieſt ſpring *. They bring one †, or very rarely two young at a time; feed on ſea-plants, fiſh, and ſhells, which they either dig out of the ſand, or force from the rocks with their great teeth. They make uſe alſo of their teeth to aſcend the iſlands of ice, by faſtening them in the cracks, and by that means draw up their bodies.

They ſleep both on the ice and in the water, and ſnore exceſſively loud ‡.

They are harmleſs, unleſs provoked; but when wounded, or attacked, grow very fierce, and are very vindictive. When ſurpriſed upon the ice, the females firſt provide for the ſafety of the young, by flinging it into the ſea, and itſelf after it, carrying it to a ſecure diſtance, then returning with great rage to revenge the injury. They will ſometimes attempt to faſten their teeth on the boats, with an intent to ſink them, or riſe in numbers under them to overſet them; at the ſame time they ſhew all marks of rage, by roaring in a dreadful manner, and gnaſhing their teeth with great violence; if once thoroughly irritated, the whole herd will follow the boats till they loſe ſight of them. They are ſtrongly attached to each other, and will make every effort in their power, even to death, to ſet at liberty their harpooned companions ‖.

A wounded Walrus has been known to ſink to the bottom, riſe ſuddenly again, and bring up with it multitudes of others, who united in an attack on the boat from which the inſult came §.

They fling the water out of their noſtrils, as the Whale does out of its head. When chaſed hard, they commonly vomit, and fling up ſmall ſtones. Their dung is like that of a Horſe, and exceſſively fetid, eſpecially where they are found in large companies.

The tongue, which is about the ſize of a Cow's, may be eaten, if boiled freſh; but if kept, ſoon runs into oil. The teeth uſed to be applied to all the purpoſes of ivory; but the animals are now killed

Uſes.

---

* *Faun. Greenl.* 4.　　† *Barentz*, 4.　　‡ *Martens*, 109.　　‖ *Martens*, 110.
§ *Phipps's voy.* 57.

　　　　　　　　only

only for the fake of the oil. Seamen make rings of the briftles of the whifkers, which they wear as prefervatives againft the cramp. The *French* coach-makers have made traces for coaches of the fkins, which are faid to be ftrong and elaftic *. The *Ruffians* formerly ufed the bone of the penis pulverifed, as a remedy againft the ftone †. *Bartholinus* ‡ recommends it, infufed in ale, in fits of the ftrangury. The *Greenlanders* eat the flefh and lard, and ufe the laft in their lamps. Of the fkin they make ftraps. They fplit the tendons into thread; and ufe the teeth to head their daits, or to make pegs in their boats.

Their only enemies, befides mankind, are the Polar Bears, with whom they have dreadful conflicts. Their feuds probably arife from the occupancy of the fame piece of ice. The Walrus is ufually victorious, through the fuperior advantage of its vaft teeth ‖. The effects of the battle are very evident; for it is not often that the hunters find a beaft with two entire tufks §.

" The Walrus, or Sea Cow, as it is called by the *Americans*," fays Lord *Shuldham* ¶, " is a native of the *Magdalene* iflands, St. *John's*, " and *Anticofti*, in the gulph of St. *Laurence*. They refort, very " early in the fpring, to the former of thefe places, which feems by " nature particularly adapted to the nature of the animals, abound- " ing with *clams* (efcallops) of a very large fize; and the moft " convenient landing-places, called *Echoueries*. Here they crawl up " in great numbers, and remain fometimes for fourteen days together " without food, when the weather is fair; but on the firft appear- " ance of rain, they retreat to the water with great precipitation.

---

* *De Buffon.*   † *Worm. Muf.* 290.
‡ As quoted in Mufeum Regium *Hafniæ*, &c. pars. i. fect. iii. 9.
‖ *Egede*, 83.   § *Crantz*, i. 126.   ¶ *Phil. Tranf.* lxv. part. i. 249.
The *French* call them *Vaches Marines. Charlevoix*, v. 216. That voyager fays, that the *Englifh* had once a fifhery of thefe animals on the *Ifle de Sable*, a fmall ifland fouth of *Cape Breton*; but it turned out to no advantage.

" They

" They are, when out of the water, very unwieldy, and move with
" great difficulty. They weigh from fifteen hundred to two thou-
" fand pounds, producing, according to their fize, from one to two
" barrels of oil, which is boiled out of the fat between the fkin and
" the flefh. Immediately on their arrival, the females calve, and
" engender again in two months after; fo that they carry their young
" about nine months. They never have more than two at a time,
" and feldom more than one.

" The *Echoueries* * are formed principally by nature, being a gradual
" flope of foft rock, with which the *Magdalene* iflands abound, about
" eighty or a hundred yards wide at the water-fide, and fpreading fo
" as to contain, near the fummit, a very large number of thefe ani-
" mals. Here they are fuffered to come on fhore, and amufe them-
" felves for a confiderable time, till they acquire a degree of bold-
" nefs, being at their firft landing fo exceedingly timid as to make
" it impoffible for any perfon to approach them.

" In a few weeks they affemble in great multitudes: formerly, when
" undifturbed by the *Americans*, to the amount of feven or eight
" thoufand. The form of the *Echouerie* not allowing them to re-
" main contiguous to the water, the foremoft are infenfibly pufhed
" above the flope. When they are arrived at a convenient diftance,
" the hunters, being provided with a fpear fharp on one fide, like a
" knife, with which they cut their throats, take advantage of a fide
" wind, or a breeze blowing obliquely upon the fhore, to prevent
" the animals from fmelling them, becaufe they have that fenfe in
" great perfection. Having landed, the hunters, with the affiftance
" of good dogs, trained for that purpofe, in the night-time endea-
" vour to feparate thofe which are moft advanced from the others,
" driving them different ways. This they call *making a cut*; it is
" generally looked upon to be a moft dangerous procefs, it being
" impoffible to drive them in any particular direction, and difficult
" to avoid them; but as the Walrufes, which are advanced above

---

* This word is derived from *Echouer*, to land, or run on fhore.

3   the

" the flope of the *Echouerie*, are deprived by the darknefs of the
" night from every direction to the water, they are left wandering
" about, and killed at leifure, thofe that are neareft the fhore being
" the firft victims. In this manner have been killed fifteen or fix-
" teen hundred at a *cut*.

" The people then fkin them, and take off a coat of fat which al-
" ways furrounds them, and diffolve it into oil. The fkin is cut
" into flices of two or three inches wide, and exported to *America*
" for carriage traces, and into *England* for glue. The teeth make
" an inferior fort of ivory, and is manufactured for that purpofe;
" but very foon turns yellow."

HIST. QUAD. Genus XLII.

Br. Zool. i. N° 71.—*Hiſt. Quad.* N° 375.—*Smellie.*
Kaſſigiak, *Faun. Greenl.* N° 6.—Lev. Mus.

S EAL. With a flat head and noſe: large black eyes: large
whiſkers: ſix cutting teeth in the upper jaw; four in the lower:
two canine teeth in each jaw: no external ears: hair on all parts
ſhort and thick: five toes on each foot, furniſhed with ſtrong ſharp
claws, and ſtrongly webbed: tail ſhort and flat.

Uſual length of this ſpecies, from five to ſix feet. Their color
differs; duſky, brinded, or ſpotted with white and yellow.

Inhabits all the *European* ſeas, even to the extreme north; and is
found far within the *Arctic* circle, in both *European* and *Aſiatic* ſeas.
It is continued to thoſe of *Kamtſchatka* *.

Theſe animals may be called the flocks of the *Greenlanders*, and
many other of the Arctic people. I cannot deſcribe the uſes of them
to the former more expreſſively than in the very words of Mr. *Crantz*,
a gentleman very long reſident in their chilly country.

" Seals are more needful to them than Sheep are to us, though they
" furniſh us with food and raiment; or than the cocoa-tree is to the
" *Indians*, although that preſents them not only with meat to eat, and
" covering for their bodies, but alſo houſes to dwell in, and boats to
" ſail in, ſo that in caſe of neceſſity they could live ſolely from it. The
" Seals fleſh (together with the Rein-deer, which is already grown
" pretty ſcarce) ſupplies the natives with their moſt palatable and ſub-
" ſtantial food. Their fat furniſhes them with oil for lamp-light, cham-
" ber and kitchen fire; and whoever ſees their habitations, preſently

* *Steller, in Nov. Com. Petrop.* ii. 290.

7   finds,

" finds, that if they even had a fuperfluity of wood, it would not do,
" they can ufe nothing but train in them. They alfo mollify their
" dry food, moftly fifh, in the train ; and finally, they barter it for all
" kinds of neceffaries with the factor. They can few better with
" fibres of the Seals finews than with thread or filk. Of the fkins of
" the entrails they make their windows, curtains for their tents,
" fhirts, and part of the bladders they ufe at their harpoons; and
" they make train bottles of the maw. Formerly, for want of iron,
" they made all manner of inftruments and working-tools of their
" bones. Neither is the blood wafted, but boiled with other ingre-
" dients, and eaten as foup. Of the fkin of the Seal they'ftand in
" the greateft need ; for, fuppofing the fkins of Rein-deer and birds
" would furnifh them with competent cloathing for their bodies,
" and coverings for their beds; and their flefh, together with fifh,
" with fufficient food; and provided they could drefs their meat
" with wood, and alfo new model their houfe-keeping, fo as to have
" light, and keep themfelves warm with it too ; yet. without the
" Seals fkins they would not be in a capacity of acquiring thefe
" fame Rein-deer, fowls, fifhes, and wood; becaufe they muft cover
" over with Seal-fkin both their large and fmall boats, in which they
" travel and feek their provifion. They muft alfo cut their thongs
" or ftraps out of them, make the bladders for their harpoons, and
" cover their tents with them ; without which they could not fubfift
" in fummer.

    " Therefore no man can pafs for a right *Greenlander* who cannot
" catch Seals. This is the ultimate end they afpire at, in all their
" device and labor from their childhood up. It is the only art
" (and in truth a difficult and dangerous one it is) to which they are
" trained from their infancy; by which they maintain themfelves,
" make themfelves agreeable to others, and become beneficial mem-
" bers of the community *.

* *Hift. Greenl.* i. 130.

                                                 " The

" The *Greenlanders* have three ways of catching Seals: either fing-
" ly, with the bladder; or in company, by the *clapper-hunt*; or in
" the winter on the ice: whereto may be added the fhooting them
" with a gun.

" The principal and moft common way is the taking them with
" the bladder. When the *Greenlander* fets out equipped according
" to the 7th Section, and fpies a Seal, he tries to furprife it una-
" wares, with the wind and fun in his back, that he may not be
" heard or feen by it. He tries to conceal himfelf behind a wave,
" and makes haftily, but foftly, up to it, till he comes within four,
" five, or fix fathom of it; mean while he takes the utmoft care
" that the harpoon, line, and bladder, lie in proper order. Then he
" takes hold of the oar with his left hand, and the harpoon with
" his right by the hand-board, and fo away he throws it at the
" Seal, in fuch a manner that the whole dart flies from the hand-
" board and leaves that in his hand. If the harpoon hits the mark,
" and buries itfelf deeper than the barbs, it will directly difengage
" itfelf from the bone-joint, and that from the fhaft; and alfo un-
" wind the ftring from its lodge on the *kajak*. The moment the
" Seal is pierced, the *Greenlander* muft throw the bladder, tied to
" the end of the ftring, into the water, on the fame fide as the Seal
" runs and dives; for that he does inftantly, like a dart. Then the
" *Greenlander* goes and takes up the fhaft fwimming on the water,
" and lays it in its place. The Seal often drags the bladder with it
" under water, though 'tis a confiderable impediment, on account
" of its great bignefs; but it fo wearies itfelf out with it, that it
" muft come up again in about a quarter of an hour to take breath.
" The *Greenlander* haftens to the fpot where he fees the bladder
" rife up, and fmites the Seal, as foon as it appears, with the great
" lance defcribed in the 6th Section *. This lance always comes
" out of its body again; but he throws it at the creature afrefh
" every time it comes up, till 'tis quite fpent. Then he runs the

* See the Sections referred to, and tab. v.

X

" little

" little lance into it, and kills it outright, but ſtops up the wound
" directly to preſerve the blood; and laſtly, he blows it up, like a
" bladder, betwixt ſkin and fleſh, to put it into a better capacity of
" ſwimming after him; for which purpoſe he faſtens it to the left-
" ſide of his *kajak*, or boat *.

" In this exerciſe the *Greenlander* is expoſed to the moſt and
" greateſt danger of his life; which is probably the reaſon that they
" call this hunt, or fiſhery, *kamavock*, i. e. the Extinction, *viz.* of life.
" For if the line ſhould entangle itſelf, as it eaſily may, in its ſud-
" den and violent motion; or if it ſhould catch hold of the *kajak*,
" or ſhould wind itſelf round the oar, or the hand, or even the neck,
" as it ſometimes does in windy weather; or if the Seal ſhould turn
" ſuddenly to the other ſide of the boat; it cannot be otherwiſe than
" that the *kajak* muſt be overturned by the ſtring, and drawn down
" under water. On ſuch deſperate occaſions the poor *Greenlander*
" ſtands in need of all the arts deſcribed in the former Section, to
" diſentangle himſelf from the ſtring, and to raiſe himſelf up from
" under the water ſeveral times ſucceſſively; for he will continually
" be overturning till he has quite diſengaged himſelf from the line.
" Nay, when he imagines himſelf to be out of all danger, and comes
" too near the dying Seal, it may ſtill bite him in the face or hand;
" and a female Seal that has young, inſtead of flying the field, will
" ſometimes fly at the *Greenlander* in the moſt vehement rage, and
" do him a miſchief, or bite a hole in his *kajak* that he muſt ſink.

" In this way, ſingly, they can kill none but the careleſs ſtupid
" Seal, called *Attarſoak* †. Several in company muſt purſue the
" cautious *Kaſſigiak* ‡ by the *clapper-hunt*. In the ſame manner they
" alſo ſurround and kill the *Attarſoit* ‖ in great numbers at certain
" ſeaſons of the year; for in autumn they retire into the creeks or
" inlets in ſtormy weather, as in the *Nepiſet* ſound in *Ball*'s river,
" between the main land and the iſland *Kangek*, which is full two

---

* See vol. i. 150. tab. viii.   † See N° 77. of this work.   ‡ Ditto, N° 72.
‖ Ditto, a variety of N° 77.

leagues

" leagues long, but very narrow. There the *Greenlanders* cut off their
" retreat, and frighten them under water by fhouting, clapping, and
" throwing ftones; but, as they muft come up again continually to
" draw breath, then they perfecute them again till they are tired,
" and at laft are obliged to ftay fo long above water, that they fur-
" round them, and kill them with the fourth kind of dart, defcribed
" in the 6th Section. During this hunt we have a fine opportunity
" to fee the agility of the *Greenlanders*, or, if I may call it fo, their
" huffar-like manœuvres. When the Seal rifes out of the water,
" they all fly upon it, as if they had wings, with a defperate noife;
" the poor creature is forced to dive again directly, and the moment
" he does, they difperfe again as faft as they came, and every one
" gives heed to his poft, to fee where it will ftart up again; which
" is an uncertain thing, and is commonly three quarters of a mile
" from the former fpot. If a Seal has a good broad water, three
" or four leagues each way, it can keep the fportfmen in play for a
" couple of hours, before 'tis fo fpent that they can furround and
" kill it. If the Seal, in its fright, betakes itfelf to the land for a
" retreat, 'tis welcomed with fticks and ftones by the women and
" children, and prefently pierced by the men in the rear. This is
" a very lively and a very profitable diverfion for the *Greenlanders*,
" for many times one man will have eight or ten Seals for his
" fhare.

" The third method of killing Seals upon the ice, is moftly prac-
" tifed in *Difko*, where the bays are frozen over in the winter. There
" are feveral ways of proceeding. The Seals themfelves make
" fometimes holes in the ice, where they come and draw breath;
" near fuch a hole a *Greenlander* feats himfelf on a ftool, putting
" his feet on a lower one to keep them from the cold. Now when
" the Seal comes and puts its nofe to the hole, he pierces it in-
" ftantly with his harpoon; then breaks the hole larger, and draws
" it out and kills it quite. Or a *Greenlander* lays himfelf upon his
" belly, on a kind of a fledge, near other holes, where the Seals

come

" come out upon the ice to baſk themſelves in the ſun. Near this
" great hole they make a little one, and another *Greenlander* puts
" a harpoon into it with a very long ſhaft or pole. He that lies
" upon the ice looks into the great hole, till he ſees a Seal com-
" ing under the harpoon; then he gives the other the ſignal, who
" runs the Seal through with all his might.

" If the *Greenlander* ſees a Seal lying near its hole upon the ice,
" he ſlides along upon his belly towards it, wags his head, and
" grunts like a Seal; and the poor Seal, thinking 'tis one of its in-
" nocent companions, lets him come near enough to pierce it with
" his long dart.

" When the current wears a great hole in the ice in the ſpring,
" the *Greenlanders* plant themſelves all round it, till the Seals come
" in droves to the brim to fetch breath, and then they kill them
" with their harpoons. Many alſo are killed on the ice while they
" lie ſleeping and ſnoring in the ſun *."

Uses in Kamt-
schatka.
Nature has been ſo niggardly in providing variety of proviſion for
the *Greenlanders*, that they are neceſſitated to have recourſe to ſuch
which is offered to them with a liberal hand. The *Kamtſchatkan* nations, which enjoy ſeveral animals, as well as a great and abundant
choice of fiſh, are ſo enamoured with the taſte of the fat of Seals,
that they can make no feaſt without making it one of the diſhes.
Of that both *Ruſſians* and *Kamtſchatkans* make their candles. The
latter eat the fleſh boiled, or elſe dried in the ſun. If they have a
great quantity, they preſerve it in the following manner:

They dig a pit of a requiſite depth, and pave it with ſtones; then
fill it with wood, and ſet it on fire ſo as to heat the pit to the warmth
of a ſtove. They then collect all the cinders into a heap. They
ſtrew the bottom with the green wood of alder, on which they place
ſeparately the fleſh and the fat, and put between every layer branches
of the ſame tree; when the pit is filled they cover it with ſods, ſo
that the vapour cannot eſcape. After ſome hours they take out both

* pp. 153, 4, 5, 6, 7.

fat

fat and flesh, and keep it for winter's provisions, and they may be preserved a whole year without spoiling.

The *Kamtschatkans* have a most singular ceremony. After they take the flesh from the heads of the Seals, they bring a vessel in form of a canoe, and fling into it all the sculls, crowned with certain herbs, and place them on the ground. A certain person enters the habitation with a sack filled with *Tonchitche*, sweet herbs, and a little of the bark of willow. Two of the natives then roll a great stone towards the door, and cover it with pebbles; two others take the sweet herbs and dispose them, tied in little packets. The great stone is to signify the sea-shore, the pebbles the waves, and the packets Seals. They then bring three dishes of a hash, called *Tolkoucha*; of this they make little balls, in the middle of which they stick the packets of herbs: of the willow-bark they make a little canoe, and fill it with *Tolkoucha*, and cover it with the sack. After some time, the two *Kamtschatkans* who had put the mimic Seals into the *Tolkoucha*, take the balls, and a vessel resembling a canoe, and draw it along the sand, as if it was on the sea, to convince the real Seals how agreeable it would be to them to come among the *Kamtschatkans*, who have a sea in their very *jurts*, or dwellings. And this they imagine will induce the Seals to suffer themselves to be taken in great numbers. Various other ceremonies, equally ridiculous, are practised; in one of which they *invoke the winds, which drive the Seals on their shores, to be propitious* *.

Besides the uses which are made of the flesh and fat of Seals, the skins of the largest are cut into soles for shoes. The women make their summer boots of the undressed skins, and wear them with the hair outmost. In a country which abounds so greatly in furs; very little more use is made of the skins of Seals in the article of dress than what has been mentioned †. But the *Koriaks*, the *Oloutores*, and *Tchutschi*, form with the skins canoes and vessels of different sizes, some large enough to carry thirty people.

* *Descr. Kamtschatka*, 425.　　† The same, 41, 42. 424.

Seals

Seals fwarm on all the coafts of *Kamtfchatka*, and will go up the rivers eighty *verfts* in purfuit of fifh.   They couple on the ice in *April*, and fometimes on the rocks, and even in the fea in calm weather.   The *Tunguft* give the milk of thefe animals to their children inftead of phyfic.

The Seals in this country are killed by harpooning, by fhooting, by watching the holes in the ice and knocking them on the head as they rife ; or by placing two or three ftrong nets acrofs one of the rivers which thefe animals frequent : fifty or more people affemble in canoes on each fide of the nets, while others row up and down, and with great cries frighten the Seals into them.   As foon as any are entangled, the people kill them with pikes or clubs, and drag them on fhore, and divide them equally among the hunters; fometimes a hundred are taken at a time in this manner.

The navigators obferved abundance of Seals about *Bering's* ifland, but that they decreafed in numbers as they advanced towards the ftraits ; for where the Walrufes abounded, the Seals grew more and more fcarce.

I did not obferve any Seal-fkin garments among thofe brought over by the navigators, fuch as one might have expected among the *Efquimaux* of the high latitudes they vifited, and which are fo much in ufe with thofe of *Hudfon's Bay* and *Labrador*.   That fpecies of drefs doubtlefsly was worn in the earlieft times.   Thefe people wanted their

MASSAGETÆ CLOATHED IN SEAL-SKINS.

hiftorians; but we are affured that the *Maffagetæ* * cloathed themfelves in the fkins of Seals.   They, according to *D'anville*, inhabited the country to the eaft of the *Cafpian* fea, and the lake *Aral* ; both of which waters abound with Seals.

Seals are now become a great article of commerce.   The oil from the vaft Whales is no longer equal to the demand for fupplying the magnificent profufion of lamps in and round our capital.   The chafe of thefe animals is redoubled for that purpofe; and the fkins, properly tanned, are in confiderable ufe in the manufactory of boots and fhoes.

* *Strabo*, lib. xi. 781.

Hift.

Hift. Quad. Nº 382.
Phoca Barbata, *Faun. Greenl* Nº 9.—Urkfuk. *Greenl.*
Lakktak, *Hift. Kamtfchatka*, 420.—Lɛv. Muꜱ.

SEAL. With long pellucid white whifkers with curled points : back arched : black hairs, very deciduous, and thinly difperfed over a thick fkin, which in fummer is almoft naked : teeth like the common Seal : fore feet like the human hand ; middle toe the longeft ; thumb fhort : length more than twelve feet.

The *Greenlanders* cut out of the fkin of this fpecies thongs and lines, a finger thick, for the Seal-fifhery. Its flefh is white as veal, and efteemed the moft delicate of any : has plenty of lard, but does not yield much oil. The fkins of the young are fometimes ufed to lie on.

It inhabits the high fea about *Greenland* ; is a timid fpecies, and ufually refts on the floating ice, and very feldom the fixed. Breeds in the earlieft fpring, or about the month of *March*, and brings forth a fingle young on the ice, ufually among the iflands ; for at that feafon it approaches a little nearer to the land. The great old ones fwim very flowly.

In the feas of the north of *Scotland* is found a Seal twelve feet long. A gentleman of my acquaintance fhot one of that fize on the coaft of *Sutherland* ; but made no particular remarks on it. A young one, feven feet and half long, was fhewn in *London* fome years ago, which had not arrived at maturity enough even to have fcarcely any teeth* : yet the common Seals have them complete before they attain the fize of fix feet, their utmoft growth.

A fpecies larger than an Ox, found in the *Kamtfchatkan* feas from 56 to 64 north latitude, called by the natives *Lachtak†*. They weighed

* *Ph. Tranf. Abr.* ix. 74. tab. v. xlvii. 120.   † *Nov. Com. Petrop.* ii. 290.

eight

eight hundred pounds : were eaten by *Bering*'s crew; but their flesh was found to be very loathsome*. The cubs are quite black.

STELLER has left behind him accounts of other Seals found in those wild seas; but his descriptions are so imperfect as to render it impossible to ascertain the species. He speaks in his MSS. of a middle-sized kind, universally and most elegantly spotted; another, black with brown spots, and the belly of a yellowish white, and as large as a yearling Ox; a third species, black, and with a particular formation of the hinder legs; and a fourth, of a yellowish color, with a great circle on it of the color of cherries †.

74. ROUGH.   *Hist. Quad.* N° 383.
                   Phoca Fœtida, *Faun. Greenl.* N° 8.—Neitseck *Greenl. Crantz,* i.

SEAL. With a short nose, and short round head : teeth like the common Seal : body almost of an elliptical form, covered with lard almost to the hind feet : hairs closely set together, soft, long, and somewhat erect, with curled wool intermixed : color dusky, streaked with white ; sometimes varies to white, with a dusky dorsal line.

Does not exceed four feet in length.

Never frequents the high seas, but keeps on the fixed ice in the remote bays near the frozen land; and when old never forsakes its haunts. Couples in *June*; brings forth in *January*, on the fixed ice, which is its proper element. In that it has a hole for the benefit of fishing; near that it remains usually solitary, rarely in pairs. Is very incautious, and often sleeps on the surface of the water, yielding itself a prey to the Eagle. Feeds on small fish, shrimps, and the like. The uses of the skin, tendons, and lard, the same with those of other Seals. The flesh is red, and fœtid, especially that of the males, which is nauseated by even the *Greenlanders.*

* *Muller's voy.* 60.   † *Dr. Pallas,* and *Descr. Kamtschatka,* 420.

The

The Seal-hunters in *Newfoundland* have a large kind, which they call the *Square Phipper*, and say weighs five hundred pounds. Its coat is like that of a Water-dog; so that it seems by the length of hair to be allied to this; but the vast difference in size forbids us from pronouncing it to be the same species.

75. LEPORINE.

Hiſt. Quad. Nᵒ 381.

Phoca Leporina, *Lepechin, Act. Acad. Petrop.* pars i. 264. tab viii. ix.— *Hiſt. Quad.* Nᵒ 381.

SEAL. With hair of an uniform dirty white color, with a tinge of yellow, but never spotted; hairs erect, and interwoven; soft as that of a Hare, especially the young: head long: upper lip swelling and thick: whiskers very strong and thick, ranged in fifteen rows, covering the whole front of the lip, so as to make it appear bearded: eyes blue, pupil black: teeth strong; four cutting teeth above, the same below *: fore feet short, and ending abrupt: the membranes of the hind feet even, and not waved: tail short and thick; its length four inches two lines.

SIZE.

Length of this species, from nose to tip of the tail, is six feet six; its greatest circumference five feet two. The cubs are milk white.

This kind inhabits the *White Sea* during summer, and ascends and descends the mouths of rivers †.with the tide in quest of prey. It is also found on the coasts of *Iceland*, and within the Polar circle from *Spitzbergen* to *Tchutki Noſs*, and from thence southward about *Kamtſchatka*.

Like the others, it is killed for its fat and skin. The last is cut into pieces, and used for straps and reins. The skins of the young, which are remarkably white, are dyed with black, and used to face caps, in imitation of Beavers skins; but the hairs are much stiffer, and do not soon drop off.

* *Mr. Lepechin* compares the number of the teeth to that of another kind (our *Harp Seal*) which, he says, has only four teeth in the lower jaw.

† The same.

Y *Hiſt.*

Hiſt. Quad. N° 384.
Phoca Leonina, *Faun. Greenl.* N° 5.

SEAL. With four cutting teeth above, four below: fore feet like the human; the thumb long: the membranes on the hind feet extend beyond the claws: on the forehead of the male is a thick folded ſkin, ridged half the way up, which it can inflate and draw down like a cap, to defend its eyes againſt ſtorms, waves, ſtones, and ſand. The females and young have only the rudiment of this guard. It has two ſpecies of hair; the longeſt white, the ſhorteſt thick, black, and woolly, which gives it a beautiful grey color.

It grows to the length of eight feet. The *Greenlanders* call it. *Neitſek-ſoak* *, or the Great *Neitſek*. It inhabits only the ſouthern parts of their country, where it inhabits the high ſeas; but in *April, May,* and *June,* comes nearer to the land. Is polygamous; copulates with its body erect. Brings forth in *April* one young upon the ice. Keeps much on the great fragments, where it ſleeps in an unguarded way. Bites hard: barks, and whines: grows very fierce on being wounded; but will weep on being ſurprized by the hunter. Fight among themſelves, and inflict deep wounds. Feed on all kinds of greater fiſh. The ſkins of the young form the moſt elegant dreſſes for the women. The men cover their great boats with thoſe of the old; they alſo cover their houſes with them, and when they grow old convert them into ſacks. They uſe the teeth to head hunting-ſpears. Of the gullet and inteſtines they make the ſea-dreſſes. The ſtomach is made into a fiſhing-buoy.

It is alſo found in *Newfoundland.* Our Seal-hunters name it the *Hooded* Seal, and pretend they cannot kill it till they remove that integument. The *Germans* call it *Klap-Mutz,* from its covering its face as if with a cap.

The moſt dreaded enemy which this ſpecies has in *Greenland,* is the *Phyſeter Microps;* on the very ſight of which it takes to the ice,

* *Crantz,* i. 25.

and

and quietly expects its fate *. The *Greenlanders* therefore detest this species of Whale, not only on account of the havock it makes among the Seals, but because it frightens them away from the bays †.

It is entirely different from the LEONINE SEAL, or from that of the South-sea, called the BOTTLE-NOSE.

Hist. Quad. N° 385.

Phoca Oceanica, *Krylatca Russis, Lepechin, Act. Acad. Petrop.* pars i. 259. tab. vi. vii.

Phoca Greenlandica, *Faun. Greenl.* N° 7.—Atak *Greenl.* Atarsoak, *Crantz,* i. 124.

77. HARP.

SEAL. With a round head: high forehead: nose short: large black eyes: whiskers disposed in ten rows of hairs: four cutting teeth in the upper jaw, the two middlemost the longest; four also in the lower, less sharp than the others: two canine teeth in each jaw: six grinders in each jaw, each three-pointed: hairs short: skin thick and strong.

Head, nose, and chin, of a deep chesnut color, nearly black; rest of the body of a dirty white, or light grey: on the top of the shoulders is a large mark of the same color; with the head bifurcated, each fork extending downwards along the sides half way the length of the body. This mark is always constant; but there are besides a few irregular spots incidental to the old ones.

The female has only two, retractile, teats; and brings only one young at a time. The cub, the first year, is of a bright ash-color, whitish beneath, and marked in all parts with multitudes of small black spots, at which period they are called by the *Russians* White Seals. In the next year they begin to be spotted; from that period the females continue unchanged in color. The males at full age, which Mr. *Crantz* says is their fifth year, attain their distinguishing spot, and are called by the *Greenlanders Attarsoak* ‡; by the *Russians, Krylatka,* or winged.

* *Faun. Groenl.* p. 9.   † The same, p. 45.   ‡ *Crantz,* i. 124.

This

This inhabits the fame countries with the *Rough* and *Leporine Seal*; but loves the coldeft parts of the coaft. Continues on the loofe ice of *Nova Zembla* the whole year; and is feen only in the winter in the *White Sea*, on the floating ice carried from the northern feas. It brings forth its young about the end of *April*, and after fuckling it a fufficient time departs with the firft ice into the *Frozen Ocean*. The young remains behind for fome time, then follows its parent with the ice which is loofed from the fhore *.

It abounds in *Greenland* and about *Spitzbergen*, efpecially in the bottoms of the deep bays. Migrates in *Greenland* twice in the year: in *March*, and returns in *May*; in *June*, and returns in *September*. Couples in *July*, and brings forth towards the end of *March* or begining of *April:* has one young, rarely two, which it fuckles on fragments of ice far from land. It never afcends the fixed ice; but lives and fleeps on the floating iflands in great herds. Swims in great numbers, having one for a leader, which feems to watch for the fecurity of the whole. Eats its prey with its head above water. Swims in various ways; on its belly, back, and fide, and often whirls about as if in frolick. Frequently fleeps on the furface of the water. Is very incautious. Has great dread of the *Phyfeter Microps*, which forces it towards the fhore. It is often furrounded by troops of hunters, who compel it even to land, where it is eafily killed.

It is found alfo about *Kamtfchatka*, being the third fpecies mentioned by *Steller*.

Size. It grows to the length of nine feet. The meafurements of one defcribed by Mr. *Lepechin* are as follow—The length, from the nofe to the tip of the tail, was fix feet: the length of the tail five inches three lines: the girth of the thickeft part of the body four feet eight.

Uses. The fkin is ufed to cover trunks; that of the young, taken in the ifle of *Solovki*, on the weft fide of the *White Sea*, is made into boots, and is excellent for keeping out water. The *Greenlanders*, in dreffing the fkins, curry off the hair, and leave fome fat on the infide to ren-

---

* *Act. Acad. Petrop.* pars 263.

3

der

der them thicker. With thefe they cover their boats, and with the undreffed fkins their tents; and, when they can get no other, make ufe of them for cloathing.

The oil extracted from the blubber of this Seal is far the moft valuable, being fweet, and fo free from greaves as to yield a greater quantity than any other fpecies. The flefh is black.

The *Newfoundland* Seal-hunters call it the *Harp*, or *Heart* Seal, and name the marks on the fides the faddle. They fpeak too of a brown fort, which they call *Bedlemer*, and believe to be the young of the former.

Hiſt Quad N° 380. fig. at p. 513.

78. RUBBON.

SEAL. With very fhort briftly hair, of an uniform gloffy color, almoſt black: the whole back and fides comprehended within a narrow regular ſtripe of pale yellow.

It is to Dr. *Pallas* I owe the knowlege of this fpecies. He received only part of the fkin, which feemed to have been the back and fides. The length was four feet, the breadth two feet three; fo it muſt have belonged to a large fpecies. It was taken off the *Kuril* iflands.

Hiſt. Quad. N° 387.
Kot *Ruſſis* Gentilibus ad Sinum *Penchinicum, Tarlatſchega, Nov. Com. Petrop.* ii. 331. tab xv.
Sea Wolf *, *Pernety*, Engl. Tr. 187. tab. xvi.—*Ulloa's voy.* i. 226.
Chat Marin, *Hiſt. Kamtſchatka*, 433.

79. URSINE.

SEAL. With a high forehead: nofe projecting like that of a dog: black irides: finaragdine pupil: whifkers compofed of triangular hairs, thinly fcattered: noftrils oval, divided by a *feptum*: lips thick; their infide red, and ferrated.

* The *French* generic name for the Seal is *Loup Marin* and the *Spaniſh*, *Lobo Marino.*

In

TEETH.

In the upper jaw four bifurcated cutting teeth; on each fide of thefe a very fharp canine tooth bending inwards; beyond thefe another, which, in battle, the animal ftrikes with, as Boars do with their tufks. Inftead of grinders, in each upper jaw are fix fharp teeth refembling canine, and very flightly exerted. In the lower jaw four cutting teeth, and canine like thofe in the upper; and on each fide ten others in the place of grinders. When the mouth is clofed all the teeth lock into each other.

TONGUE, EARS.

The tongue rough and bifid: the ears fhort, fmall, and fharp-pointed, hairy on the outfide, fmooth and polifhed within.

FORE LEGS.

Fore legs two feet long, not immerfed in the body, like thofe of other Seals, but refemble thofe of common quadrupeds. The feet are furnifhed with five toes, with the rudiments of nails; but thefe are fo entirely covered with a naked fkin, as to be as much concealed as a hand is with a mitten. The animal ftands on thefe legs with the utmoft firmnefs; yet the feet feem but a fhapelefs mafs.

HIND LEGS.

The hind legs are twenty-two inches long, and fituated like thofe of Seals; but are capable of being brought forward, fo that the animal makes ufe of them to fcratch its head: on each are five toes,

TAIL.

connected by a large web; and are a foot broad. The tail is only two inches long.

BODY.

The body is of a conoid fhape. The length of a large one is about eight feet; the circumference near the fhoulders is five feet, near the tail twenty inches. The weight eight hundred pounds.

FEMALE.

The female is far inferior in fize to the male: it has two teats, placed far behind.

COLOR.

The whole animal is covered with long and rough hair, of a blackifh color; that of the old is tipt with grey; and on the neck of the males is a little longer and erect: beneath the hair is a foft fur of a bay color. The females are cinereous. The fkin is thick and ftrong.

PLACE.

Thefe animals are found in amazing multitudes on the iflands between

tween *Kamtfchatka* and *America* *; but are fcarcely known to land on the *Afiatic* fhore: nor are they ever taken except in the three *Kurilian* iflands, and from thence in the *Bobrowoie More*, or *Beaver Sea*, as far as the *Kronofki* headland, off the river *Kamtfchatka*, which comprehends only from 50 to 56 north latitude. It is obfervable that they never double the fouthern cape of the peninfula, or are found on the weftern fide in the *Penfchinfka* fea: but their great refort has been obferved to be to *Bering*'s iflands. They are as regularly mi- MIGRATORY. gratory as birds of paffage. They firft appear off the three *Kurili* iflands and *Kamtfchatka* in the earlieft fpring. They arrive exceffively fat; and there is not one female which does not come pregnant. Such which are then taken are opened, the young taken out and fkinned. They are found in *Bering*'s ifland only on the weftern fhore, being the part oppofite to *Afia*, where they firft appear on their migration from the fouth. They continue on fhore three months, during which time the females bring forth. Excepting their employ of fuck- ling their young, they pafs their time in total inactivity. The males LONG SLEEP AND fink into the moft profound indolence, and deep fleep; nor are they FASTING. ever roufed, except by fome great provocation, arifing from an inva- fion of their place, or a jealoufy of their females. During the whole time they neither eat nor drink. *Steller* diffected numbers, without finding the left appearance of food in their ftomachs.

They live in families. Every male is furrounded by a feraglio of LIVE IN FAMI- from eight to fifty miftreffes; thefe he guards with the jealoufy of LIES. an eaftern monarch. Each family keeps feparate from the others, notwithftanding they lie by thoufands on the fhore. Every family, with the unmarried and the young, amount to about a hundred and twenty. They alfo fwim in tribes when they take to the fea.

---

* They fay that the *Sea-Cat*, or *Siwutcha*, is found in thofe iflands; but *Siwutcha* is the name given by the *Kamtfchatkans* and *Kurilians* to the *Leonine* Seal only. *Northern Archipelago*, &c. by *Von Stæhlen*. Printed for *Heydinger*, 1774, p. 34.

The

AFFECTION TO-
WARDS THEIR
YOUNG.

The males shew great affection towards their young, and equal tyranny towards the females. The former are fierce in the protection of their offspring; and should any one attempt to take their cub, will stand on the defensive, while the female carries it away in her mouth. Should she happen to drop it, the male instantly quits its enemy, falls on her, and beats her against the stones till he leaves her for dead. As soon as she recovers, she crawls to his feet in the most suppliant manner, and washes them with her tears; he at the same time brutally insults her misery, stalking about in the most insolent manner. But if the young is entirely carried off, he melts into the greatest affliction, likewise sheds tears, and shews every mark of deep sorrow. It is probable that as the female brings only one, or at most two cubs, he feels his misfortune the more sensibly.

Those animals which are destitute of females, through age or impotence, or are deserted by them, withdraw themselves from society, and grow excessively splenetic, peevish, and quarrelsome; are very furious, and so attached to their antient stations, as to prefer death to the loss of them. They are enormously fat, and emit a most nauseous and rank smell. If they perceive another animal approach its seat, they are instantly roused from their indolence, snap at the

CONFLICTS.

encroacher, and give battle. During the fight they insensibly intrude on the station of their neighbor. This creates new offence; so that at length the civil discord spreads through the whole shore, attended with hideous growls, their note of war. They are very tenacious of life, and will live a fortnight after receiving such wounds as would soon destroy any other animal.

CAUSES OF THEM.

The particular causes of disputes among these irascible beasts are the following :—The first and greatest is, when an attempt is made to seduce any of their mistresses, or a young female of the family : a battle is the immediate consequence of the insult. The unhappy vanquished instantly loses his whole seraglio, who desert him for the victorious hero.

The

The invasion of the station of another, gives rise to fresh conflicts; and the third cause is the interfering in the disputes of others. The battles they wage are very tremendous; the wounds they inflict very deep, like the cut of a sabre. At the conclusion of an engagement they fling themselves into the sea to wash off the blood.

Besides their notes of war, they have several others. When they **NOTES.** lie on shore, and are diverting themselves, they low like a Cow. After victory they chirp like a Cricket. On a defeat, or after receiving a wound, mew like a Cat.

Common Seals, and Sea Otters, stand in great awe of these ani- **DREAD THE LEO-** mals, and shun their haunts. They again are in equal awe of the **NINE SEAL.** Leonine Seals, and do not care to begin a quarrel in their sight, dreading the intervention of such formidable arbitrators; who like-wise possess the first place on the shore.

The great and old animals are in no fear of mankind, unless they **FEAR NOT MAN-** are suddenly surprized by a loud shout, when they will hurry by **KIND.** thousands into the sea, swim about, and stare at the novelty of their disturbers.

When they come out of the water, they shake themselves, and smooth their hair with their hind feet: apply their lips to those of the females, as if they meant to kiss them: lie down and bask in the sun with their hind legs up, which they wag as a Dog does its tail. Sometimes they lie on their back, sometimes roll themselves up in-to a ball, and fall asleep. Their sleep is never so sound but they are awoke by the left alarm; for their sense of hearing, and also that of smelling, is most exquisite.

They copulate, *more humano*, in *July*, and bring forth in the *June* **COPULATION.** following; so they go with young eleven months. The cubs are **GESTATION.** as sportive as puppies; have mock fights, and tumble one another on the ground. The male parent looks on with a sort of compla-cency, parts them, licks and kisses them, and seems to take a greater affection to the victor than to the others.

SWIFT SWIMMERS. They fwim with amazing fwiftnefs and ftrength, even at the rate of feven or eight miles an hour, and often on their back. They dive well, and continue a great while under water. If wounded in that element, they will feize on the boat, carry it with them with great impetuofity, and often will fink it.

When they wifh to afcend the rocks, they fix their fore feet on them, arch their backs, and then draw themfelves up.

CAPTURE. The *Kamtfchatkans* take them by harpooning, for they never land on their fhore. To the harpoon is faftened a long line, by which they draw the animal to the boat after it is fpent with fatigue; but in the chafe, the hunters are very fearful of too near an approach, leaft the animal fhould faften on and fink their veffel.

USES. The ufes of them are not great. The flefh of the old males is rank and naufeous; that of the females is faid to refemble lamb; of the young ones roafted, a fucking pig. The fkins of the young, cut out of the bellies of the dams, are efteemed for cloathing, and are fold for about three fhillings and four pence each; thofe of the old for only four fhillings.

RE-MIGRATION. Their re-migration is in the month of *September*, when they depart exceffively lean, and take their young with them. On their return, they again pafs near the fame parts of *Kamtfchatka* which they did in the fpring. Their winter retreats are quite unknown; it is probable that they are the iflands between the *Kurili* and *Japan*, of which we have fome brief accounts, under the name of *Compagnie Land*, *States Land*, and *Jefo Gafima*, which were difcovered by *Martin Uriel* in 1642 *. It is certain that by his account the natives employ themfelves in the capture of Seals †. Sailors do not give themfelves the trouble of obferving the nice diftinction of fpecific marks, we are therefore at liberty to conjecture thofe which he faw

---

* He failed from the eaft fide of *Japan* in the fhip *Caftricom*, vifited the ifle of *Jefo*, and difcovered the iflands which he called *States Land* and *Company Land*, the laft not very remote from the moft fouthern *Kurili* ifland. *Recueil de voy. au Nord*, iv. 1.

† The fame, 12.

to be our animals, efpecially as we can fix on no more convenient place for their winter quarters. They arrive along the fhores of the *Kurili* iflands, and part of thofe of *Kamtfchatka*, from the fouth. They land and inhabit only the weftern fide of *Bering*'s ifle, which faces *Kamtfchatka*; and when they return in *September*, their route is due fouth, pointing towards the difcoveries of *Uriel*. Had they migrated from the fouth-eaft as well as the fouth-weft, every ifle, and every fide of every ifle, would have been filled with them; nor fhould we have found (as we do) fuch a conftant and local refidence.

Before I quit this article I muft obferve, that there feems to be in the feas of *Jefo Gafimo* another fpecies of Seal, perhaps our little Seal, N° 386. *Hift. Quad.* The account indeed is but obfcure, which I muft give as related by *Charlevoix* in his compilations refpecting that ifland. "The natives," fays he, "make ufe of an oil "to drink, drawn from a fort of fifh, a fmall hairy creature with "four feet." If this account is true, it ferves to point out the fartheft known refidence of this genus, on this fide of the northern hemifphere.

Finally, the *Urfine* Seals are found in the fouthern hemifphere, even from under the line, in the ifle of *Gallipagos* †, to *New Georgia* ‡, in fouth latitude 54. 15. and weft longitude 37. 15. In the intermediate parts, they are met with in *New Zeland* ||, in the ifle of *Juan Fernandez*, and its neighbor *Maffa Fuera*, and probably along the coafts of *Chili* to *Terra del Fuego*, and *Staten Land*. In *Juan Fernandez*, *Staten Land*, and new *Georgia* ¶, they fwarm; as they do at the northern extremity of this vaft ocean. Thofe of the fouthern hemifphere have alfo their feafons of migration. *Alexander Selkirk*, who paffed three lonely years on the ifle of *Juan Fernandez*, remarks

URSINE SEAL IN THE SOUTHERN HEMISPHERE.

---

† *Woodes Rogers's voy.* 265. He fays that they are neither fo numerous there, nor is their fur fo fine as thofe on *Juan Fernandez*, which is faid to be extremely foft and delicate.

‡ *Cook's voy.* ii. 213. || *Cook*, i. 72. 86. *Forfter's Obf.* 189.   ¶ *Anfon's voy.* 122. *Cook*, ii. 194. 213.

Z 2

that

that they come afhore in *June*, and ftay till *September* *. Captain *Cook* found them again, in their place of remigration, in equal abundance, on *Staten Land* and *New Georgia*, in the months of *December* and *January* †; and Don *Pernety* ‡ found them on the *Falkland* iflands, in the month of *February*.

According to the *Greenlanders*, this fpecies inhabits the fouthern parts of their country. They call it *Auvekæjak*. That it is very fierce, and tears to pieces whatfoever it meets; that it lives on land as well as in water, fwims moft impetuoufly, and is dreaded by the hunters ‖.

80 LEONINE.

Hift. Quad. N° 389.
Beftia Marina, *Kurillis, Kamtfchadalis* et *Ruffis, Kurillico* nomine *Siwutfcha* dicta. *Nov. Com. Petrop.* ii. 360.
Lion Marin, *Hift. Kamtfchatka*, 428.

SEAL. With a large head: nofe turning up like that of a pug Dog: eyes large; pupil fmaragdine: the greater angle of each as if ftained with cinnabar color. In the upper jaw four fmall cutting teeth; the exterior on each fide remote, and at fome diftance from thefe are two large canine teeth: in the lower jaw four fmall cutting teeth, and the canine: the grinders fmall and obtufe; four on each fide above, and five below: ears conic and erect: feet exactly like thofe of the *Urfine* Seal.

Along the neck of the male is a mane of ftiff curled hair; and the whole neck is covered with long waved hairs, fuch as diftinguifh a Lion; the reft of the animal cloathed with fhort reddifh hairs: thofe of the female are of the color of ochre; the young of a much deeper. The old animals grow grey with age.

* *Selkirk's* account in *W. Rogers's voy.* 136.   † ii. 194. 213.   ‡ His veyage, *Engl. Tr.* 187.   ‖ *Faun. Greenl.* p. 6.

The

The weight of a large male beaſt is ſixteen hundred pounds. Length of the males is ſometimes fourteen, or even eighteen feet *. The females are very diſproportionably leſſer, not exceeding eight feet.

Inhabits the eaſtern coaſts of *Kamtſchatka*, from cape *Kronozki* as low as cape *Lapatka* and the *Kurili* iſlands, and even as far as *Matſmai*, which probably is the ſame with *Jeſo Gaſima*. Near *Matſmai* Captain *Spanberg* obſerved a certain iſland of a moſt pictureſque form, bordered with rocks reſembling buildings, and ſwarming with theſe animals, to which he gave the name of the *Palace of the Sea Lions* †. Like the *Urſine* Seals, they are not found on the weſtern ſide of the peninſula. They abound, in the months of *June*, *July*, *Auguſt*, and *September*, on *Bering*'s iſland, which they inhabit for the ſake of quiet parturition and ſuckling their young. *Steller* alſo ſaw them in abundance in *July* on the coaſts of *America*.

They do not migrate like the former; but only change the place of reſidence, having winter and ſummer ſtations ‡. They live chiefly on rocky ſhores, or lofty rocks in the ſea, which ſeem to have been torn away from the land by the violence of ſome earthquake ‖. Theſe they climb, and by their dreadful roaring are of uſe in foggy weather to warn navigators to avoid deſtruction.

They copulate in the months of *Auguſt* and *September*; go ten months, and bring only one at a time. The parents ſhew them little affection, often tread them to death through careleſsneſs, and will ſuffer them to be killed before them without concern or reſentment. The cubs are not ſportive, like other young animals, but are almoſt always aſleep. Both male and female take them to ſea to learn them to ſwim; when wearied, they will climb on the back of their dam; but the male often puſhes them off, to habituate them to the

---

* *Narborough*, 31. *Penroſe Falkland Iſles*, 28. *Pernetti, voy. Malouines*, 240. By his confounding the names of this and the Bottle-noſe Seal, N° 288. *Hiſt. Quad.* he led me into a miſtake about the length of this.

† *Deſcr. Kamtſchatka*, 433. ‡ *Nov. Com. Petrop.* ii. 365. ‖ *Muller's voy.* 60.

exerciſe.

exercife. The *Ruffians* were wont to fling the cubs into the water, and they always fwam back to fhore.

The males treat the females with great refpect, and are very fond of their careffes. They are polygamous, but content themfelves with fewer wives than the former, having only from two to four apiece.

FEAR MANKIND; The males have a terrible afpect, yet they take to flight on the firft appearance of a human creature; and if they are difturbed from their fleep, feem feized with great horrors, figh deeply in their attempts to go away, fall into vaft confufion, tumble down, and tremble in fuch a manner as fcarcely to be able to ufe their limbs. But if they are reduced to a ftrait, fo as not poffibly to effect an efcape, they grow defperate, turn on their enemy with great fury and noife, and even put the moft valiant to flight.

UNLESS HABITU-
ATED. By ufe they lofe their fear of men. *Steller* once lived for fix days in a hovel amidft their chief quarters, and found them foon reconciled to the fight of him. They would obferve what he was doing with great calmnefs, lie down oppofite to him, and fuffer him to feize on their cubs. He had an opportunity of feeing their conflicts about their females; and once faw a duel between two males, which lafted three days, and one of them received above a hundred wounds. The Urfine Seals never interfered, but got out of the way as faft as poffible. They even fuffered the cubs of the former to fport with them without offering them the leaft injury.

NOTES. This fpecies has many of the fame actions with the former, in fwimming, walking, lying, and fcratching itfelf. The old bellow like Bulls; the young bleat like Sheep. *Steller* fays, that from their notes he feemed like a ruftic amidft his herds. The males had a ftrong fmell, but were not near fo fetid as the Urfine fort.

FOOD. Their food is fifh, the leffer Seals, Sea Otters, and other marine animals. During the months of *June* and *July* the old males almoft entirely abftain from eating, indulge in indolence and fleep, and become exceffively emaciated.

The

The voyagers made ufe of them to fubfift on, and thought the flefh of the young very favoury. The feet turned into jelly on being dreffed, and in their fituation were efteemed great delicacies. The fat was not oily; that of the young refembled the fuet of mutton, and was as delicious as marrow. The fkin was ufeful for ftraps, fhoes, and boots.

The *Kamtfchatkans* efteem the chafe of thefe animals a generous diverfion, and hold the man in higheft honor, in proportion to the number he has killed. Even thefe heroes are very cautious when they attack one of the animals on fhore: they watch an opportunity when they find it afleep, approach it againft the wind, ftrike their harpoon, faftened to a long thong, into its breaft, while their comrades faften one end to a ftake, and that done, he takes to his heels with the utmoft precipitation. They effect his deftruction at a diftance, by fhooting him with arrows, or flinging their lances into him; and when exhaufted, they venture to come near enough to knock him on the head with clubs.

When they difcover one on the lonely rocks in the fea, they fhoot it with poifoned arrows: unable to endure the pain of the wound, heightened by the falt-water, which it plunges into on the firft receiving it, it fwims on fhore in the greateft agony. If they find a good opportunity, they transfix it with their weapons; if not, they leave it to die of the poifon, which it infallibly does in twenty-four hours, and in the moft dreadful agony *.

They efteem it a great difgrace to leave any of their game behind: and this point of honor they often obferve, even to their own deftruction; for it happens that when they go in fearch of thefe animals to the ifle of *Alait*, which lies fome miles fouth-weft of *Lapatka* promontory, they obferve this principle fo religioufly, as to overload their boats fo much, as to fend them and their booty to the bottom; for they fcorn to fave themfelves, at the expence of throwing overboard any part †.

* *Defcr. Kamtfchatka*, 377.   † *Nov. Com. Petrop.* ii. 302.

This

This species has been difcovered very low in the fouthern hemi-fphere; but, I believe, not on the weftern fide. Sir *John Narbo-rough* * met with them on an ifland off *Port Defire*, in lat. 47. 48. Sir *Richard Hawkins* † found them on *Pinguin* ifle, within the fecond *Narrow* of the ftreights of *Magellan*. They abound in the *Falkland Iflands* ‡; and were again difcovered by Captain *Cook* on the *New Year's Iflands*, off the weft coaft of *Staten Land* ‖. In thofe fouthern latitudes they bring forth their young in the middle of our winter, the feafon in which our late circumnavigators § vifited thofe diftant parts.

---

* *Voy.* 31.   † *Voy.* 75.   ‡ *Pernety's voy.* 188. tab. xvi.

‖ *Cook*, ii. 194. 203. The months in which thefe animals were obferved by the navigators, were *January* and *February*; but by Sir *J. Narborough*, in the ftreights of *Magellan*, about the 4th of *March*, O. S.

§ *Forfter's voy.* ii. 514.

HIST. QUAD. Genus XLIII.

Hift. Quad. N° 390.
Morſkaia Korowa, *Ruſſorum. Nov. Com. Petrop.* ii. 294.
Vaches Marines, *Defcr. Kamtfchatka,* 446.

MANATI. With a ſmall oblong ſquariſh head, hanging down: mouth ſmall: lips doubled, forming an outward and inward lip: about the junction of the jaws a ſet of white tubular briſtles, as thick as a pigeon's quil, which ſerve as ſtrainers to permit the running out of the water, and to retain the food: the lips covered with ſtrong briſtles, which ſerve inſtead of teeth to crop the ſtrong roots of marine plants: no teeth, but in each jaw a flat white oblong bone with an undulated ſurface, which being placed above and below, performs the uſe of grinders to comminute the food.

Noſtrils placed at the end of the noſe, and lined with briſtles: no ears, only in their place a ſmall orifice.

Eyes very ſmall, not larger than thoſe of a Sheep, hardly viſible through the little round holes in the ſkin; the irides black; the pupil livid: tongue pointed and ſmall.

The whole animal is of great deformity: the neck thick, and its union with the head ſcarcely diſcernible: the two feet, or rather fins, are fixed near the ſhoulders; are only twenty-ſix inches long; are deſtitute of toes, or nails, but terminate in a ſort of hoof, concave beneath, lined with briſtles, and fitted for digging in ſand.

The outward ſkin is black, rugged, and knotty, like the bark of an aged oak: without any hair; an inch thick, and ſo hard as ſcarcely to be cut with an ax; and when cut, appears in the inſide like ebony. From the nape to the tail it is marked with circular wrinkles riſing into knots, and ſharp points on the ſide. This ſkin covers the whole

A a   body

body like a cruft, and is of fingular ufe to the animal during winter, in protecting it againft the ice, under which it often feeds, or againft the fharp-pointed rocks, againft which it is often dafhed by the wintry ftorms. It is alfo an equal guard againft the fummer heats; for this animal does not, like moft other marine creatures, feed at the bottom, but with part of the body expofed, as well to the rays of the fun as to the piercing cold of the froft. In fact, this integument is fo effential to its prefervation, that *Steller* has obferved feveral dead on the fhore, which he believes were killed by the accidental privation of it. The color of this fkin, when wet, is dufky, when dried, quite black.

Tail.

The tail is horizontally flat; black, and ending in a ftiff fin, compofed of laminæ like whale-bone, terminating with fibres near nine inches long. It is flightly forked; but both ends are of equal lengths, like the tail of a Whale.

It has two teats placed exactly on the breaft. The milk is thick and fweet, not unlike that of a Ewe. Thefe animals copulate *more humano*, and in the feafon of courtfhip fport long in the fea; the female feigning to fhun the embraces of the male, who purfues her through all the mazes of her flight.

The body, from the fhoulders to the navel, is very thick; from thence to the tail grows gradually more flender. The belly is very large; and, by reafon of the quantity of entrails, very tumid.

Size.

Thefe animals grow to the length of twenty-eight feet. The meafurements of one fomewhat leffer, as given by Mr. *Steller*, are as follow:

The length, from the nofe to the end of the tail, twenty-four feet and a half: from the nofe to the fhoulders, or fetting-on of the fins, four feet four. The circumference of the head, above the noftrils, two feet feven; above the ears, four feet: at the nape of the neck, near feven feet: at the fhoulders, twelve: about the belly, above twenty: near the tail, only four feet eight: the extent of the tail, from point to point, fix feet and a half.

The.

The weight of a large one is eight thousand pounds.

Inhabits the shores of *Bering*'s and the other islands which intervene between the two continents. They never appear off *Kamtschatka*, unless blown ashore by tempests, as they sometimes are about the bay of *Awatscha*. The natives style them *Kapustnik*, or cabbage-eaters, from their food. This genus has not been discovered in any other part of the northern hemisphere. That which inhabits the eastern side of *South America*, and some part of *Africa*, is of a different species. For the latter I can testify, from having seen one from *Senegal*. Its body was quite smooth; its tail swelled out in the middle, and sloped towards the end, which was rounded * To support my other opinion, I can call in the faithful *Dampier*; who describes the body as perfectly smooth † : had it that striking integument which the species in question has, it could not have escaped his notice. Let me also add, that the size of those which that able seaman observed, did not exceed ten or twelve feet; nor the weight of the largest reach that of twelve hundred pounds ‡.

I suspect that this species extends to *Mindanao*, for one kind is certainly found there ‖. It is met with much farther south; for I discover, in the collection of Sir JOSEPH BANKS, a sketch of one taken near *Diego Rodriguez*, vulgarly called *Diego Rais*, an isle to the east of *Mauritius*; and it may possibly have found its way through some northern inlet to the seas of *Greenland*; for Mr. *Fabricius* once discovered in that country the head of one, half consumed, with teeth exactly agreeing with those of this species §.

These animals frequent the shallow and sandy parts of the shores, and near the mouths of the small rivers of the island of *Bering*, seemingly pleased with the sweet water. They go in herds : the old keep behind and drive their young before them : and some keep on their sides, by way of protection. On the rising of the tide they

---

* A figure of this species is given in *De Buffon*, xii. tab. lvii. and in *Schreber*, ii. tab. lxxx.

† *Voy.* i. 33.   ‡ Ibid.   ‖ *Dampier*, i. 321.   § *Faun. Greenl.* p. 6.

approach

approach the fhores, and are fo tame as to fuffer themfelves to be ftroked: if they are roughly treated, they move towards the fea; but foon forget the injury, and return.

They live in families near one another: each confifts of a male and female, a half-grown young, and a new-born one. The families often unite, fo as to form vaft droves. They are monogamous. They bring forth a fingle young, but have no particular time of parturition; but chiefly, as *Steller* imagines, about *autumn*.

They are moft innocent and harmlefs in their manners, and moft ftrongly attached to one another. When one is hooked, the whole herd will attempt its refcue : fome will ftrive to overfet the boat, by going beneath it; others will fling themfelves on the rope of the hook and prefs it down, in order to break it; and others again will make the utmoft efforts to force the inftrument out of its wounded companion.

Their conjugal affection is moft exemplary : a male, after ufing all its endeavours to releafe its mate which had been ftruck, purfued it to the very edge of the water; no blows could force it away. As long as the deceafed female continued in the water, he perfifted in his attendance; and even for three days after fhe was drawn on fhore, and even cut up and carried away, was obferved to remain, as if in expectation of her return.

They are moft voracious creatures, and feed with their head under water, quite inattentive of the boats, or any thing that paffes about them; moving and fwimming gently after one another, with much of their back above water. A fpecies of loufe harbours in the roughnefs of their coats, which the Gulls pick out, fitting on them as Crows do on Hogs and Sheep. Every now and then they lift their nofe out of the water to take breath, and make a noife like the fnorting of Horfes. When the tide retires, they fwim away along with it; but fometimes the young are left afhore till the return of the water : otherwife they never quit that element : fo that in nature, as well as form, they approach the cetaceous animals, and are the link between Seals and them.

They

They were taken on *Bering*'s iſle by a great hook faſtened to a long rope. Four or five people took it with them in a boat, and rowed amidſt a herd. The ſtrongeſt man took the inſtrument, ſtruck it into the neareſt animal; which done, thirty people on ſhore ſeized the rope, and with great difficulty drew it on ſhore. The poor creature makes the ſtrongeſt reſiſtance, aſſiſted by its faithful companions. It will cling with its feet to the rocks till it leaves the ſkin behind; and often great fragments of the cruſty integument fly off before it can be landed. It is an animal full of blood; ſo that it ſpouts in amazing quantities from the orifice of the wound.

They have no voice; only, when wounded, emit a deep ſigh.

They have the ſenſes of ſight and hearing very imperfect; or at leſt neglect the uſe of them.

They are not migratory; for they were ſeen about *Bering*'s iſland the whole of the ſad ten months which Mr. *Steller* paſſed there after his ſhipwreck.

In the ſummer they were very fat; in the winter ſo lean that the ribs might be counted.

The ſkin is uſed, by the inhabitants about the promontory *Tchuktchi*, to cover their boats. The fat, which covers the whole body like a thick blubber, was thought to be as good and ſweet as *May*-butter: that of the young, like hogs-lard. The fleſh of the old, when well boiled, reſembled beef: that of the young, veal. The fleſh will not refuſe ſalt. The crew preſerved ſeveral caſks, full, which was found of excellent ſervice in their eſcape from their horrible confinement *.

To this article muſt be added an imperfect deſcription of a marine animal ſeen by Mr. *Steller* on the coaſt of *America*, which he calls a *Sea Ape*. The head appeared like that of a Dog, with ſharp and upright ears, large eyes, and with both lips bearded: the body round and conoid; the thickeſt part near the head: the tail forked;

* *Muller's voy.* 62. *Nov. Com. Petrop.* ii. 329.

the

the upper lobe the longeſt: the body covered with thick hair, grey on the back, reddiſh on the belly. It ſeemed deſtitute of feet.

It was extremely wanton, and played a multitude of monkey-tricks. It ſometimes ſwam on one ſide, ſometimes on the other ſide of the ſhip, and gazed at it with great admiration. It made ſo near an approach to the veſſel, as almoſt to be touched with a pole; but if any body moved, it inſtantly retired. It would often ſtand erect for a conſiderable ſpace, with one-third of its body above water; then dart beneath the ſhip, and appear on the other ſide; and repeat the ſame thirty times together. It would frequently ariſe with a ſea-plant, not unlike the Bottle-gourd, toſs it up, and catch it in its mouth, playing with it numberleſs fantaſtic tricks *.

On animals of this ſpecies the fable of the *Sirens* might very well be founded.

SEA
BELUGA.

I ſhall conclude this article with a recantation of what I ſay in the 357th page of my Synopſis, relating to the *Beluga*; which I now find was collected, by the author I cite, from the reports of *Coſſacks*, and ignorant fiſhermen. The animal proves at laſt to be one of the cetaceous tribe, of the genus of *Dolphin*, and of a ſpecies called by the *Germans Wit-Fiſch*, and by the *Ruſſians Beluga* †; both ſignifying White fiſh: but to this the laſt add *Morſkaia*, or *of the ſea*, by way of diſtinguiſhing it from a ſpecies of Sturgeon ſo named. It is common in all the *Arctic* ſeas; and forms an article of commerce, being taken on account of its blubber. They are numerous in the gulph of St. *Lawrence*; and go with the tide as high as *Quebec*. There are fiſheries for them, and the common *Porpeſſe*, in that river. A conſiderable quantity of oil is extracted; and of their ſkins is made a ſort of Morocco leather, thin, yet ſtrong enough to reſiſt a muſquet-ball ‡. They are frequent in the *Dwina* and the *Oby*; and go in ſmall families from five to ten, and advance pretty far up the rivers in purſuit of fiſh. They are uſually caught in nets; but are ſome-

* *Hiſt Kamtſchatka*, 136.
† *Pallas, Itin.* iii. 84. tab. iv. *Crantz Greenl.* i. 114. *Purchas's Pilgrims*, iii. 549.
‡ *Charlevoix*, v. 217.

times

times harpooned. They bring only one young at a time, which is dufky; but grow white as they advance in age; the change firft commencing on the belly. They are apt to follow boats, as if they were tamed; and appear extremely beautiful, by reafon of their refplendent whitenefs *.

It being a fpecies very little known, and never well engraven, I fhall give a brief defcription, and adjoin an engraving taken from an excellent drawing communicated to me by Dr. *Pallas*.

The head is fhort: nofe blunt: fpiracle fmall, of the form of a crefcent: eyes very minute: mouth fmall: in each fide of each jaw are nine teeth, fhort, and rather blunt; thofe of the upper jaw are bent, and hollowed, fitted to receive the teeth of the lower jaw when the mouth is clofed: pectoral fins nearly of an oval form: beneath the fkin may be felt the bones of five fingers, which terminate at the edge of the fin in five very fenfible projections. This brings it into the next of rank in the order of beings with the *Manati*. The tail is divided into two lobes, which lie horizontally, but do not fork, except a little at their bafe. The body is oblong, and rather flender, tapering from the back (which is a little elevated) to the tail. It is quite deftitute of the dorfal fin.

DESCRIPTION.

Its length is from twelve to eighteen feet. It makes great ufe of its tail in fwimming; for it bends that part under it, as a Lobfter does its tail, and works it with fuch force as to dart along with the rapidity of an arrow.

SIZE.

A full account of the fifh of the Whale kind, feen by the Reverend Dr. *Borlafe* † between the *Land's End* and the *Scilly* iflands, is a *defideratum* in the *Britifh* Natural Hiftory. He defcribes them as being from twelve to fifteen feet long; fome were milk-white, others brown, others fpotted. They are called *Thornbacks*, from a fharp and broad fin on the back. This deftroys my fufpicion of their being of the above fpecies.

* *Faun. Groenl.* 51.   † *Obf. Scilly Iflands,* 3.

IV   WINGED,

IV. WINGED.

BAT.   *HIST. QUAD.*   GENUS XLIV.

82. NEW YORK.   *Hift. Quad.* Nº 403.—LEV. MUS.

BAT. With the head like that of a Moufe: top of the nofe a little bifid: ears broad, fhort, and rounded: in each jaw two canine teeth: no cutting teeth: tail very long, inclofed in the membrane, which is of a triangular form: the wings thin, naked, and dufky: bones of the hind legs very flender.

Head, body, and upper part of the membrane inclofing the tail, covered with very long hair of a bright tawny color, paleft on the head, beginning of the back, and the belly: at the bafe of each wing is a white fpot.

Length from nofe to tail two inches and a half; tail, one inch eight-tenths: extent of the wings, ten inches and a half.

Inhabits the province of *New York*; and difcovered by Dr. *Forfter* * n *New Zealand,* in the *South Seas.*

83. LONG-HAIRED.   *Mr. Clayton, in Ph. Tranf. Abridg.* iii. 594.

BAT. With long ftraggling hairs, and great ears. The above is all the account we have of this fpecies; which is faid to be an inhabitant of *Virginia.*

Mr. *Lawfon* fays, that the common Bat is found in *Carolina* †.

* *Obfervations, &c.* 189.   † *Hift. Carolina.* 125.

Hift.

Hiſt. Quad. Nº 407.—Great Bat, *Br. Zool.* i. Nº 38.   84. NOCTULE

BAT. With the noſe ſlightly bilobated : ears ſmall and rounded : on the chin a ſmall wart : body of a cinereous red.

Extent of wings fifteen inches : body between two and three in length : tail, one inch ſeven-tenths.

Brought from *Hudſon's Bay* in ſpirits. I ſaw it only in the bottle ; but it appeared to be this ſpecies.

A. COMMON BAT, *Hiſt. Quad.* Nº 411.—*Br. Zool.* i. Nº 41.—LIV. MUS.

THIS ſpecies is found in *Iceland*, as I was informed by the late Mr. *Fleiſcher*, which is the moſt northernly reſidence of this genus. In *Aſia* I can trace them no farther eaſtward than about the river *Argun*, beyond lake *Baikal*.

B b   CLASS

Printed in the United States
By Bookmasters